P9-CQJ-239

Praise for POWER TRIP

"*Power Trip* offers a panoramic view of our energy crisis, exploring past, present, and future with hope, passion, and humor. Whether you are liberal or conservative, expert or novice, young or old, you'll find adventure and insight in this book."

—Robert F. Kennedy Jr., president of Waterkeeper Alliance

"Energy is the most important story in the world bar none, and no one has ever told it with more verve than Amanda Little. If you want to know how the world works, and why it may not work much longer, this is the book you need." —Bill McKibben, author of
Fight Global Warming Now:
The Handbook for Taking Action in Your Community

"Lively, engaging, and most thought-provoking, *Power Trip* takes us on a journey through the very wide world of energy, from its colorful past to its high-tech future. Little answers the questions that perplex many—and, so importantly, identifies the key questions that only the future will answer." —Daniel Yergin, Pulitzer Prize–winning author of
The Prize: the Epic Quest for Oil, Money & Power

"A fast, fun, and gripping story—one that's both candid and unflinching in its approach. Little represents the best of a new, young perspective, a new voice of green." — Robert Redford

"A wonderfully illuminating voyage. Little charts a fresh path outside the usual doctrinaire accounts on energy. Her intelligence and enthusiasm will change the way you think about the future."

— Steven Johnson, bestselling author of
The Ghost Map and *The Invention of Air*

"It's hard to imagine a book about energy that would appeal as much to a business executive as it would to an eco-activist—or, for that matter, to a soccer mom, a farmer, a politician or a student. Here it is. This provocative story about America's love affair with energy is a must-read for everyone."   —Jim Rogers, chairman and CEO of Duke Energy

"This fascinating book should now be a must-read." —Gwyneth Paltrow

"Charming, fun, and deeply informed, *Power Trip* is a great way to get a handle on our energy and environmental future. Little talks to the key players across America, digs into the reasons we have so many problems, and finds hope for a better world ahead."
     —Jim Cooper, congressman from Tennessee

"Ambitious and highly readable. . . . [Little] expertly ties together disparate strains of history to make her case. . . . Jargon-free and written with a fine eye for detail—one of the best books on America's energy crisis to emerge in recent years."   —*Kirkus Reviews* (starred review)

"Amanda Little [is] faster, funnier, smarter and more tenacious than anyone else. . . . Little has a big voice and tells her story with style."
     —Iain Finlayson, *The Times* (UK)

"Greens like author Little want America to wean itself from oil, but rather than shout righteously from the mountaintop, she traveled the fruited plain to meet people immersed in practicalities of transitioning from a carbon- to a renewable-energy economy. . . . Writing jaunty descriptions of her interviewees, Little will appeal to energy-minded readers who don't want another anti-oil screed or a pro-green policy manual but, rather, a locally-oriented, human-interest introduction to businesspeople and government officials busy with doing rather than talking when it comes to the future of energy."   —*Booklist*

"Little examines the role fossil fuels play in her own life, and out of her self-examination emerges a thoughtful book that gives energy neophytes an accessible way to learn about fossil fuels and their fallacies."
     —*Publishers Weekly*

"Combine the historical intrigue of Jared Diamond, the journalistic flare of Tom Wolfe, and the passionate advocacy of Rachel Carson—and you get *Power Trip*. Amanda Little's multifaceted approach makes this the one book about our energy past and future that everyone should read and all will enjoy."

—Andrew Shapiro, founder and president of GreenOrder

"[A] sweeping account." —*Seed* magazine

"[A] wide-ranging, ambitious story of energy. . . . Like a physician diagnosing a patient. . . . Little has found true insight into the past, present, and future interaction of culture and energy. She shares what she has learned with a mixture of humor, gravitas, and optimism that is both rare and welcome in environmental reporting. . . . It is [an] understanding—not just of Americans' collective guilt in burning energy as casually as we breathe air, but that things and people are often not as simple they seem—that makes *Power Trip* a fine and enlightening read." —Chris Scott, *Chapter 16*

"Tell[s] a more human side of the energy story."

—Laura Fitzpatrick, Time.com

"Little maps America's energy landscape, traveling to deep-sea oil rigs, NASCAR speedways, Google's twenty-first-century campus, and the Pentagon's DefCon-haunted warrens. Her findings: American ingenuity created cheap fossil fuel, and that same ingenuity is spurring monumental change." —*Outside* magazine

"To truly understand energy consumption, our dependency on fossil fuels and how it touches virtually every product we rely upon, Little takes us on a voyage of discovery. . . . Business visionaries, environmental leaders, and concerned citizens alike will find *Power Trip* illuminating, and its ideas inspiring."

—J.P. Morgan/Barnes & Noble, winter 2009 reading list

"A well-written, engaging account of how cheap and abundant energy is the long-running constructive theme in the United States' rise to world dominance. . . . Even know-it-all greenies will get some new info out of it." —Matthew McDermott, Treehugger.com

"Ambitious and engaging . . . grabs readers from page 1 with lively profiles of just about everyone touched by oil."
— *Monterey County Herald MediaNews*

"Far from simply relaying abstract theories of peak oil or alternative energy technologies, Little . . . paints a human face on each innovation—eco-friendly or not. Her rapier wit slices through discussions of the 'love affair between the petroleum industry and politics' . . . Still, the predominant attitude remains optimistic. Little leaves readers hopeful that the same ingenuity that bound America to oil will eventually liberate us from it." —Lynne Peeples, *Audubon* magazine

"A fascinating journey. . . . Packed with insight and wit. . . . You need to read this book." —Shea Gunther, Mother Nature Network

"Engrossing . . . a form of reporting that isn't done enough."
—Greening of Oil

"Ambitious and engaging . . . grabs readers from page 1 with lively profiles of just about everyone touched by oil from NASCAR fans and their gas-guzzling cars to the plastic surgeon who uses synthetic (i.e. oil-derived) body parts to remake clients into their dreams come true."
—*San Jose Mercury News*

"Amanda Little tries talking our energy-addicted nation down from a bad *Power Trip*." —*Vanity Fair*

HEIDI ROSS

About the Author

AMANDA LITTLE has published widely on energy, technology, and the environment. Her columns on green politics and innovation have appeared in Grist.org, Salon.com, and *Outside* magazine. Her articles have been published in the *New York Times Magazine, Vanity Fair, Rolling Stone, Wired, New York Magazine, O, The Oprah Magazine, In Style, Men's Journal,* and the *Washington Post.* She is the recipient of the Jane Bagley Lehman Award for excellence in environmental journalism. She lives with her husband and daughter in Nashville, Tennessee.

Power Trip

HARPER ⬤ PERENNIAL

NEW YORK • LONDON • TORONTO • SYDNEY • NEW DELHI • AUCKLAND

POWER TRIP

The Story of
America's Love Affair
with Energy

Amanda Little

HARPER ● PERENNIAL

A hardcover edition of this book was published in 2009 by Harper, an imprint of HarperCollins Publishers.

Photo Credits

Pages viii–ix	Photo by Brian Harkin/*New York Times*/Redux
Page 1	Photo by the Texas Energy Museum/Newsmakers/Getty Images
Page 2	Photo by Mickey Driver
Page 44	Photo by Kenn Mann/USAF/Getty Images
Page 80	Photo by Rusty Jarrett/Getty Images for NASCAR
Page 110	Photo by Heidi Ross
Page 146	Photo by Robert W. Kelley//Time Life Pictures/Getty Images
Page 178	Photo by Heidi Ross
Page 204	Photo by Gina LeVay
Page 241	Photo copyright 2009 W. T. Pfeffer
Page 242	Photo by Robert Nickelsberg/Getty Images
Page 282	Photo by KAREN BLEIER/AFP/Getty Images
Page 316	Cook + Fox Architects
Page 348	Photo by Shadia Fayne Wood

I would like to thank *Wired*, Grist.org, *Outside*, the *New York Times*, *Vanity Fair*, and *Men's Journal* for permission to include in *Power Trip* excerpts from articles I wrote for their publications. I would also like to thank Cole Porter Musical and Literary Property Trusts for permission to use an excerpt from "You're the Top!"

HarperCollins books may be purchased for educational, business, or sales promotional use. For information please write: Special Markets Department, HarperCollins Publishers, 10 East 53rd Street, New York, NY 10022.

FIRST HARPER PERENNIAL EDITION PUBLISHED 2010.

Designed by Jennifer Daddio/Bookmark Design & Media Inc.

Library of Congress Cataloging-in-Publication Data is available upon request.

ISBN 978-0-06-135326-0

10 11 12 13 14   OV/RRD   10 9 8 7 6 5 4 3 2 1

For CARTER
trip leader, power source

Contents

One

LIFE, LIBERTY, AND THE PURSUIT OF OIL

The Story of the American Century

Two

GREENER PASTURES

The Dawn of the New Energy Era

Foreword to the Paperback Edition

Sometimes the least remarkable events trigger history's most remarkable moments. That was the case on April 20, 2010, when a nameless rubber component eroded within BP's drilling machinery in the Gulf of Mexico, causing failures that sparked a gigantic explosion on the Deepwater Horizon rig and left its well bleeding oil uncontrollably. The result was the single worst environmental disaster in U.S. history—an event that, more than any other in the last century, revealed the extreme but hidden risks of America's oil usage. It also shed unexpected light on our national identity—both our greatest weaknesses as a country, and our greatest strengths.

Weeks of media coverage showed the gradual destruction of a marine ecosystem the size of Wyoming, as the crude oil killed vast swaths of coral reefs and threatened hundreds of bird, fish, marine mammal, and plant species. The region's multibillion-dollar seafood industry was crippled indefinitely, putting thousands of shrimpers, oystermen, and fishermen out of work. Tourism along the Gulf was also devastated, with hundreds of miles of tainted shorelines.

Americans monitored the event like it was gory reality TV, watching twenty-four-hour real time video of oil hemorrhaging into the Gulf at a rate of roughly five hundred gallons per minute. Outrage toward BP's executives flowed with equal force: How could they have overlooked key safety measures in order to fast-track oil production? How could they have the means to drill the deepest offshore well on record—boring over 35,000 feet into the earth—and not know how to plug a leak?

But while we bridled at the greed and incompetence of BP and the oil industry in general, few of us acknowledged our own roles, as consumers, in the catastrophe. The event exposed our collective ignorance about what it takes to satisfy our nation's appetite for oil. Even today, few of us understand how large that appetite truly is. In a single day, Americans consume roughly 750 million gallons of oil—about 10 times more than the total estimated volume of crude that has spewed into the Gulf of Mexico so far.

Our hunger for oil, like our fondness for fast food, has spawned a kind of obesity epidemic—but one that doesn't manifest itself in visible pounds of flesh. The very fact that we generally can't *see* the consequences of our oil consumption has created a fantasy of sorts—that we can live energy-lavish lifestyles without experiencing any negative effects. The Gulf spill, if only temporarily, punctured the myth: images of thick brown scum floating like a funeral shroud over thousands of square miles of ocean, coating the corpses of egrets and dolphins, gave an emotional texture to a substance that remains a mystery to most of us.

Even though I had spent nearly a decade reporting on energy and environmental policy, I'd never personally witnessed the production of crude oil or coal until a few years ago. I understood, in theory, that these resources feed virtually every aspect of our economy—transportation, electricity, plastics, pharmaceuticals, farming, computing, military, and so much more—but I knew very little in practice about where and how the fuels are produced, and how they became so prevalent in our lives.

In search of answers to these mysteries, I traveled to a deepwater oil rig in the Gulf of Mexico known as the "Cajun Express" in the spring of 2007. That voyage was the first step of a two-year journey through America's energy landscape—a journey that changed my understanding of who I am, both on a personal level and as a citizen of the United

States. I learned many sobering truths about just how much the strength and survival of our nation depend on fossil fuels, and about the risks that come with that dependence. But I also witnessed much cause for optimism in looking towards the future. *Power Trip* is the story of what I discovered and why I remain hopeful.

Eight months after the book's release, both my concerns and my hopes for America's energy future have intensified. As long as we are a nation tied to fossil fuels, we will be vulnerable to colossal operations like the Deepwater Horizon and the tens of thousands of other rigs run by companies like BP, Chevron, and ExxonMobil that are scattered across the world's oceans. The industry will continue going further offshore and inserting drills deeper into sea beds to tap new resources, and, in turn, face ever-greater risks.

It's not these risks *per se* that concern me; I believe they can be managed with proper oversight. It's our reluctance as a nation to address the root cause of these risks: our relentless demand for oil. In the wake of the oil spill (which, as I write, is ongoing), few political leaders have called on Americans to cut their oil use. Instead, many are arguing to prohibit drilling on America's coasts. The worst thing we could do, in fact, is to push the problem out of view.

In America and throughout the world, land-based oil reserves are dwindling, forcing the industry to look to offshore regions for major new oil discoveries. If our coastlines are closed to oil development, and our demand continues unabated, the risks of offshore production will simply shift to the coastlines of Nigeria, South Korea, Kazakhstan, Angola, and other nations where environmental standards are extremely lenient, if they exist at all. Moreover, in order to secure the oil resources in these volatile nations, we will have to send American troops at great financial and human cost. We also face the growing threat of global warming, which is worsened by the burning of fossil fuels.

Perhaps only when we see oil rigs crowded along our coastlines, and experience the human consequences of oil extraction and consumption on our home turf, will we mobilize the shift toward energy efficiency and clean alternatives on a grand scale.

I believe that we *are* capable of solutions on a grand scale. The journey I took through America's energy past and present—exploring the first oil wells, the birth of electricity, the first mass-produced automobiles, the emergence of the plastics industry, the modern military, modern cities, and modern medicine—convinced me of our staggering potential as a nation to solve problems and innovate. But transformative change can happen only when mainstream America embraces it as absolutely necessary. The question that persists today is: what will it take to convince Americans that shifting away from fossil fuels is absolutely necessary?

One week after the explosion on the Deepwater Horizon, I was struck by a blog headline mocking the absurd risks we're taking to produce oil. The Department of the Interior had just approved the first massive offshore wind project in the United States off the coast of Cape Cod, Massachusetts—a move that promised to pave the way for huge growth in the U.S. wind power industry. The headline read, "BREAKING NEWS: Large Air Spill at Wind Farm. No Threats Reported. Some Claim to Enjoy the Breeze."

Clean energy on a grand scale remains an outlandish thought. But it's no more outlandish than the extraordinary industrial growth of the twentieth century would have seemed three generations ago—growth in roadways, skyways, industry, and agriculture fed by the unforeseen discovery of cheap and abundant fossil fuels.

Understanding this past is essential to building a better future. Understanding the true promise and potential of current innovations in renewable energy, electric cars, green buildings, sophisticated plastics, and the smart grid can give us the courage to face our past and its costs and consequences.

The Gulf oil disaster can also serve as a powerful lesson to America and persuade us that it is now absolutely necessary to shift to an efficient, innovative, sustainable economy, so that our children and their children will be able to walk along clean coastlines and enjoy the breeze.

— *Nashville, Tennessee, June 28, 2010*

Confessions of a Petroleum Addict

The trouble started on an August afternoon in a remote field in northern Ohio, miles from any town large enough to be marked on a standard road atlas. The field was empty except for scattered deciduous trees—maple, poplar, oak—thick with late-summer leaves. The ground was scrubby and parched. A nearby river rolled lazily in the summer heat. The only trace of humanity hung above the trees—an electrical cable known as the Harding-Chamberlin Line, carrying 345,000 volts of power.

By three o'clock the air temperature had risen to 90 degrees, and the cable itself had reached nearly 200 degrees Fahrenheit—roughly twice its average temperature. The aluminum core of the 3-inch-thick wire was expanding with the heat and beginning to sag.

Five hundred miles due east of that meadow I was sitting at my desk in New York City when, at 4:09 p.m., my computer suddenly shut down. The lights, music, and air-conditioning died. I heard a strange lurching sound as the elevator in my building froze with passengers trapped on board. I rushed to the window along with my officemates and was amazed to see traffic snarling to a halt up the entire length of Broadway as street signals went black. The Verizon landlines were dead and our cell phones had no signals. We hurried down eleven flights of stairs, into streets already thickening with crowds of evacuees. Storefronts, groceries, and cafés were darkened. Subway stations were emptying of travelers as word spread that the trains had no power and hundreds of people were stuck underground. It was 2003, and like most New Yorkers, we initially jumped to the same conclusion—another terrorist attack.

What had in fact happened to us, and to a majority of the residents of the metropolitan areas of New York, Newark, Baltimore, Cleveland, Detroit, and Toronto, was a blackout—larger than any other blackout in recorded history. One of the greatest achievements in industrial engineering, the 93,600 miles of electrical cable known as the Eastern Interconnection, had been brought to its knees. All because of unseen events in that distant Ohio meadow where an overloaded wire had drooped into high tree branches and short-circuited, triggering a massive cascade effect throughout the aging power grid.

As night fell, I walked up to Times Square to see its flashing billboards snuffed out, leaving the commercial El Dorado quaint and sheepish. I passed the main post office building and Bryant Park, where thousands of stranded commuters were sprawled in a mass slumber party, using their suit jackets and briefcases as pillows. Candlelight flickered in apartment windows, and I looked up past the walls of darkened buildings at a sky so brilliant with stars I could make out the soft haze of the Milky Way and the faint pulses of orbiting satellites.

Before-and-after satellite images of the event tell the story. In the first picture there is a thick streak of foamy white across the northeastern portion of the United States and southeastern Canada. In the second is just a scattering of faint droplets, the rest absorbed into the blackness of space. Fifty million Americans were without power.

Up to that point, I had spent most of my brief career as a journalist trying to gain a better understanding of the causes of just such events—an understanding, more broadly, of the strengths and vulnerabilities of America's energy landscape. The twenty-four-hour blackout made me realize how little I actually did know, and how much I still had left to learn.

Just out of college in 1997, I had started out as a technology reporter, swept up in the exuberance of the dawning digital age—when stock prices jumped from 60¢ to $60 overnight, and business plans scrawled on cocktail napkins could get six-figure backing. I went on to write "Urban Upgrade," a column in the *Village Voice* about how digital technology was transforming New York City—from Wall Street's trading floor to the billboards of Times Square—into an intelligent, networked, high-speed metropolis.

The more I learned over time about New York's electricity grid, the more shocked I was to discover that the torrent of pixels and megabytes newly pulsing through the city was sustained by an antiquated grid no smarter than a plumbing system—and much harder to repair. Moreover, this system was powered by the decidedly unfuturistic force of fossil fuels (specifically, coal and natural gas). I began to wonder: as cell phones, PDAs, ATMs, iPods, laptops, and flat-screen televisions proliferated, how long could this brittle grid hold up, and what kind of impact would these new pressures have on the environment?

My interest in America's energy dependence redoubled after the events of September 11, 2001, this time focusing on a different aspect of our fossil fuel usage—the oil that powers our cars, trucks, buses, ships, and airplanes. That morning I was riding my bike over the Brooklyn Bridge into Manhattan when, at 8:46 a.m., I watched the North Tower of the World Trade Center explode suddenly, inexplicably, into flames. All movement—cars, bikes, pedestrians—froze. From where I stood on the crest of the bridge I saw, in the foreground, the orange plume of fire flaring skyward from the building. (It was ignited, we would soon learn, by 11,000 gallons of jet fuel from the tanks of American Airlines flight 11.) Dwarfed in the background was a dim fleck of light wavering from the Statue of Liberty. This attack and its tragic consequences would not have happened, as news coverage following the event made clear, without our presence in the Middle East—a presence closely tied to our reliance on the oil reserves heavily concentrated in that region.

The months after September 11 revealed further evidence of vulnerability and change in America's energy system. Petroleum prices soared in response to the attack, as speculators feared interruptions to the flow of oil between the Middle East and the United States. Meanwhile, a group of two thousand scientists who constitute the Intergovernmental Panel on Climate Change came out with a landmark (and widely ignored) report declaring that global warming was accelerating faster than ever predicted—a phenomenon largely driven by our use of fossil fuels, which release carbon dioxide and other greenhouse gases into the atmosphere when burned. Then Enron, one of the world's leading producers of electricity and natural gas, collapsed into bankruptcy amid revelations of widespread corporate fraud.

The focus of my reporting pinballed from the digital revolution to what seemed a bigger, more urgent shift—the wholesale rebuilding of our energy landscape. Fancying myself an amateur detective, I started traveling throughout the country, from Ashland, Oregon, to Tampa Bay, Florida, to write about the architects and early adopters of emerging energy technologies that could provide alternatives to fossil fuels: solar, wind, geothermal, biofuels, and hybrid-electric cars such as the Toyota Prius. I began studying and writing about the legislation that was being drafted (and blocked) to push these innovations into the mainstream. I began criticizing the federal government's failure to take action on climate change and its unwillingness to encourage the development of clean, efficient, next-generation energy technologies.

But when the August 2003 blackout hit, I realized one major blind spot in my understanding of energy. Nothing I'd learned in my reporting had quite prepared me for the feeling of utter helplessness and paralysis that a blackout of that scale would cause. It was the first time, for me and for millions of Americans, that the story of energy was conveyed in human terms. Here I was crisscrossing the country, chasing after innovators and wagging fingers at the government, but I'd completely neglected to examine the role of energy in my own life. One morning I began with a seemingly simple task: I took a much smaller and quieter, but for me equally momentous, tour around my office. My aim was to count the things in my midst that were, in one way or another, tied to fossil fuels.

Oil, coal, and natural gas—the three most common forms of fossil fuels—were all formed over a period of millions of years from the remains of plants and animals (primarily tiny aquatic organisms) that were exposed to the combined effects of time, compression, and temperature. Oil accounts for roughly half of our nation's total fossil fuel usage. What it provides, by and large, is movement—America's transportation sector is almost entirely dependent on petroleum (a term used interchangeably with *oil*, referring to its raw, unrefined state). Petroleum is chemically complex and can be refined not just into gasoline, kerosene, and motor oil but also into the petrochemicals that are the basic building blocks of a vast range of consumer products, from plastic bags to bulletproof vests.

What petroleum doesn't provide is electricity, which accounts for the other half of America's fossil fuel use. Electricity generation can be broken down into three main sources: coal (about 50 percent), natural gas (20 percent), and nuclear (20 percent). Natural gas is essentially petroleum that has been slow-cooked over time into a gaseous form.

Despite their many different applications, fossil fuels have a common purpose. My Merriam-Webster dictionary defines *energy* as "the ability to do work," and fossil fuels have been doing America's industrial work for more than a century. When I refer to America's "energy landscape," I'm talking about the whole picture—the combination of oil, coal, and natural gas that feeds the intricate organism of modern society. That feeds our own lives—my own life, as I realized that morning while conducting my amateur fossil fuel audit.

Since nearly all plastics, polymers, inks, paints, fertilizers, and pesticides are made from petrochemicals, and all products are delivered to market by trucks, trains, ships, and airplanes, there was virtually nothing in my office—my body included—that wasn't there because of fossil fuels.

There I sat at a desk made of Formica (a plastic), wearing a sweatshirt made of fleece (a polymer) over yoga pants made from Lycra (ditto), sipping coffee shipped from Zimbabwe, eating an apple trucked from Washington, surrounded by walls covered with oil-derived paints, jotting notes in petroleum-derived ink, typing words on a petrochemical keyboard into a computer powered by coal plants. Even the supposedly guilt-free whole-grain cereal I had for breakfast and the veggie burger I ate for lunch came from crops treated with oil-derived fertilizers. My purse yielded another trove of specimens: capsules of Extra-Strength Tylenol made from acetaminophen (a substance, like many commercial pain relievers, that is refined from petroleum); glossy magazines and a packet of photographs printed with petrochemicals; mascara, lip balm, eyeliner, and perfume that, like most cosmetics, have key components derived from oil.

I had understood this intellectually before—that the energy landscape encompasses not just our endless acres of oil fields, coal mines, gas stations, and highways, as well as the vast network of copper wires that feeds electricity to our homes and offices. It's also the cornfields in America's heartland, the battlefields of Iraq, and the medical labs that produce penicillin, novocaine, chemotherapy drugs, and many other treatments and cures. It's the cosmetics shelves and glossy magazine racks in our drugstores. It's the constantly humming, behind-the-scenes network of ships, planes, trains, and trucks that transport products to our store shelves. It's even our own bodies, which we routinely drape in synthetic fabrics like spandex and nylon, and feed with crops that were fertilized by fossil fuels.

What I hadn't fully managed to grasp was the intimate and invisible omnipresence of fossil fuels in my own life—the plastic sutures that stitched up my split lip when I was seven, the photographic CAT scan images that evaluated my concussion after an accident when I was twenty-seven. Once I connected the dots between so many seemingly disparate elements of my life—my car, my clothes, my e-mail, my makeup, my burger, even my health—I saw an energy landscape far more vast and complex than I'd ever imagined.

I also realized that this thing I'd thought was a four-letter word (oil) was actually the source of many creature comforts I use and love—and many survival tools I need. It seemed almost miraculous. Never had I so fully grasped the immense versatility of fossil fuels on a personal level and their greater relevance in the economy at large.

Energy, I realized that morning, is *everything*. It's life, liberty, the pursuit of happiness—and our very survival. But if fossil fuels are a part of everything we do, how do we go about removing them from the picture? How can we kick America's addiction to fossil fuels, given its sheer magnitude?

What I'd been chasing ever since my first efforts at reporting, as I tried to make sense of the power grid, September 11, the 2003 blackout, and the role of fossil fuels in daily life, was connections—the ways in which energy connects us all, beyond our homes, our cities, our state and national borders. Energy is the thread from which our modern lives dangle, but it is an invisible thread—pumped through underwater oil pipelines, coursing through unseen cables in remote meadows, and tucked away in basement fuse boxes, just as the veins are hidden beneath our skin.

It is common knowledge now that America's energy-dependent economy is facing radical change. We are in the midst of economic, geopolitical, and environmental turmoil—a triple threat that is deeply rooted in our use of fossil fuels. Many countries face these issues, but Americans stand to lose more than the people of any other nation given our formidable appetite for energy. We use roughly 25 barrels of oil per person annually; Europeans, by comparison, use about 17, and the citizens of Japan use 14.

Nevertheless, we are reminded as frequently by ExxonMobil commercials as we are by White House officials and eco-activists that our nation has begun the shift away from ancient energy sources toward cleaner, homegrown sources of power and fuel. We regularly hear menacing threats and utopian promises—on the one hand, global warming is well on its way to producing irreversible coastal floods and mega-drought, and we are in the throes of an "energy crisis"; on the other, we are in "an era of change," heading toward a "green revolution." We are told that the global recession we've faced in recent years is an opportunity to reengineer our industries and infrastructure with green technology. "We will rebuild, we will recover," President Obama has said, "and the United States of America will emerge stronger than before." He has vowed to "lay a new foundation for economic growth by beginning a new era of energy exploration in America." We are promised a future in which solar panels will glitter across rooftops, wind turbines will whirl across prairies, super-efficient cars will glide silently along our roads, and new clean industries will provide millions of jobs to a "green-collar" workforce that will repair our aging infrastructure and revive our struggling economy.

For most of us, however, America's energy landscape still feels like distant and impersonal terrain. We wonder how those grand threats and promises will translate into action. Very little reporting tells America's energy story in human terms. Books on energy tend to be dense and technical; they often examine the economic, scientific, and political aspects of energy, but they rarely explain what these changes mean in our lives in the most practical, personal sense. They complain about the mess we're in, but few explain how we got ourselves here and, in simple terms, how we can climb our way out.

The father of a friend of mine, who is now a successful businessman, defined his approach to problem solving in terms he learned through a painful experience as a boy growing up on a farm in Ohio. When a cow gets stuck in a ditch, first you have to get the cow out of the ditch. Second, you have to figure out how the cow got into the ditch. Third, you have to figure out how to stop the cow from getting into the ditch in the future.

I want, like a majority of Americans today, to get myself out of the ditch of fossil fuel dependence. But to do it right, I—and we—need to understand the roots of the problem, to understand how, during the twentieth century, fossil fuels became so thoroughly woven into the fabric of our lives. We need to recognize what our options are going forward—how America as a whole could build an actual, factual "green" future, free from fossil fuels.

This book searches America's past for clues to understand our future. It tracks the meteoric growth of our superpower thanks to fossil fuels. It describes how cheap fuel and electricity built our sprawling cities, unequalled military might, and major industries—from automobiles and agriculture to plastics and computing—and seeped into the fibers of our daily lives. It examines how the oil and coal that built us up now threaten our ruin, tainted as they are with political strife and corruption, and pollutants that kindle environmental chaos. It also looks forward—to the transfusion of clean and renewable power sources that are beginning to course through the copper veins and combustion-engine heart of our nation.

The story of America is, in sum, the story of a power trip; to understand it, I had to go on my own. In January 2007, I set out to explore the most extreme frontiers of our energy landscape—from its deepest wells to its tallest towers. I wanted to pull at the threads of connection between fossil fuels and everyday American life and see what places they led me to, however strange or unexpected. They led me, as it turned out, to some very strange spots, from deep-sea oil rigs to Kansas cornfields, NASCAR tracks to high-priced plastic surgeons, dank city manholes to Texas wind farms, Pentagon offices to my local produce aisle.

I saw places where the old energy system is flaming out and the new system—smart, efficient, whisper-quiet, and high-performance—is sparking to life. I interviewed architects of the oil-guzzling twentieth-century economy—great and innovative in their own right—and pioneers of tomorrow's new machine.

My goal as I describe this journey is not to cast judgment on what has gone wrong in America's energy landscape—as I have said, I'm guilty myself of buying into and even relishing it. Instead, I want simply to understand this landscape, and to celebrate its successes for all their unintended consequences. It was, after all, American ingenuity that led us down the path of fossil fuel dependence. So I set out to discover how that same ingenuity can change our future course. The following pages are a chronicle of both journeys through the fast-changing energy frontier—America's, and my own.

One

LIFE, LIBERTY, AND THE PURSUIT OF OIL

The Story of the American Century

J.R.: There's nothing realer than oil, that's for sure.

Sue Ellen: Not to you darlin', except perhaps money.

J.R.: Same thing, honey, same thing.

—*Dallas*, season two

Over a Barrel

THE BOOM AND BUST OF AMERICA'S DOMESTIC OIL EMPIRE

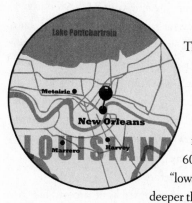

The oil field known as "Jack" is located 175 miles off the coast of Louisiana, below 7,200 feet of water and another 30,000 feet under the seabed, occupying a geological layer formed in the Cenozoic Era more than 60 million years ago. This layer—the "lower tertiary"—lies beneath waters far deeper than those surrounding any other Gulf of Mexico oil discovery, which is one reason why many in the industry initially dismissed it as too remote to exploit. But in 2006, Chevron defied the odds when its engineers drilled a test well at Jack and discovered that oil could flow from this ancient sediment at profitable rates. Their success opened up a new drilling frontier—a monster oil patch holding between 3 billion and 15 billion barrels of crude. It was hailed as the largest discovery in the United States since 1968—a discovery potentially big enough to boost national oil reserves up to 50 percent.

Since then, global oil companies have been pouring billions of dollars into these so-called ultradeep waters of the Gulf in pursuit of the region's buried treasure. Jack is among a cluster of nearly a dozen new fields there—including "Blind Faith," "Great White," and "Cascade"—that companies are now tapping in waters from 4,000 to 8,000 feet deep and in sedimentary rock extending between 1 and 6 miles below the seabed.

Coaxing oil from such great depths poses unprecedented risks for oil drilling—and that's why I decided to visit the area. I wanted to witness firsthand the world's most extreme drilling territory, the Mount Everest of oil frontiers, where the industry has to tackle the tallest odds and gravest circumstances to eke out new discoveries before global petroleum production peaks and begins to decline.

I set out at dawn on an April morning in a Sikorsky S-76 helicopter. The sky above the New Orleans heliport was a pea-soup green, thick with rain and pitchfork lightning. I was traveling with a Chevron executive and three of his staffers, all of us wearing regulation jumpsuits, hard hats, and steel-toed boots. The chopper lurched and shuddered in the squalls, but my travel companions nodded to the pilot to press on—this was typical weather for the Louisiana coast, and routine flying conditions. We hurtled over the bayou's emerald marshlands, patterned like marbled paper with coiling blue inlets and flecks of white from puttering shrimp boats. Soon the marsh gave way to the Gulf of Mexico's open waters and the storm lifted. I relaxed my grip on the edge of my seat as a smooth two-hour voyage stretched out before us.

"Isn't this transcendent?" Paul Siegele shouted as he pressed his nose to the window. The early morning sun glinted off a colony of metal structures pocking the surface of the sea. Siegele, the director of Chevron's offshore drilling division, identified the objects below with the geeky verve of a birder: a miniature oil rig known as a mono pod, a drill ship nearly as big as the *Titanic*, and circular, tiered platforms scattered like steel chandeliers that fell from the storm-shaken clouds.

A lanky six foot three, Siegele has none of the cowboy swagger you might expect from a top oil executive. He's earnest and quick to smile, with a mild, professorial manner and a boyish mop of brown hair. Siegele was an aspiring artist as an undergraduate at California Lutheran University, until his studies of rock as a sculptural medium sparked his interest in geology. Oil drilling, he says, is not unlike sculpture: "It's about precision—guiding tools into the earth artfully, not just blindly hammering at rock."

The Gulf yields 25 percent of all U.S. oil production, and is home to more than 3,700 production platforms, most of them located in relatively shallow waters of less than 2,000 feet. Many geologists believe that the ultradeep regions of the Gulf—those covered by waters greater than 4,000 feet—hold more untapped oil reserves than any other parts of the Western world. Today, offshore rigs are capable of operating in 10,000 feet of water and boring through more than 35,000 feet of seabed (twice the depth they could manage a decade ago). One rig sits atop each field, thrusting its tentacles into up to a dozen wells throughout the bed. The rig pulls up oil and then pumps it back to onshore refineries via underwater pipelines.

Chevron is one of the largest leaseholders in the deep-sea Gulf, which means it has much to gain from these waters—but also much to lose. Three out of four exploration wells in this area come up dry—nerve-wracking odds when the wells cost $100 million apiece, or as much as twenty times what they cost on land. And even if you hit pay dirt, there's no guarantee of profit: in the past decade, Chevron has abandoned nearly a quarter of the successful wells it has drilled because they wouldn't flow at profitable rates.

Our specific destination was an offshore rig known as the "Cajun Express"—a massive rectangular ship with onboard drilling equipment. It was the Cajun's 5-mile-long drill that months earlier had burrowed the legendary discovery well at Jack. That day, the rig had motored over to a nearby field known as "Tahiti," where Siegele would be overseeing final preparations for another drilling endeavor. (Field names are given by the geologists who discover and study them. There are plenty of fields in the deep-sea Gulf with names that carry more gravitas—"Thunder Horse" and "Atlantis," for instance; "Jack" was named in honor of its founder's lifelong mentor, and "Tahiti" by a geologist who loved the Polynesian islands.)

The activity we were about to witness on the Cajun would help determine the fate of this extreme deep-sea frontier: Would the risks outweigh the rewards? Would Chevron be able to scale this drilling Everest?

Truthfully, there's no single Everest of oil drilling, given how many unthinkable extremes—environmental, technological, political—the industry and its workers face today. In the Chukchi Sea of the Arctic Ocean, for instance, drillers have to wear heated bodysuits to survive winter temperatures averaging 60 degrees below zero; the region descends into near-complete darkness for four months of the year. In the Kashagan field of Kazakhstan, workers have to wear gas masks to protect against the high concentration of poisonous hydrogen sulfide present in the oil they're extracting. On Sakhalin Island, located in the Sea of Okhotsk off the coast of Russia, pipelines are placed in the vicinity of active fault lines, exposing workers to the ongoing threat of explosions and other disruptions caused by earthquakes. This remote territory is also home to brown bears and bandits, so security guards must stand watch. Add to that the more routine hazards of the job: five workers died in a single year on Sakhalin from accidents such as high falls. The oil wells, pipelines, and refineries of Chad, Iraq, and Iran, meanwhile, are vulnerable to terrorist attacks.

But in the minds of oil executives, one risk outweighs all others: cost. In this category, no drilling frontier is more extreme than the world's deepest seas. And no seas are more costly to exploit than the Gulf of Mexico, where the undersea sediment is particularly difficult to map and penetrate and the oil is embedded so deep underground.

Chevron's discovery at Jack was like spring in the autumn of the oil industry. Many of the world's largest fields today—from Saudi Arabia's Ghawar to Alaska's Prudhoe Bay—are dwindling, verging on retirement. Nearly 70 percent of the world's oil comes from fewer than eight hundred fields with an average age of forty years. Of those, only twenty-five are supergiants—a term industry insiders use to refer to fields that contain more than 5 billion barrels. Amazingly, only nine supergiant fields have been discovered worldwide since 1970, according to the International Energy Agency. Despite increasing exploration activity, the rate of supergiant discoveries has been slowing over time: Of these nine, six were discovered in the 1970s, two in the 1980s, and one in the 1990s. (Discoveries in the Gulf's ultradeep waters have not been included in the tally, as the region is still being mapped.)

Add to that the political pressures associated with many petroleum-rich areas: three-quarters of the world's oil reserves are controlled by nations such as Saudi Arabia, Iran, and Nigeria, which have governments that are either at odds with the United States or vulnerable to corruption and conflict. Managing drilling operations in these regions can be dangerous, costly, and unpredictable. Even those oil-producing countries that are U.S.-friendly, such as Brazil and Mexico, require American oil companies to pay high taxes to produce their crude abroad.

Taken together, these constraints on U.S. oil producers make domestic discoveries in the deep-sea Gulf momentous. But there's a hitch—a few, in fact. Just because the oil is there doesn't mean it can be reliably and affordably pumped up from extreme depths and into refineries and gas stations. The deeper and farther offshore you go to find oil, the bigger the technological and financial hurdles: temperatures "down hole" get ever hotter and the pressures more intense, the seas get rougher, and the likelihood of placing the drill in the right location gets more and more remote.

In short, penetrating fields like Jack and Tahiti is as much an exercise in brute mechanical force as it is an act of extremely delicate surgery—or careful artistry, as Siegele sees it—the likes of which my travel companions and I were about to witness.

INTO THE DEEP

From my helicopter window, the Cajun looked like a child's toy—a multicolored Erector Set floating on a buoy. But once we landed and I stepped out into the salty, sunny Gulf air, the rig gave an entirely different impression, awesomely vast and imposing. Looming above us like an elephant above ants was a massive hydraulic drill encased in a 250-foot cage of steel scaffolding. The rest of the hulking industrial curios on the platform looked miniature by comparison. Siegele explained these objects and their functions as he walked me past two huge red cranes; six smokestacks releasing exhaust from the rig's diesel generators; a robotic submarine that oversees drilling activity on the seafloor; mountainous piles of metal pipes used to tap the dormant oil bed; and steel holding tanks for the sediment, mud, and thick black crude that soon would be pulled up from below.

We entered the boxy three-story cement building that houses the dorm rooms and offices. So austere were the surroundings—and so far removed from civilization—that I found myself heartened by the daily, familiar details of a Snickers wrapper crumpled on the floor, a dust bunny underneath a desk, and a family snapshot tacked to an office wall— evidence that people actually do live and work on this floating city.

"It isn't the *Queen Elizabeth*," Siegele told me, "but we've got what we need." The cafeteria was a grim, prisonlike chamber of gray linoleum and stainless steel, supplying a diet of rib-sticking but tasty fare: bratwurst, cheese fries, Frito pie, and twice-baked potatoes were the items piled on my lunch plate, for instance. The living quarters, which taken together house up to 150 workers, are each the size of a walk-in closet, crammed with two cot-sized Murphy beds. I poked my head into one room, finding that it held little trace of its occupants except for a wooden crucifix and a *Sports Illustrated* swimsuit centerfold Scotch-taped to the wall. These are temporary dwellings—most of the occupants work two weeks of each month, going ashore in between.

While the Cajun did have an Internet café, a gym, and a movie theater (starkly furnished venues that look more like conference rooms than recreation areas), these luxuries are rarely used. Few of the men (the rig workers I met were invariably men) have the energy for entertainment after working twelve-hour shifts on the drilling floor. There's not much contact with family on the job: cell phones don't work this far offshore, so workers have only the options of e-mailing (by satellite Internet connection) or calling from a community phone. And while the sapphire ocean views are beautiful, especially when painted with the pale light of dawn and fiesta-colored sunsets, the workers don't indulge in recreational swimming. I found out why when I saw a lone dorsal fin circle the platform—this is shark territory.

But not one Cajun Express worker I spoke to complained about the unforgiving environment. As global demand for oil increases and supplies become scarcer, oil industry profits in recent years have never been higher—and there are generous salaries to show for it. Entry-level tool pushers make about $60,000 a year and high-level geologists and engineers can make in the middle six figures. There's also the guaranteed Rocky Balboa–sized testosterone rush of this type of work: "This is the best big-boy toy you'll ever find," said Chevron's public affairs manager Mickey Driver, patting a railing on the platform. "There's more horsepower beneath this puppy than in all the engines of the Indy 500."

Rising from the concrete floor and up through the bottoms of my boots was a strange and subtly apocalyptic vibration. "The thrusters," said Siegele, noticing my puzzlement. Thrusters, he told me, are gigantic engines at each corner of the platform relentlessly pushing and pulling against the ocean currents. Picture yourself standing in shallow waters at a beach and incessantly shifting your weight to stay balanced as the waves surge and the tides ebb and flow. Thrusters do an extreme version of this in order to keep the rig "on station," meaning within six inches in any direction of the drill's charted entry point into the seabed below. Anchors can't be used to moor drilling vessels at these depths—the motion of the ocean would strain even the strongest of moorings, and rigs need to be able to motor to safety in the event of a hurricane.

The thruster solution is ingenious, but it carries an astonishing energy burden: these 9,500-horsepower engines use a combined total of 27 megawatts of power when running at full capacity—enough to power about twenty-one thousand homes. The generators that power the thrusters and keep the lights on, the electric drill turning, and the computers humming in this village at sea require about 40,000 gallons of diesel per day. It's roughly the amount of fuel that 13,300 Hummers consume in a typical day of driving.

PRESSURE POINTS

You have to burn fossil fuels to harvest them—that's a reality in any drilling scenario—but the ratio of energy invested to energy gained gets slimmer as the drilling conditions get more extreme. (By "energy invested" I'm referring to all fossil fuels used to discover, drill, pump, and refine the oil and transport it to market.) During the glory days of U.S. oil production in the 1930s, an investment of 1 barrel of oil would yield a return of about 100 barrels. By 1970, when oil deposits had become scarcer and more difficult to extract and refine, the ratio had shrunk by more than half: 40 barrels of oil gained for every 1 barrel invested. By 2005, as the industry faced ever-greater limits, the ratio had diminished still further: about 14 to 1. Returns will continue to diminish, some experts argue, until we reach a 1:1 ratio; and that would spell the end of the petroleum era.

As I watched the Cajun in action, I began to understand why extreme drilling conditions can be so treacherous and demanding. It's an expensive fuel-intensive process by itself to grind a drill into the farthest reaches of the earth; it's an even bigger challenge to overcome the inevitable barriers and delays that occur along the way, draining more fuel and resources as the project wears on. That morning, workers on the Cajun Express had begun scraping clean the 5-mile drill hole so that perforating guns could be dropped down to the base of the well. There the guns would be triggered, releasing a spray of buckshot to loosen the sediment and stimulate the flow of oil into the well. If these highly sensitive instruments encounter unexpected obstacles on the way down, they can fire prematurely and this can permanently cripple the well. The well therefore has to be thoroughly cleared first by a tool known as a junk basket—an 8-inch-wide hunk of iron that's forced up and down the entire 5-mile length of the hole, removing loose earth, rocks, and other possible barriers.

Halfway through our visit, Siegele took me to the rig's control room—a small glassed-in chamber that contains a thronelike chair and a desk with a red joystick that operates the drill. I could hear the *clank BOOM clank*

BOOM of the drill's robotic arm sounding rhythmically as it gripped, positioned, and screwed together 90-foot sections of pipe to plunge the junk basket ever deeper into the hole. Minutes later, Siegele got some bad news. "The junk basket is stuck way down there on some debris," reported Ron Byrd, a weather-beaten Chevron employee who has captained Gulf rigs for more than thirty years. Siegele winced almost imperceptibly. "Just a little bump in the road," he muttered when I pressed him for details. Technically, it was a million-dollar bump. The crew would have to spend the next forty-eight hours fishing the jammed cleaning tool out of the hole, halting all other activity on a rig that costs over $500,000 a day to lease, fuel, and operate. But this is chump change to Siegele, with his annual budget of more than $1 billion. "If snags like this didn't happen so frequently, you'd probably let them get to you," Siegele told me, sucking in a breath of salt air. "But you can't do these kinds of wells without stuff breaking—it comes with the territory."

It's one of many hazards that come with the territory. Take, for instance, loop currents. These mighty flows of water propelled by the Gulf Stream can threaten to bend or snap the drill shaft as it plunges toward the seafloor, and have to be vigilantly monitored for any directional shifts. The rig's electrical system is also highly vulnerable—if a fuse blew, the thrusters would seize up, and the drill shaft would have to be severed. Still another challenge is guiding the drill on its optimal course down through 30,000 feet of sediment—a challenge akin to "flying above New York City in a jumbo jet, aiming a baseball at the pitcher's mound in Yankee Stadium, and hitting it dead center," said Siegele. The margin of error as the drill enters the seafloor is only about a meter in any direction. Any farther, and chances go up that you'll hit a fault line or air pocket that will throw the whole well off.

Charting the course of the drill is an implausibly difficult task of its own. "We're pretty much shooting in the dark," said Siegele. Chevron runs its offshore drilling operations out of a gleaming Houston skyscraper that's the shape of twin cylinders, resembling the nose of a double-barreled shotgun aimed skyward. The company devotes billions of dollars annually to mapping out the subsea landscape of its deepwater fields on high-tech equipment at this location, but there's a limit to what these maps can show.

Geologists work in cavernous visualization rooms with floor-to-ceiling monitors and computers that have the processing power of "a PlayStation the size of an eighteen-wheeler," as one engineer described it to me. The computers crunch seismic data that are then translated into maps of ancient sediment. To collect the data, geologists deploy ships that cruise above deep-sea prospects and pop off air guns—underwater cannons that emit gigantic burps of air into the ocean, bouncing sound waves off the underwater rock formations. Aquatic microphones tethered to the vessel record the response.

Gathering seismic data for subsea oilfields in the Gulf of Mexico is far trickier than in other offshore drilling regions. The sediment beneath the Gulf has a salt layer that's as massive and ragged as the Swiss Alps; this layer acts like a fun house mirror for sound waves, deflecting and distorting them in ways that other sediments don't. So Siegele's team had to trigger multiple air guns at once while microphones took hundreds of thousands of recordings simultaneously. The vast constellation of data points enabled Chevron's seismologists to unscramble the salt layer's distortions. Still, the maps were largely inscrutable. "Reading these maps is like looking through a wall of thick glass brick," as one geologist told me, "and trying to count the eyelashes of a person on the other side."

The maps also can't predict how hard it will be to extract the crude. You might think of oil as situated in big pools under layers of rock. But it's actually embedded in the rock, like water in a sponge. "When you drive the drill down you're going into porous rock that can be either kind of squishy or kind of rigid," Siegele explained. Squishy is better, but as rocks age in deeper terrain, they typically become tighter—meaning less productive. You also confront more debris that can clog the well shaft: in other words, instead of sucking up the oil in one big swig like a soft milkshake, it's as though chunks of ice and strawberry get stuck in the straw. That's why, when I visited the Cajun, teams of geologists were standing by to analyze the rocks and mud that got pulled up by the junk basket, hoping to gather a better understanding of the conditions deep below.

Temperature and pressure also pose risks to drilling activities, so engineers must vigilantly scan the computer readouts that monitor these conditions as machinery travels down through the sediment, crossing geological layers that range from hard bedrock to sand to empty voids. The rapid pressure changes between these layers routinely disturb equipment. At the well bottom, there is enough pressure to implode a human head—or more pertinently, to crack iron casings. And the closer you get to the earth's core, the hotter the rocks become. At 20,000 feet below seabed, the oil is hot enough to boil an egg. At 30,000 feet, the oil can reach over 400 degrees Fahrenheit, hot enough to cook off into natural gas or carbon dioxide. Meanwhile, the water at the bottom of the deep sea is at near freezing temperatures, creating a dangerous contrast as the oil is pulled up.

Any one of these factors—loop currents, faulty drill placement, electrical glitches, rock porosity, pressure and temperature changes—could delay operations for days, weeks, even months. At more than half a million dollars a day, the operating costs add up on deep-sea rigs like the Cajun. Hurricanes, too, pose an ominous threat. In 2005, the year of Katrina, Chevron had to carry out seven emergency evacuations. BP's legendary Gulf of Mexico platform Thunder Horse suffered a $250 million blow when a hurricane tore a tiny hole in its hull that eventually sank half the rig, requiring a stem-to-stern reconstruction.

FINAL FRONTIER

Given the challenges that plague ultradeep drilling, it's sobering to think that this frontier holds the oil industry's best hope for finding new petroleum reserves. "The odds are incredibly low that we're going to hit some fabulous new discovery on land," Matthew Simmons, a leading investor and industry analyst, told me. "Everybody's looking to the deep sea for big new finds." To an outsider, it was at once impressive and baffling to watch engineers burrow 5 miles into the earth for oil. "It has all the audacity and technological complexity of launching a space shuttle," as Simmons put it. I found the enterprise doggedly ambitious, but also seemingly desperate—like an addict forcing a syringe into the earth's innermost veins.

Siegele himself admitted that "there's no guarantee that the rewards in this field will outweigh the risks." After my visit, in fact, an even greater snag than the one I'd witnessed occurred on the Tahiti field's production platform: an incorrectly soldered mooring would cause a year-long setback that cost Chevron over a hundred million dollars, by a conservative estimate. But the sunken treasure was worth it: the company proceeded with repairs despite the high cost and began to pump oil from that platform by mid-2009.

One question persisted in my mind: if an energy company is going to throw a billion dollars into something untested and possibly doomed to failure, wouldn't it make more sense to invest in the inexhaustible, greener technologies that will likely replace fossil fuels? None of the Chevron employees I spoke with seemed concerned that their industry may be fast approaching obsolescence. "Do you heat your home? Do you fly on planes? Do you drive a car?" Siegele challenged me. "What do you think makes that heat and moves those jets?"

Siegele was right. Even as innovators have been producing breakthroughs in clean cars, green buildings, and renewable energy and efficiency, the Department of Energy projects that American oil demand will hold steady—not decline—in the decades ahead. American oil demand on the whole has been holding steady in recent years, not declining. And even if America were to slash its oil consumption, industrial growth in China and India is pushing global petroleum demand ever higher. The *New York Times* has reported that the global demand for energy could triple by midcentury. "So long as people need oil," Siegele told me, "we'll find a way to supply it." In other words, the oil industry will go to whatever lengths (literally and otherwise) it must to get oil so long as consumer demand persists and the oil is there for the taking.

But how much oil *is* there for the taking, and how long will supplies last? It depends on who you ask. The moment when the global economy reaches "peak oil" will be the most significant tipping point of the twenty-first century—the point in time when the world's oil producers can no longer increase their supply, and the industry enters "terminal decline." Though "peak oil" is a confusing term, it can be pictured simply as the peak or high point on a graph of production over time. It doesn't mean that we've run out. It means that the world's oil fields will be producing less and less each year. After this peak, the falling-off of oil supplies will in turn bear directly on the basic demand-supply curve of Economics 101: when supply declines and demand stays steady (or rises), prices will rise. A mere 4 percent shortfall in oil production, for instance, could lead to a 177 percent increase in the price of oil.

It's true that oil could stay cheap if demand dropped faster than supply. We saw that happen when the economy slumped in 2008. Industrial activity slowed, curbing the flow of fuel into commercial trucks, bulldozers, cargo trains, buses, airplanes, and ships. For this and other reasons, demand for oil plunged, causing crude prices to fall from an all-time high of $147 per barrel in July 2008 to $40 per barrel in November 2008. Over the long term, however, demand will inevitably outstrip supply as the global population continues to expand and industrial growth trends upward in the developing world.

Peak oil happened long ago within most industrialized countries, including the United States. Our domestic oil production peaked in 1970 after decades of meteoric growth. That's why America—the world's biggest oil producer for most of the twentieth century—now contributes roughly 10 percent of the global supply, and imports about two thirds of its petroleum from overseas. The question remains: when will our suppliers hit *their* peaks?

No one can predict with total certainty the geological limits of the petroleum era. Just look at the range of opinions among top experts. One of the world's leading oil industry analysts, Daniel Yergin, who was awarded a Pulitzer for his book *The Prize: The Epic Quest for Oil, Money & Power,* told me that oil supplies "may reach a plateau . . . perhaps in two to three decades." Meanwhile, former industry executives and geologists such as T. Boone Pickens and Kenneth Deffeyes insist the peak has already occurred, and supplies will only continue to decline from here on out. How can these savvy insiders disagree by thirty years? In part, because the data on global reserves are largely unknown: as much as 90 percent of the world's oil is owned by government-run or privately held oil companies, and they tend to keep as closely guarded secrets information about the size of their reserves. Estimating the total volume of the world's remaining oil involves a great deal of guesswork.

Siegele and his engineers are hardly worried about the long-term threat of oil supply shortages. Technological breakthroughs have, decade after decade, revived the perpetually doomed oil industry: petroleum reserves often seemed too remote or too expensive to exploit over the last century, yet engineers invariably managed to come up with better, cheaper drilling tools. "Predicting peak oil," Siegele told me, "is almost like predicting peak technology"—an exercise that to him seems inherently small-minded, even absurd.

Siegele's comment reminded me of something the fictional oil baron J.R. Ewing said on the TV show *Dallas*. (I confess I watched six seasons of the series back-to-back, justifying a guilty pleasure as "research" during the reporting of this book.) When J.R.'s younger brother Bobby announced plans to start a solar energy company to prepare for a world without oil—this was in the late 1970s, when America was still reeling from the Arab oil embargo—J.R. scoffed: "We've been running out of oil since the day we first drilled it." His clear implication: peak oil is simply a mirage.

In fact, the timing of peak oil may soon become irrelevant. Political and environmental forces could end our oil addiction long before supplies run dry. "The Stone Age did not end for lack of stone," as Sheikh Ahmed Zaki Yamani, a former oil minister to Saudi Arabia, famously said, "and the Oil Age will end long before the world runs out of oil."

The drilling technology itself may become too costly, for one thing. As the supergiant oil fields dwindle, we are getting our supplies from an ever greater number of smaller fields. "We've had to drill more and more wells to keep production steady," Matthew Simmons told me. Much of the current stock of drilling equipment is aging and in need of replacement.

There are, furthermore, external costs to America's oil dependence that could hobble the industry well before oil reserves vanish. The military costs of defending our petroleum interests abroad are tremendous—and growing. Just to keep U.S. forces on the ground worldwide to protect the country's energy supplies costs U.S. taxpayers some $100 billion a year, according to the National Priorities Project, a nonprofit organization that analyzes federal data.

Likewise, the climate impacts of burning fossil fuels are becoming increasingly costly. In the state of California alone, "$2.5 trillion of real estate assets are at risk from extreme weather events, sea level rise and wildfires expected to result from climate change over the course of a century," according to a recent report from the University of California, Berkeley. Add to that the growing impact of warming trends on the state's and the nation's water supplies, road and transportation networks, tourism industries, agriculture, and public health.

In the coming decades, the double whammy of climate change and military spending could sufficiently drive up the price of petroleum to make green alternatives such as electric cars and wind power look increasingly attractive and affordable—pushing petroleum out of the market long before supplies run out.

Siegele brushed aside such concerns, stating matter-of-factly that the game of oil diplomacy has been running for the better part of a century, and the United States has gotten out of military predicaments far thornier than those it faces today. As for global warming, he believes technology will triumph here, too: we'll find a way to scrub carbon from the atmosphere, making fossil fuels climate-friendly.

Just as we can't be certain how much oil is left, it's also too early to predict what new technology breakthroughs will win out in the long run. We can be certain about some things, though. Looking back at history, it is clear that the U.S. oil industry is in a drastically different place today than when it was born just a century ago, when great pools of petroleum bubbled up unbidden from Pennsylvania streams, Oklahoma prairies, and Texas's Golden Triangle. When lucky prospectors could just about pop a straw in the ground and release a gusher. I knew about this golden era only from the mythic heroes it spawned—like Jed Clampett of *The*

Beverly Hillbillies, sitcom's fluky Appalachian mountaineer who struck oil with a stray hunting bullet and made instant millions; Jett Rink, James Dean's character in the movie *Giant*, the down-and-out ranch hand who stepped on a patch of soft Texas soil, saw glistening black fluid pool in his footprint, and became a prodigal oil tycoon; and Daniel Plainview, the character played by Daniel Day-Lewis in 2008's *There Will Be Blood*, who built a petroleum empire from a ramshackle California ranch he'd bought for pocket change.

After my trip to the Cajun Express, I felt the need for more than a fictional knowledge of the glory days of American oil. I would never fully understand the work of Paul Siegele and his fellow petroleum hunters— or the magnitude of the challenges they face—without exploring the roots of their industry and the combination of luck, gumption, and sheer geological abundance that created it. How, I wanted to know, did it come to this—to scenarios as remote and arduous as ultradeep drilling? How did a resource that is now so hard to come by in America become the basis of our economic survival?

BLACK GOLD RUSH

A 60-foot-tall pink granite obelisk fringed by manicured grass and tubs of red geraniums rises up from the town square of Beaumont, Texas, population 110,000. Its inscription reads: "On this spot on the tenth day of the twentieth century, a new era in civilization began." The day was January 10, 1901, and the spot—until then a scrubby knoll known as Spindletop—yielded a mammoth gusher that tripled U.S. oil production overnight. It was because of this one gusher, some historians have argued, that petroleum became the dominant fuel of the twentieth century.

In reality, as I learned from the history books I scoured at my local library, oil prospecting had begun well before the find at Spindletop—in the mid-1800s, when the petroleum by-product kerosene was discovered as an illuminant that burned brighter, cleaner, longer, and more safely than whale oil, and was far cheaper to produce. "It is the light of the age," read an advertisement written in 1860, just months after Edwin Drake tapped America's first oil field in Titusville, Pennsylvania. "Its light is no moonshine, but something nearer to the clear, strong, brilliant light of day, to which darkness is no party."

Instantly, demand for the lantern fuel began to escalate. Droves of zealous prospectors began digging wells and sinking shafts throughout Pennsylvania—wells that came to be known as wildcat wells (and those who dug them as wildcatters) after a term for speculative ventures that originated in the early 1800s. The moniker was particularly apt in this case because of the bobcats and mountain lions that roared around prospectors at night in the dense Pennsylvania woods. It stuck as shorthand for any exploratory well in previously untapped terrain as prospectors expanded their efforts into West Virginia, Ohio, and New York. Public demand for the new lantern fuel soared—not just in the United States but throughout Europe, Russia, and Asia. One man, John D. Rockefeller, presided over those ballooning markets. His mission, as author Daniel Yergin described it, was to deliver "the gift of 'new light' to the world of darkness"—a gift that, for the moment, was supplied exclusively by oil wells in the American Northeast.

But the oil game changed radically in 1901. For years, leading geologists had insisted that no petroleum could be found at Spindletop. Two rogue wildcatters disagreed—Anthony Lucas, a salt miner from Louisiana, and Pattillo Higgins, a local mechanic and amateur geologist. Higgins was the willful son of a gunsmith; at the age of seventeen he lost his left arm in a shootout with a sheriff. A self-starter, Higgins tried running logging and brickmaking businesses before he taught himself petroleum geology from secondhand textbooks and fell in love with oil. Lucas was a Croatian immigrant with a degree in mechanical engineering. Handsome and stout, with a square jaw and deep-set eyes, he mined gold and salt in Colorado and Louisiana, where he developed a theory that salt deposits could be indicators of big oil fields lying deep below.

As early as the eighteenth century, scientists had theorized that petroleum and other fossil fuels (including coal and natural gas) originated, as Russian scientist Mikhail Lomonosov wrote in the mid-1700s, from "tiny bodies of animals buried in the sediments which, under the influence of increased temperature and pressure acting during an unimaginably long period of time, transform." These "tiny bodies" were predominantly of marine plankton: both Lucas and Higgins were ahead of the curve in connecting salt deposits with the contours of former seas and the precious buried residue of marine life.

The idea to drill at Spindletop first came to Higgins when he took his Baptist Sunday school class on a picnic to Spindletop Hill—a bleak plateau thick with rock salt on the edge of the sleepy rice mill town of Beaumont. It was a strange spot for lunch: pools of foul-smelling water and sulfurous gases oozed from the land, and wild bulls were known to wander the hills. But Higgins knew the place from his geology studies and had a trick up his sleeve: he jabbed his cane into the ground, and as the sulfurous vapor escaped he struck a match, creating an instantaneous bloom of fire that dazzled the kids.

Convinced that oil lay beneath the gas seep, Higgins subsequently tried to sink the first exploratory well himself in 1896 but quickly ran out of funds. He ran an ad in the local newspaper seeking investors, and Lucas answered, curious about the location's rock salt terrain. The men formed a partnership and over the following year drilled multiple test wells with no luck, again draining their coffers. Higgins bowed out but Lucas persisted, obtaining funds from Pennsylvania-based oil investors and contributing his own limited savings—at one point even selling all his furniture to keep the project alive.

Lucas commissioned the Hamill brothers, jovial do-it-yourself engineers from Corsicana, Texas, to construct a more powerful drill. Following his guidelines, they jury-rigged the Spindletop derrick from hand-cut lumber and powered it with a boiler fed by firewood. Lucas kept the drilling operation going for nearly a year, grinding down into the earth with this primitive drill 500 feet, then 700, then 1,300. Suddenly he struck the payload.

At 10:30 a.m. on January 10, 1901, the dry earth began to shudder. Mud gurgled up to the mouth of the well, giving way to a thunder-ous blast of gas that launched hundreds of feet of thick steel pipe and hunks of bedrock into the air. Rocks the size of cannonballs rained down. The oil bed had been penetrated, and pressure from its rocks and sur-rounding gases had triggered the explosion. The debris was followed by oil—shooting up over the top of the derrick in a stream 150 feet high. As a crowd gathered to watch, the geyser coated the onlookers with a mist of inky, pungent liquid the consistency of corn syrup: black gold.

The following day, the *Dallas Morning News* reported that the "people of Beaumont, of every sort and condition, are in a feverish state of excite-ment. . . . The throng on the streets appears to be childishly happy and grown men are going about smiling and bowing to each other like school girls." The news quickly spread to the national media. "There is wild excitement throughout Southeast Texas over the oil strike," the *New York Times* proclaimed on the front page of its January 13, 1901, edition. "The well has a flow of over 18,000 barrels every twenty-four hours. It is said to be the greatest oil strike in the history of that industry."

The pressure within the field's oil-bearing rocks was so intense that the well could not be capped for days, and no tools could be found to quell the current. By January 14, a headline in the *Dallas Morning News* read, "Want It Stopped: Reward Offered to Any One Who Will Control the Flow." Reporters estimated that more than 60,000 barrels of oil had gone to waste, despite the fact that every vehicle in town had been mo-bilized to ferry buckets of crude to holding tanks. Fear spread that the petroleum-drenched soil would catch fire.

After many sleepless nights, the Hamill brothers—the same engi-neers whom Lucas had originally hired to drill the well—managed to clamp an iron T-joint and pressure valve to the top of the derrick. The "weary and oil-saturated Hamills still had enough strength and sense of humor left to line up and bow to their audience," wrote Beaumont histo-rian and oil prospector Michel Halbouty.

Halbouty was himself a prime example of the swashbuckling breed of wildcatter that came to be uniquely associated with Texas. A lifelong oil entrepreneur who died in 2004 at the age of ninety-five, Halbouty told me at a 2003 energy conference that the day of Spindletop's discovery "was the day the United States became a world power." At the time, I looked sideways at Halbouty's claim—he was a Beaumont native, after all, and struck me as someone who might exaggerate the impact of his hometown. An exuberant man with snow-white hair and a silver handlebar mustache, Halbouty was the son of a grocer who began his oil career as a boy carting ice water to workers at Spindletop. He went on to discover dozens of oil and gas fields from Texas to Alaska, building a fortune of millions—one of thousands of Americans whose lives were changed by the Beaumont gusher.

Soon after Spindletop was uncorked, investors, wildcatters, and thrill seekers rushed to the Texas fields from places as far away as California, Illinois, and New York. Within months, Beaumont's population had leapt from 9,000 to 50,000, and land prices had surged by a factor of thousands. The town itself had become a forest of wooden derricks shaped like church steeples; iron drill pumps continually lifted and bowed their heads like religious supplicants. The first well drilled (known as the Lucas well) could alone produce half the total U.S. output at the time—as much oil as 37,000 eastern wells, and twice the production of Pennsylvania, the leading oil state.

The discovery had global implications. "Within a year," wrote one awed geologist, "Texas oil was burning in Germany, England, Cuba, Mexico, New York and Philadelphia." By January 1902, 440 gushers had been tapped in the area.

OIL AND TROUBLE

"The frenzy has penetrated every mind and the blood has coursed like fire through the hearts of everyone who owned a bit of [Beaumont] land," reported the *Dallas Morning News* on January 17, 1901. "The mind has been stunned with the possibilities of great wealth and men have been carried further from sound business principles than they care to admit."

Plenty of men struck gold at Spindletop, but many more lost their shirts. Pattillo Higgins, for one, never saw a dime of profit off the well he had discovered and in which he had invested his fortune and sweat. Even Anthony Lucas abandoned his efforts at Spindletop within two years of tapping the gusher; the investors who had funded his efforts had offered him a mere 12 percent ownership of the oil produced, and quickly bought him out.

Like all oil boomtowns, Beaumont fell victim to land speculation, overproduction, and general disrepair. "Now every bubble of gas, every oil-coated pool in all the region is prized as a possible indication of vast oil deposits beneath," reported the *Galveston Daily News* on February 3, 1901. "Consequently there is anxiety to tie up the land with leases. Prospectors are scanning the maps and running about the country in search of owners." One farmer sold his land for $20,000 to a buyer who within minutes had turned it around to another investor for $50,000. At the height of the frenzy, the price of a single acre within Beaumont was as much as $900,000 ($22 million in today's dollars). The disenchanted began calling the area "Swindletop."

Boomtowns such as Beaumont had all the chaos of a carnival, teeming with prostitution, heavy drinking, theft, and public brawls. Drilling was governed by the Rule of Capture, a British law "by which oil was regarded in legal terms like a wild animal—who ever caught it first could keep it," wrote historian Anthony Sampson. The producer or driller was entitled to as much oil as he could extract, and had no obligation to pay community taxes or government royalties.

Growth was haphazard, poorly planned. Trees were clear-cut from hillsides to provide wood for derricks and shack housing, the soil beneath them eroding in avalanches of mud during heavy rains. Relentless mule, wagon, and foot traffic turned dirt roads into impassable swamps. The mire had to be covered with wood planks so that roustabouts wouldn't drown. Here's one author's description of a boomtown known as Oil City that captures the squalor surrounding the wildcatter's existence:

> Oil City, with its one long crooked and bottomless street. Oil City, with its dirty houses, greasy plank sidewalks, and fathomless mud.

Oil City, where horsemen ford [sic] the street in four to five feet of liquid filth, and inhabitants wear knee-boots as part of indoor equipment. Where weary travelers consider themselves blessed if they can secure their claim to six feet of floor for the night. . . .

Air reeks with oil. The mud is oil, the rocks hugged by the narrow street perspire oil. The water shines with rainbow hues of oil. Oil boats, loaded with oil, throng the oily stream, and oil men with oily hands fasten oily ropes around the oily snubbing posts. Oily derricks stand among the houses . . . and the citizens are busy boring in their back yards, in waste lots, or wherever a derrick can be erected.

Beyond the grime and disorder of these conditions, the problem with such a reckless approach to drilling was that it inefficiently depleted underground reserves. The Spindletop field produced 3.5 million barrels of oil in 1901; the following year production soared to 17.5 million barrels; in 1903 it dipped to 8.6 million barrels. By February 1904, production was down to just 10,000 barrels a day, and the supply dried up over the next four years. "The cow was milked too hard," Spindletop founder Anthony Lucas commented during a 1904 Beaumont visit, "and moreover she was not milked intelligently."

After his big discovery, Lucas had patented a long list of devices he'd developed at Spindletop, many of which are used to this day—including a wellhead to contain high-pressure reserves, and specific valve designs and blowout-prevention methods. He went on to serve as a successful consulting engineer throughout Europe, Russia, and Mexico. He focused his later career on devising controlled drilling methods, looking disapprovingly at production binges and the hasty, inefficient exploitation of oil resources.

Nowhere are boom-and-bust economic cycles more dramatic than in the oil industry, with its rapid swings between wild abundance and scarcity. The story of Spindletop bears this out in full color: first the massive field was discovered, then oil fever set in and brought with it overzealous drilling, then the tremendous supply this yielded far outpaced demand, in turn creating a glut in the market that caused oil prices to plummet.

When Spindletop was first discovered in 1901, Lucas sold his crude for $1 a barrel, but by the time the field was producing at its peak volume, overproduction had sunk the price to 3¢ a barrel—cheaper than the drinking water that was carted out to the field's workers. When the oil bed began to dry up a year later and demand caught up to supply, prices then predictably began to soar.

Spindletop had one significant distinction from the majority of active fields of its era that worked both in its economic favor and against it: its oil was a lightweight sulfurous variety of petroleum that was not well suited for using as a lubricant or burning in lamps—the two biggest markets for oil at that time. Spindletop's crude was, however, compatible as a fuel for tanker ships, locomotives, and generators—machines that were mostly coal-fired at the time. It also worked to fuel the automobile—a new European invention that had debuted in the American market just a few years before the Spindletop find.

"It is the most fortunate thing that could have happened in connection with this well—that it is not illuminating oil," J. H. Galey, a part owner of the Lucas well, told the *Dallas Morning News* soon after the discovery. "[T]here will be a good market for fuel oil when the country has had time to adjust itself to using liquid oil. The railroads will use it, every factory will make steam with it, and the steamships can carry much more power in oil than they can in coal." What Spindletop produced, in other words, was a sudden volume of cheap fuel that had nowhere to go but toward transportation. "Spindletop transformed the fuel of light into the fuel of engines," as wildcatter Michel Halbouty told me. Over the subsequent decades, the new cheap engine fuel would set dozens of other mechanized industries in motion, in turn helping to build a young democracy into an industrial and economic powerhouse.

In the meantime, there was much work to do in Beaumont. By 1902, hundreds of oil companies had been chartered to manage the economic risks associated with petroleum production and steer the whirlwind of activity. The building of refineries became just as important as the drilling of wells—crude was worthless until it could be processed. Wildcatters, investors, and refiners brokered casual partnerships—many over whiskeys at local saloons—that in some cases grew into corporations we still know today. Exxon (formerly Humble Oil), Texaco (formerly the Texas Company), and Gulf Oil all had their beginnings in Beaumont. Gulf Oil, for instance, was an outgrowth of Anthony Lucas's legendary first gusher: William Mellon, one of the investors who had backed Lucas and then bought him out, also purchased many other successful wells in the area and established a refinery business, and the sum of these parts became Gulf.

The Texas Company, meanwhile, was founded by Joseph Cullinan, known as "Buckskin Joe" for his tough, unyielding persona. A former employee of Standard Oil's pipeline division, Cullinan raced to the scene of Spindletop after the discovery and soon became one of Beaumont's most successful oilmen. As he snatched up valuable leases in Beaumont, Cullinan also began building storage facilities 20 miles outside of town, giving him a big advantage over the majority of wildcatters who overlooked the need for infrastructure. He also built a pipeline from Texas to Oklahoma, establishing a major artery of southwestern petroleum distribution. By the end of 1901, he had begun consolidating his various oil producing and distributing operations into the Texas Company.

Mellon and Cullinan were among the many independent oil producers who grew out of Spindletop—brash, confident risk takers with keen instincts and the will to follow them. They quickly became formidable economic forces to reckon with—specifically, for Standard Oil to reckon with. By 1902, an estimated $235 million had been invested in the Texas oil boom alone. Standard Oil, the titan of the Northeast, was valued at less than half that: $100 million.

THE FIRST TYCOON

Looming two thousand miles north of the Texas bacchanal, at his Manhattan headquarters at 26 Broadway, was John D. Rockefeller. A pious, bespectacled, and impossibly austere man, Rockefeller had extraordinary reserves of restraint and self-control in an industry with barely a trace of either.

Rockefeller's entrepreneurial instincts took hold in high school, when he dropped out his senior year to study banking, bookkeeping, and law at a professional school near his family's home in Cleveland, Ohio. Three years later, with a partner and $1,000 of his own savings, he set up a company that packaged and distributed regional crops and meats. Just as Rockefeller was growing his young company, Edwin Drake discovered oil less than 150 miles away in Pennsylvania. As he watched demand for the illumination fuel escalate, Rockefeller became convinced of its enormous commercial potential. He acquired his first refinery in 1863, at the age of twenty-four, and had achieved near-total dominion over the production and distribution of oil in the United States by the age of forty.

A compulsively orderly and fastidious man, Rockefeller was never interested in the grimy, frenzied oil-prospecting side of the industry; it was making the end product and selling it to customers that intrigued him. He realized early on that it was much more profitable to let the wildcatters take the risks—let them grapple with the inevitable fluctuations and uncertainties of production—while he maintained tight control of the refining, marketing, and distribution of petroleum goods. Rather than fall victim to the fear and havoc created by boom-bust cycles, he found a way to benefit from them. Cheap oil prices offered him a chance to increase his stake in the market.

I sought out some insight on Rockefeller's business acumen from Daniel Yergin, who invited me to his house. The walls were lined with history books, oil paintings, and tchotchkes from his many visits with industrial leaders of Russia, Asia, and the Middle East. Though Yergin spends much of his time analyzing reams of industry data, he takes a raconteur's approach to the subject of oil, conveying its details as though reciting verses about mythic heroes and their deeds. "Rockefeller *believed* in oil," he told me, clenching a fist for emphasis. "He knew in his gut that the market moved in cycles. So whether prices were high or low, whether there was flood or shortage, he was unfazed by short-term fluc-tuations and relentlessly focused on the future. Any drop in the price of oil was not a reason for despair but an opportunity to *buy*."

Rockefeller's mission was to buy up not just crude but also his com-petitors. Putting the competition out of business and acquiring their re-fineries, pipelines, and delivery fleets, as Rockefeller saw it, generated a positive feedback loop: the greater his control over refining and distri-bution, the more he could control the price of oil, therefore the better his position to topple more of the competition and amass a still-larger share of the market. When Rockefeller founded Standard Oil in 1870, there were some 250 refinery operations in the Northeast, Yergin ex-plained, and by 1880 there were just a handful—and more than 85 per-cent of that market was under his control. Rockefeller's reach extended far beyond refining: he aimed to commandeer all stages of the flow of oil, from the moment it surged from a derrick to its processing, packaging, and transportation to store shelves and gas pumps.

Standard Oil quickly became the most recognizable brand-name product in America. "That notion of this light fuel being *standard*—being reliable, consistent, and safe wherever you used it—was really a novel concept at the time," Yergin explained. "Rockefeller created a national product in a country that had never had national products. We now think of brand names as a part of our lives, but they didn't really exist until the Standard brand emerged."

For all his success in spreading light and unlocking the power of the hydrocarbon, Rockefeller exhibited very little of the wildcatter's exuber-ance for his product. He was as obsessed with controlling his words and emotions as he was with controlling volatile markets. A devout Baptist and lifelong Sunday school teacher, Rockefeller and his wife never drank, rarely socialized, and raised their children with the belief that pastimes such as square dancing were frivolous if not depraved.

"Do not many of us who fail to achieve big things," he once conjec-tured, "fail because we lack concentration, the art of concentrating the mind on the thing to be done at the proper time and to the exclusion of everything else?" Rockefeller made it a rule to speak rarely, if ever, in meetings—letting his staff and competitors do the talking. "A man of words and not deeds is like a garden full of weeds," went one of his credos. Another was summed up in a curious nursery rhyme he kept on his desk:

A wise old owl lived in an oak
The more he listened the less he spoke
The less he spoke the more he heard
Why aren't we all like that old bird?

One Standard Oil staff member said of his dour boss, "He is the most unemotional man I have ever known." An industry competitor commented, "I guess he's 140 years old, for he must have been 100 years old when he was born." But behind his expressionless veneer, the titan wrestled with anxiety: "I am eating celery which I understand to be very good for nervous difficulty," he once wrote to his mother. A fitness fanatic, Rockefeller kept in his office an unusual rubber-and-wood exercise machine that he used daily to unwind.

He had good reason to be anxious. First, there was the mystery of the earth's hidden resources—the extent of their abundance was beyond even Rockefeller's control. As early as the 1920s—a decade that saw tremendous growth in U.S. oil production—geologists were predicting that U.S. petroleum supplies were about to run out. "The question at the heart of the oil struggle has always been: how much does the earth have to offer?" Yergin told me. "Skepticism over crude supplies is a recurrent malady of the business—it is in the first decade of the twenty-first century, just as it was in 1920 and after World War II and again in the 1970s."

Rockefeller also had a deeper, ethical angst to contend with: he famously used cutthroat tactics to increase his control over domestic and international markets. He undercut prices so drastically that competitors had to shut down or sell out to Standard Oil. He commandeered the market for the barrels used to distribute oil so his competitors would have no way of packaging their product for customers. He negotiated an alliance with the railroads that transported oil between refineries and markets to get special, clandestine rates far below those afforded his competitors. He reportedly planted spies within competitors' backroom negotiations to get advance warning on any new deals in the making.

Rockefeller expressed no outward remorse over crushing smaller companies in his pursuit of control. Instead, he celebrated it: "The American Beauty rose," he famously reasoned in 1905, "can be produced in all its splendor only by sacrificing the early buds that grow up around it." He described his tactics not as predatory but rather as an act of public service—eliminating what he considered to be the waste and inefficiency endemic to free markets, and ensuring reliable flows of oil and profits that were insulated from excessive volatility, overproduction, and periods of scarcity. Rockefeller believed that he had a divine duty to impose order on the chaos of the oil markets, and "that the Almighty had buried the oil in the earth for a purpose," wrote Ron Chernow in his Rockefeller biography, *Titan*. The executive was known to have calmed his employees' anxieties about market volatility with the assurance "The Lord will provide."

For all his ruthlessness and eccentricity, Rockefeller was arguably good for America in the way that oil itself was good for America—both brought tremendous power and versatility to the American economy. Never before had a commodity been so widely distributed, so reliable, so profitable, and so meticulously managed. Rockefeller's methods of vertical integration, price stabilization, efficiency, and economies of scale to this day inform the business strategies of the world's most successful companies.

"What makes him problematic—and why he continues to inspire such ambivalent reactions—is that his good side was every bit as good as his bad side was bad," wrote Chernow. Rockefeller also contributed roughly $550 million (the equivalent of about $12 billion today) in charitable donations to support education, medical research, and other philanthropic causes. "Seldom has history produced such a contradictory figure."

The media and the public of his day were primarily concerned with

Rockefeller's bad side. While the federal government took its time in cracking the whip on Standard Oil's more abusive tactics, the media flew into an investigative frenzy in the early 1900s. Leading the pack were journalists known as muckrakers—watchdogs who sought to expose behind-the-scenes misconduct within government and industry. Ida Tarbell, the daughter of a Pennsylvania wildcatter, spent more than six years bird-dogging Standard Oil, gaining the trust of its executives, plumbing its accounting books, and for the first time publicly exposing its inner workings. Her articles were published in a series of cover features in *McClure's Magazine* from 1902 to 1904 and were then compiled into a tell-all bestseller, *The History of the Standard Oil Company*.

What is most noteworthy about Tarbell's writings, beyond their shrewd analysis and detailed evidence of business corruption, are the more intimate passages in which the author paints a portrait of Rockefeller himself—a man she described as "the victim of a money-passion which blinds him to every other consideration in life":

> To know every detail of the oil trade, to be able to reach at any moment its remotest point, to control even its weakest factor—this was John D. Rockefeller's ideal of doing business. It seemed to be an intellectual necessity for him to be able to direct the course of any particular gallon of oil from the moment it gushed from the earth until it went into the lamp of a housewife. There must be nothing—*nothing* in his great machine he did not know to be working right.

Tarbell saw Rockefeller—his greed, obsession with order, and chokehold on the competition—as the embodiment of all that was wrong with America, whereas independent wildcatters (like her father) embodied all that was right. "Life ran swift and ruddy and joyous in these men," she wrote adoringly. "There was nothing too good for them, nothing they did not hope and dare." It was a blatantly biased judgment, but nevertheless it voiced the opinion of many Americans: while Rockefeller represented caution and distrust, the wildcatter represented boldness and hope.

Yet Rockefeller and the wildcatters also shared strong commonalities. Above all, they shared the insatiable lust for more—the feeling that the industry must stop at nothing to find ever more distant, deeper, greater pockets of reserves. William Mellon, the Gulf Oil executive who invested in Spindletop, explained the common thread this way: "For a great many . . . the oil business was more like an epic card game, in which the excitement was worth more than great stacks of chips. None of us was disposed to stop, take his money out of the wells, and go home. Each well, whether successful or unsuccessful, provided stimulus to drill another."

BUSTED

The wildcatters and Standard Oil were on a collision course, and the battle between them would be fought in the halls of Congress and the Supreme Court. This struggle would transform the diverse, chaotic oil business into the modern contours we know today as Chevron, Exxon, and other industry leaders. Howls of protest from independent oil companies against Standard Oil grew louder as the behemoth grew in might. The independents argued that Standard was blocking competition and controlling prices with cutthroat tactics that forced them to sell out. President Theodore Roosevelt allied himself with these independent companies, publicly denouncing Rockefeller and his colleagues as morally corrupt: "The methods by which the Standard Oil people . . . have achieved great fortunes," he argued, "can only be justified by the advocacy of a system of morality which can also justify every form of . . . violence, corruption and fraud, from murder to bribery and ballot-box stuffing in politics."

Roosevelt's administration launched an investigation into Standard Oil's business practices on the grounds that it was violating the Sherman Antitrust Act—a federal statute outlawing corporate monopolies or trusts. A trust is a megacompany that aggregates an array of smaller companies, controlling them under a single leadership. The law, still in force today, forbids such conglomerates on the grounds that they can stifle competition, innovation, and economic progress, hurting small businesses and disadvantaging consumers.

This conflict between Capitol Hill and Standard Oil trust escalated into a public spectacle that was equal parts Watergate and *The Magnificent Seven*, and the public lined up for ringside seats. "[It] became a kind of morality play for any individual who felt threatened by the new industrial order; a kind of commercial equivalent to the Western, as the last glimpse of a heroic age," wrote historian Anthony Sampson. "The trust-busters saw themselves as defending the very core of democracy."

Rockefeller was characteristically unfazed by the investigation, and cast the charges against his company as a challenge to the nation's economic structure as a whole: "It must in good time be perceived by all that the centralized corporation is a necessity of progress. There has been substantial basis for popular suspicion . . . but it is poor logic to find against the whole idea of corporations because of these few failures."

The Justice Department sued Standard Oil for antitrust violations in 1909 and, after 444 witnesses gave testimony, the company was found guilty; Rockefeller and his executives were slapped with the maximum penalty. The gray eminence himself was playing golf when news of the decision reached him. He read the messengered letter, slid it into his pocket, and said stonily, "Well, gentlemen, shall we proceed?" When his colleagues pressed him, he disclosed the damage: a $29 million fine ($638 million in today's dollars). Standard appealed the decision, and the case went to the Supreme Court. On a sunny day in May 1911, the Supreme Court upheld the earlier verdict and ordered the monopoly to splinter into more than a dozen independent companies.

"This was an extraordinary task," Yergin explained to me. "Here you had a company that spanned every state in the Union, operated in multiple countries, refined more than three-quarters of America's oil, exported four-fifths of its kerosene, sold the railroads nearly all of their lubricating oil, and had a massive transportation business. How are you going to divide that up?" Standard's executives decided to split up regionally. The largest of its subsidiaries was Standard Oil of New Jersey, which acquired Spindletop-born Humble Oil and later became Exxon. Standard Oil of California later became Chevron, subsequently merging with Buckskin Joe's Texaco. Standard Oil of New York later became Mobil. Standard Oil of Ohio later became part of BP. Standard Oil of Indiana became Amoco. These subsidiaries eventually comprised six of the legendary "Seven Sisters" that dominated global oil production throughout the twentieth century.

As these spin-off companies competed to produce oil products better, faster, more cheaply, and more efficiently, they pushed each other to innovate. Mobil, for instance, developed a breakthrough method for refining oil into gasoline, which increased by 45 percent the amount of product it could get for every barrel of crude. By 1917, the profits of the spin-off companies had collectively shot up to double, then triple the profits of Standard Oil before the dissolution. Rockefeller, in turn, had nearly quadrupled his wealth and become the richest man in the world.

FATEFUL PLUNGE

An even more transformative event was looming on the near horizon. World War I would have a profound effect on the growth and influence of the young oil industry. The war presented America's oil executives with an opportunity to gain protection, stability, and security by expanding their influence into politics.

This was the first major war of the Industrial Age—a "war that was fought between men and machines," as Yergin put it. Close to 13 million people lost their lives. Never before had petroleum-powered battleships and tanks been on the front lines of battlefields. They replaced the horses, trains, and slower ships that had moved troops in wars past. The defining decision for this mechanized war—a decision often credited with the Allies' victory—was in fact made in 1911, the same year as the trust bust, before the war began. Winston Churchill, serving in the admiralty, made the difficult choice to transform his entire naval fleet from coal-powered engines to oil, a new and riskier fuel but one that promised the great strategic benefits of faster ships and more efficient use of manpower. This momentous decision—he termed it the "fateful plunge"—ultimately helped clinch the Allies' victory, as their ships outmaneuvered the coal-fired fleet of the German Reich.

Oil also won big victories on land. Yergin told me the story of the "Paris taxi armada": Germans were amassing troops in the hills to the north of Paris in 1914, far outnumbering the Allied soldiers who were dispersed to the south. There was no train system to transport the Allies northward, and traveling on foot would never get them to the front in time to repel the assault on Paris. French general Joseph Gallieni had an idea: in a matter of hours he rounded up six hundred Parisian taxicabs and commissioned them to shuttle six thousand French reserve infantry troops to the battlefield—paying each driver his standard meter fare after a swift victory.

Military successes such as these amounted to dollar signs in the eyes of American oil executives, whose fields and refineries were at the time supplying the vast majority of the Allies' oil. In 1915, when President Woodrow Wilson was making plans to join England and France in the war, he summoned to a strategy meeting the chief executives of America's top oil companies, asking what they would need from the government to ensure that a cheap and abundant supply of oil flowed from U.S. oil wells to European battlefields. This represented a U-turn in government-industry relations: not five years after the dissolution of Standard Oil's trust, its new chief executives were in Washington deciding how they could cooperate—amongst themselves, and with their federal disciplinarians.

Wilson created a Petroleum War Service Committee and tapped A. C. Bedford, president of Standard Oil of New Jersey (later Exxon), to lead it. Bedford, a cutthroat executive who had worked at Standard Oil for more than thirty-five years, saw an opportunity that would extend well into the postwar future. "He and his committee of fellow oil company presidents proposed that supplies be pooled and efforts coordinated to produce all possible oil," wrote historian Carl Solberg. "Overnight, the industry, with President Wilson's support, was doing what the Sherman Antitrust Act forbade."

It didn't stop there. Industry leaders made a plea for a substantial tax break wrapped in the battle flag of patriotism. Bedford made the case that "if America were to have enough oil to win the war, Congress must create greater inducements to produce the stuff," Solberg wrote. A so-called depletion allowance was established, eventually granting a 27 percent tax deduction for any investment made by U.S. petroleum companies to discover new oil.

As the war came to an end in late 1918, oil industry leaders formed the American Petroleum Institute, and installed Bedford as president. This was the nation's first organization designed to protect the industry's interests and guide its influence in public policy making. At the top of the Institute's agenda: extending the wartime tax breaks and opening western public lands to prospectors.

In the postwar years, U.S. oil companies advanced steadily into overseas markets. Already the industry had established itself as America's first multinational enterprise—remember Rockefeller's initial mission to bring "new light" to the world's darkest corners—but World War I accelerated that trend. Thanks to the celebrated success of trucks, automobiles, and oil-powered ships during the war, the Age of Mechanization shifted into high gear. The mass production of gas-powered vehicles for the war (more than 160,000 were constructed for Allied troops) dramatically reduced the cost of these conveniences for mainstream consumers in Europe and at home.

The Standard spin-offs quickly expanded their operations throughout Western Europe, Russia, and Asia. Soon they had more foreign customers than domestic. The United States was, in today's terms, the Saudi Arabia of the world, a country whose vast plains, deserts, and ranch lands supplied the majority of the world's oil—and would continue to do so for several decades.

"After the Great War," Yergin told me, "there was a surge of consumption as automobiles began to crowd city streets and a new lifestyle based on cheap oil took hold. That's when Americans began to perceive low-cost travel, cheap electricity, and affordable heating almost as a right conferred by our democracy." In 1923, when English writers Davenport and Cooke visited the United States during its postwar oil heyday, they reported:

> Travel but a little in the country and you will gain the impression that the modernism of the United States flowed from its oil wells. . . . The oil-tank car is as ubiquitous on his railroads as the coal-truck on ours. The oil-tin litters his waste places. His wayside is dotted with the petrol pump and at night illuminated oil "filling stations" make his streets more beautiful. A network of oil pipe-lines underlies his country, more extensive than the network of railways overlying ours. . . . Does not the American partly live in oil? Certainly he cannot move without it. Every tenth man owns an automobile, and the rest are saving up to buy one.

The cozy relationship that had been established between Washington and Big Oil during World War I grew more complicated in its aftermath when questionable public-private ties were exposed. As early as 1910, Congress gave the president authority to set aside strategic oil fields in California and Wyoming for emergency use in the event that military oil supplies ran low. This provision would lead to one of the more notorious moments in American political history—the Teapot Dome scandal.

In 1921, Albert Fall, President Warren Harding's secretary of the interior, who oversaw millions of acres of public land, was caught taking bribe money from oilmen. Fall, who had worked as a chuck wagon cook, a hard-rock miner, a rancher, and a lawyer before he won political leadership, had secretly leased more than 100,000 acres of land located inside of the military oil reserves in eastern Wyoming (named Teapot Dome after a boulder in the shape of a teapot that overlooked the oil field) to prominent eastern oil prospectors. He had done so in exchange for personal gifts and loans totaling more than $5 million in today's dollars.

When Fall was investigated, Congress was hard pressed to find a lawyer to represent him who didn't have a conflict of interest with the oil industry. "Just think; America has 110 million population, 90 percent of them lawyers," remarked the popular humorist Will Rogers, "yet we can't find two of them who have not worked at some time or another for an oil company. There has been at least one lawyer for each barrel that ever came out of the ground."

In response to the scandal, rumblings arose about the oil industry's influence inside Congress itself: "[I can] name a dozen senators spotted with this flood of oil," said Senator George Moses, a New Hampshire Republican. "Those yet unnamed are greater in number."

GREASING PALMS

The love affair between the petroleum industry and politics endured long after Albert Fall was imprisoned and replaced with a more responsible executor of public land. In the 1940s, for instance, eighteen oil executives were mysteriously forgiven sizable federal penalties for manipulating the price of petroleum distribution over their pipelines; according to some reports at the time, they dodged their fines by bribing members of the Democratic Party with contributions totaling about $13 million in today's dollars.

That same decade, as the oil industry was beginning to explore offshore oil fields, debate raged over who owned the shallow coastal areas known as the tidelands. One side argued that drilling in this region should be controlled through tight federal restrictions to preserve the oil for military use and protect the nation's fisheries. The other argued that it should be managed by states and more loosely (and favorably) leased to private companies. In 1945, President Harry Truman declared the tidelands federal land, infuriating the industry, which battled on to reverse this decision. During his 1952 presidential campaign, Dwight D. Eisenhower, who received large contributions from the oil industry, pledged to support state control of the tidelands. He honored the promise soon after he was elected. "Eisenhower's smashing personal victory," wrote historian Robert Engler, "was quickly followed by a fulsome display of gratitude to its oil supporters."

Indeed, at the presidential level, even the most respected administrations of the twentieth century—*especially* the most respected administrations, including Franklin D. Roosevelt's, John F. Kennedy's, and Ronald Reagan's—have brokered and nurtured relationships with oil producers both domestic and international that openly abetted America's appetite for oil. "The trouble with the country is that you can't win an election without the oil bloc," Roosevelt famously said, "and you can't govern with it."

In the last century, the federal government leased tens of millions of acres of public lands for oil drilling, upholding a philosophy that affordable, free-flowing energy was a matter of public well-being—a resource that functioned as a public service as much as it did a private commodity. For close to a hundred years, this resource has also been seen as the linchpin of America's national security.

America's relationship to petroleum was irrevocably altered in 1970 when the country hit peak oil. From that point forward the U.S. could no longer hope to provide this public service entirely for itself, but had to buy ever-increasing amounts from outside its borders—putting our national security increasingly into the hands of political allies and enemies alike. In the 1970s America went from being the world's premier source of oil to being a net importer of oil. The shock of the Arab oil embargo in 1973 exposed this vulnerability for the first time to the American public. The embargo—along with rising concern in the 1970s over pollution caused by drilling and burning crude—sparked widespread public criticism of the marriage of oil and politics in America.

The industry has adapted. Today the American Petroleum Institute is one of the most powerful business lobbies in U.S. politics, representing a membership of more than four hundred companies and investing between $3 million and $5 million annually in recent years on lobbying efforts. Most notably, the API has led the fight to open up the Arctic National Wildlife Refuge (ANWR) and the Outer Continental Shelf to oil drilling, opposed federal requirements for greenhouse gas reduction, and fought to secure and prolong industry-specific tax breaks.

In 2001, President George W. Bush selected as his chief of staff for the White House Council on Environmental Quality a former API lobbyist, Philip A. Cooney. Cooney, who had been the climate team leader at API, came under fire in 2005 when the *New York Times* reported that he had been altering government reports to raise doubts about the effects of greenhouse gas emissions on global warming. Days after the *Times* published its report, Cooney resigned. He went on to work at ExxonMobil.

No administration has been so overt in its ties to the petroleum industry as the George W. Bush administration, which employed more than fifty Oval Office staff members who had previously worked for oil companies, named the CEOs of the nation's biggest oil companies to the task force that shaped the administration's energy and war policies, and opened up a record amount of public land to development, including vast areas of the Gulf of Mexico, home to Chevron's Cajun Express. Though I was a critic of Bush's energy policies, after my study of history I began to see that the administration's collaborations with Big Oil was a fairly predictable (which is not to say optimal) extension of political patterns that had been in place throughout the entire twentieth century.

For all that has shifted in America's relationship to oil over the last century, what is most surprising is how much of the basic dynamics of petroleum policy and prospecting have stayed the same. At the core, the industry has been fueled by an abiding optimism: "Oil drilling isn't for the pessimist," Michel Halbouty said, "or even for the realist. You've got to be an optimist. You've got to believe no matter how many dry holes you drill, the next one is going to hit."

In one of my favorite scenes from the TV show *Dallas*, J.R. Ewing describes his passion for oil to his disenchanted wife Sue Ellen:

J.R.: There's nothing realer than oil, that's for sure.
Sue Ellen: Not to you darlin', except perhaps money.
J.R.: Same thing, honey, same thing.

Today, America's obsession with drilling well after deeper well has eclipsed our ability to scale back demand by developing more enduring energy sources that could take the place of oil. This obsession has, in essence, taken us back to the roots of our domestic oil production. As technology evolved over the decades between the 1920s and 1950s, drillers continued to grind down to ever greater and more pressurized depths—from 1,000 feet to 2,500 feet and then 8,000 feet—unleashing deeper pockets of reserves. In recent decades, prospectors have begun double-dipping in already exploited wells—even in some surrounding Spindletop. They are now venturing as low as 40,000 feet in the hope of opening up still deeper treasures.

INDEPENDENCE DAY

Against the softness of the interminable blue seas, the Cajun Express with its landscape of iron, steel, and cement was the unmistakable mark of human enterprise, appearing as improbable—as unnatural—as a rose garden in the Mojave Desert, a Hyatt Hotel in Antarctica, or a flag planted on the moon. Here, it seemed, was another wilderness conquered.

At the end of my visit, Paul Siegele took me, via a jury-rigged elevator (a coffin-sized plastic box attached to a forklift), to the crown of the rig, a harrowing widow's walk suspended at the top of the drill's 250-foot scaffolding. The body of the rig below looked like the loneliest place on earth—a tiny, solitary circuit board floating in a boundless blue sea. Then, out in the distance, I spotted fleets of trawlers the size of thumbnails setting off seismic guns in search of the next big deep-sea prospect. "A decade ago, I never even dreamed we'd get here," marveled Siegele. "And a decade from now, this moonscape could be populated with rigs as far as the eye can see."

Though the image to me was jarring, Siegele's scenario does seem increasingly likely. Both opponents and proponents of domestic drilling in areas such as the Gulf of Mexico and ANWR share a common conviction that it is necessary to free the United States from dependence on foreign oil. Of the 85 million barrels of petroleum consumed daily in the world, America consumes 21 million—nearly a quarter. Our net imports of petroleum are about 12 million barrels a day. Even as we invent ever smarter, more efficient buildings, appliances, and cars, and even as we develop cleaner, renewable energy sources, the transition from this prodigious oil usage to a new energy landscape will be gradual.

By any measure, America is in no position to drill its way to energy independence. Our proven domestic reserves stand at 21 billion barrels—enough, at our current levels of consumption, to meet our needs for roughly 1,000 days if we stopped importing any oil. There are another 697 million barrels in the Strategic Petroleum Reserve (an emergency fuel stockpile the Department of Energy maintains in underground caverns along the Gulf Coast, to be used in the event of a sudden shortage or spike in oil prices); but that would only give us another 34 days of supply. Allowing drilling in the long-protected areas of the outer continental shelf could potentially expand reserves by 18 billion barrels—giving us at best another 860 days of supply. ANWR has an estimated 7.7 billion barrels—another 372 days. In total (and ignoring the time needed to tap and test new wells, and our limited refining capabilities), these new frontiers would give us fewer than 2,500 days of supplies—less than seven years.

Today, America is still indulging energy-lavish habits that it formed more than eighty years ago, in the domestic oil boom following World War I. And now, as the nation's homegrown oil supplies become ever harder to come by, we are faced with tremendous costs—not just of drilling for oil at the ends of the earth, but of protecting our access to cheap and abundant supplies from overseas, particularly from the Persian Gulf. How much do we taxpayers actually pay, I wondered, to conduct U.S. diplomacy in petroleum-rich nations? What are the moral, economic, and political costs of relying on foreign producers to feed America's oil addiction? And what, on a more practical level, does it take to fuel the military charged with protecting these supplies?

My men can eat their waist belts, but my tanks need gas.

—General George S. Patton

2

War and Grease

HOW OIL BUILT AND SUSTAINS
A MILITARY SUPERPOWER

In the summer of 2006, Marine Corps Major General Richard Zilmer sent the Pentagon an unusual "Priority 1" request for emergency battle-field supplies. Stationed at a temporary base in Fallujah, Zilmer was commanding a force of 30,000 troops responsible for protecting Al Anbar, the vast territory in western Iraq bordering Saudi Arabia, Jordan, and Syria. Heavily armed insurgents were hammering the region, and Al Qaeda was quickly gathering recruits. Zilmer's beleaguered soldiers were running low on fuel for the diesel generators powering their barracks—fuel that cooled their tents in the 115-degree weather, refrigerated and cooked their food, and kept the communication lines open. The general, however, was wary of trucking in backup supplies during a time of so much turmoil. The U.S. fuel convoys that chugged along the back roads of Iraq every day—long lines of eighteen-wheelers hauling armored vats of gas—were among the insurgents' prime targets.

This had been a growing problem since the Pentagon had begun staging its 2003 invasion of Iraq. In one particularly brutal incident, on April 9, 2004—a year to the day after Saddam Hussein's statue was toppled in Baghdad's Firdos Square—an American convoy of twenty-six vehicles delivering fuel to Baghdad International Airport came under fire. The mile-long convoy was transporting roughly 125,000 gallons of jet fuel to support U.S. military operations. At approximately 10:30 a.m., a large group of insurgents hidden in the grass alongside the Abu Ghraib Expressway launched its assault. Homemade bombs planted along the road began to detonate, blasting shrapnel into the sides of the trucks. Insurgents stormed the convoy with a fusillade of machine gun fire so relentless that the siding of some supply vehicles gave in. The trucks "looked like water-sprinkler systems wetting down the pavement, which was slick with the oily diesel fuel," wrote Thomas Hamill, a convoy driver who survived the assaults. "The trucks slid through like hogs on ice." Amid the chaos, insurgents launched rocket-propelled grenades and several trucks exploded, shooting flames 200 feet into the air. Truck drivers had to abandon their vehicles and escape by foot. A prolonged gun battle ensued. Only a handful of the fuel tankers made it to safety—the others were left burned and leaking on the highway, their valuable cargo looted by insurgents and civilians.

KBR (formerly Kellogg Brown & Root), then a Halliburton subsidiary and one of the companies contracted to oversee fuel supplies distribution for the U.S. military, was forced to suspend convoys around Baghdad in the days thereafter. But there were bigger losses. Two U.S. soldiers and six truck drivers had died in the ambush. Another soldier—Army Pfc. Keith Matthew Maupin, age twenty, of Batavia, Ohio—was taken hostage and later executed. Thomas Hamill, a KBR employee, was also kidnapped, but managed to flee his captors after twenty-four days. "We had a duty to deliver fuel to our troops," Hamill recounted in his memoir *Escape in Iraq*. "There was no time to worry about what might happen or about things beyond our control." In total, at least nine lives were claimed as a result of the ambush.

There were many more such attacks: between the summer of 2005 and the time of Zilmer's memo in 2006, there were some 280 reported attacks on supply convoys. In the twelve months that followed, there were over 850. Many though not all of these ambushes targeted fuel convoys. (Fuel and water represent 70 percent of all cargo carried by supply convoys in Iraq.) The insurgents were carrying out a century-old battle strategy: starve an army of fuel, and it will be immobilized. Fuel is the lifeblood of mechanized warfare. Each day in Iraq, the U.S. military uses a staggering 1.5 million gallons of fuel to power the tanks, fighter jets, Black Hawks, Humvees, hospitals, and base camps on the front lines of war. Energy supplies on the battlefield give American soldiers a huge advantage in communications, agility, and firepower. But the loss of life that April morning was a grim reminder of the hidden costs of fueling combat.

Major General Zilmer's memo, dated July 25, 2006, presented the Pentagon with an unprecedented request: "a self-sustainable energy solution," including "solar panels and wind turbines." This was the first time a frontline commander had formally requested renewable energy backup in battle. Without alternative power sources, the memo continued, U.S. forces "will remain unnecessarily exposed" and will "continue to accrue preventable . . . serious and grave casualties." Put in civilian-speak: too many of Zilmer's troops were dying in fuel convoys, and the relentless gasoline demands of the diesel generators were partly to blame.

"By reducing the need for [petroleum] at our outlying bases, we can decrease the frequency of logistics convoys on the road, thereby reducing the danger to our marines, soldiers, and sailors," the request stated. "Without this solution, personnel loss rates are likely to continue at their current rate. Continued casualty accumulation exhibits potential to jeopardize mission success." Renewable energy was not an environmental consideration for Zilmer, it was a tactical necessity—a matter of life and death, of victory or defeat. A matter of national security.

GREENER BERETS

The Pentagon is the largest consumer of petroleum in the United States. In recent years it has used between 130 million and 145 million barrels of oil annually—comprising 2 percent of America's total petroleum demand. That translates to nearly 400,000 barrels per day, roughly the total daily energy consumption of the United Arab Emirates. Over the last century, no institution has done more to propel America's rise to power than our military—or consumed more oil in the process. We have petroleum to thank for building the Department of Defense into an as-yet-unmatched fighting machine—but our troops are only as powerful as the flow of fuel that sustains them.

"And herein lies the dilemma. Oil makes this country strong; dependency makes us weak," noted Michael Klare, a professor at Hampshire College who wrote the books *Resource Wars* and *Blood and Oil*.

I was baffled and hopeful when I read about Zilmer's memo in a September 2006 issue of *USA Today*. Here was a no-nonsense Marine Corps general who has served more than thirty years in the U.S. military (not your typical tree-hugger) stationed in a country that's virtually floating on an ocean of oil (Iraq has the world's third-largest oil reserves, after Iran and Saudi Arabia) demanding clean energy solutions that only a few years earlier had been regarded as rinky-dink hippie technology suitable only for yurts and Earthships. Zilmer's plea struck me as a clear harbinger of change in America's attitudes about energy. If there was ever an opportunity to "man up" the effete image and role of solar panels, wind power, and other fossil fuel alternatives, this was it. Just think of what the Pentagon could do to fast-track alternative-energy innovations going forward—after all, it was military R & D that led to the invention of jet airplanes, helicopters, radar, remote-control mechanisms, cell phones, global positioning systems (GPS), microchips, and the Internet.

But for all the promise it augured, Zilmer's memo also carried overtones of despair that spoke to the massive challenges that come with fueling the military—one more oil-dependent today than ever before in history. The newspaper story left me wondering: How did the American military get so hooked on petroleum? How much does it really cost—in both blood and treasure—to fuel war? What would it take to transform the world's biggest and strongest military into a petroleum-free enterprise? And how did this become the primary concern of a man leading 30,000 troops?

The current imbroglio in Iraq, I would learn as I researched twentieth-century military history, is by no means the first war tied to oil. It's the culmination of decades of foreign policy and international relations that have been deeply connected to, and shaped by, petroleum. For the better part of a century, oil has not just been fueling our military equipment and shaping our battle strategies; it has also been provoking the very wars in which these machines and tactics are deployed.

OILS OF WAR

While the "fateful plunge" from coal to petroleum undoubtedly gave the Allies a leg up in World War I, by World War II oil had become more than an advantage—it was a tactical necessity. To get a full-color account of this first fully oil-dependent war, I visited David Painter, a professor of history at Georgetown University and one of the leading experts on America's energy diplomacy. Reference books consumed almost every inch of space in Painter's office—filling the floor-to-ceiling bookshelves and rising in stacks like stalagmites from the floor. The only section of visible wall in the room held a print of a Christo sketch of half a million stacked oil drums, titled *America: The Third Century*. Painter wore round horn-rimmed glasses and a necktie slightly askew, and had the introspective patience of a man who's spent a lifetime searching the past for clues to understand the present.

Painter explained that World War II was a high-tech, heavy-artillery operation, one in which Allied forces used 7 billion barrels of oil—many times more fuel than they did in World War I, when machinery was mostly powered by coal. Oil is the most convenient form of energy—it generates 40 percent more thermal capacity than coal, which means it can take vehicles greater distances at higher speeds. The major weapons systems used in World War II—long-range bombers and other aircraft, aircraft carriers, surface warships, submarines, tanks, and trucks—were fueled by oil, and most had been produced in American factories. Moreover, the United States had become the main source for the 100-octane fuel and specialty lubricants that improved the speed, power, and reliability of the most sophisticated aircraft engines. "The high-performance engines climbed faster, flew higher, and got better mileage—huge tactical advantages," said Painter. American citizens, for their part, understood the importance of oil in fueling this war, and were asked, as a patriotic duty, to carpool, ration gasoline, ban auto racing, and observe no-driving days.

What surprised me more than the sheer volume of oil consumed in World War II battles were Painter's stories of how the capture and control of oil had been, for the first time in history, a motivation for war. When the United States entered World War II, it produced two-thirds of the world's petroleum. Nearly all the oil—6 billion of the 7 billion gallons—that fueled the Allied war effort came from U.S. fields. Only the Soviet Union, among the other great powers, had any significant oil production, while Britain and France were short on domestic oil and dependent on foreign suppliers. "Oil was known as the 'master resource,'" Painter explained, "in the sense that it enables you to do so many things: energy for mining, for agriculture, for manufacturing, for home heating, for transportation—not just for direct military use."

The Germans and Japanese also had extremely limited domestic petroleum reserves and had been shut out of the major foreign oil-producing areas, leaving both nations highly vulnerable. Nazi leader Adolf Hitler understood this vulnerability—how could he build an autonomous Third Reich if it relied on foreign countries to fuel its industries? So Hitler advanced a technology that the United States is now exploring seventy-five years later: he pushed the development of synthetic fuels from coal (a resource abundant inside German borders) shortly after taking power in 1933. By the outbreak of World War II, coal-derived "synfuels" accounted for nearly half of Germany's oil needs. But the process of deriving fuel from coal was complicated and expensive, and it required huge installations of steel that could be spotted by surveillance planes and became easy targets for Allied bombers.

Germany relied mainly on the Soviet Union and Romania for its oil supplies, but Hitler quickly realized this wouldn't be enough to fuel his military machine and sustain his long-term vision of the Third Reich. He began eyeing as a potential fuel source the oil-rich fields in Russia's Caucasus Mountains. This was one of his primary motivations for Operation Barbarossa, the invasion of the Soviet Union in June 1941. When Hitler's intent became clear, Russian troops destroyed their oil fields, refining equipment, and pipelines to prevent their resources from being tapped—one of several desperate measures ordered by Joseph Stalin. "Not one step back!" Stalin told his army. "The execution of this task means the preservation of our country, the destruction of our enemy, and a guarantee of victory."

Like Germany, Japan was heavily motivated by oil. The "master resource" was a driving factor in Japan's infamous attack on U.S. forces at the Hawaiian base of Pearl Harbor. By the end of the 1930s, Japan depended on the United States to provide the vast majority of its oil needs. Even after Japan announced its "Axis" alliance with Germany and Italy, the United States continued to supply it with petroleum. (As internal documents would later reveal, Roosevelt did not want to do anything that Japan might interpret as provocative, such as banning oil, for fear that this would trigger an attack before the United States was prepared to respond.) When Japan suddenly tried to import massive amounts of 100-octane aviation fuel, Roosevelt and the State Department limited oil exports to Japan to 86 octane. "If we stopped all oil, it would simply drive the Japanese down to the Dutch East Indies," Roosevelt wrote to his secretary of state, who was urging an oil ban, "and it would mean war in the Pacific."

But Japan was already conspiring to capture its own fuel source in order to fulfill its nationalistic ambitions, which were much like Germany's. "Japan hoped to conquer all of Southeast Asia, including the oil fields of the East Indies, and open the shipping lanes between those fields and Japan," Painter told me, gesturing at a worn World War II–era map. The bombing of Pearl Harbor was an attempt to immobilize the U.S. Pacific fleet so it could not cut off Japanese forces when they moved south on the East Indies. Japanese planes launched the first bomb at 7:55 a.m. One eyewitness, U.S. Navy Commander Hubert Gano, would later recall, "Ships were sunk and burning everywhere. [Battleships] had tried to escape but ran aground at the entrance to the harbor. Entire squadrons of our airplanes were destroyed on the ground. I saw aircraft engines in puddles of molten aluminum." A dozen battleships were sunk and hundreds of U.S. planes went down. Within hours, 2,335 American soldiers and 68 civilians had been killed.

Rather than disable the United States, the attack only quickened its resolve to lead the Allies to victory—and to starve the Axis nations of essential fuel. As the Allies systematically destroyed German and Japanese tankers and trains delivering oil, and sabotaged their fuel supply routes, the Axis powers were eventually forced to design programs and battle tactics around gas shortages. Hitler's *blitzkrieg* (lightning war) attack style was a case in point: the idea was to strike quickly and forcefully before fuel supply problems could arise. The Japanese, meanwhile, were driven to pursue a desperate "pine root campaign" that brandished the slogan "Two hundred pine roots will keep a plane in the air for an hour." Civilians were frantically exhorted to dig up pine roots in the hope that the vegetation could be fermented to produce an alternative to oil. Though pine root fuel production reached 70,000 barrels per month, the entire operation was ultimately a bust—the refining process was riddled with problems, and Japanese planes could barely get off the ground with the botched alternative fuel.

"Ravenous for oil, Japan was facing defeat," Painter told me. "It was a nation running on empty." In part because of petroleum shortages, Japan resorted to its tragic strategy of kamikaze attacks, whereby suicide pilots crashed their planes into the decks of Allied ships. Such an attack was portrayed as an act of glory—a sacrifice for the good of the Japanese nation—but it also served a horribly practical purpose for a country lacking oil: if the pilots were going only one way, they would need just half the fuel.

German general Erwin Rommel summed up the desperate challenge of fueling war in a letter to his wife written as the tide of battle turned against his forces in North Africa: "Shortage of petrol!" the stoic general wrote. "It's enough to make one weep."

In a sense, General Zilmer's appeal for relief from fossil fuels is reminiscent of Rommel's lament. While the challenges the two generals faced were technically different—Zilmer was hamstrung by a flawed fuel-delivery system and Rommel by a supply shortage—both felt the tactical vulnerability that comes with relying on oil. Just as Germany and Japan relied on their foes Russia and America for fuel before World War II, now America relies on volatile nations—most notably in the Middle East—to fuel our daily demands and to sustain our military presence in this oil-rich region. It is a dependence that had its origins in those final years of World War II.

THE KINGDOM

As I tried to gain a clearer understanding of the relationship between oil and war—specifically, of the military tactics David Painter had described —one book I turned to was Sun Tzu's *The Art of War,* among the first books on military strategy and still, twenty-five hundred years after it was written, one of the most influential. Judging by the due dates stamped in the margins of the various weathered versions housed at my neighborhood library, it still remains a popular book even in the unlikely locale of Nashville, Tennessee. One sentence struck me as particularly resonant, probably because I didn't expect to find anything like it in a book that's ostensibly a how-to of combat: "In peace prepare for war; in war prepare for peace."

That phrase could be said to sum up the relationship dynamics between the United States and Saudi Arabia, I realized as I thought through my conversation with Painter. For decades America has been carefully nurturing its friendship with the Saudis while simultaneously girding for conflict with Saudi-funded groups.

The origins of the stormy U.S.-Saudi friendship can be traced back to February 14, 1945. President Franklin Delano Roosevelt had spent years shepherding the United States through the bloodiest conflict in world history. In his preparations for peace toward the end of World War II, Roosevelt made the desert kingdom on the Arabian Peninsula a top priority.

Immediately after the Yalta Conference of February 4–11, 1945, at which the Allies established guidelines for their now-certain victory in Europe, Roosevelt set sail aboard USS *Quincy* for a destination on the Suez Canal where he'd arranged to meet Abdul Aziz Ibn Saud, king of Saudi Arabia. This was to be one of the most fascinating and pivotal meetings of the twentieth century.

Roosevelt wanted to establish a collaborative relationship with the king. As early as 1943 he had proclaimed in a letter to his secretary of state that the "defense of Saudi Arabia is vital to the defense of the United States."

That same year, the American geologist Everette Lee DeGolyer had visited Saudi Arabia to assess the region's oil reserves: He concluded that "the center of gravity of the world's oil production is moving from the Gulf of Mexico–Caribbean area, to the Middle East . . . and will continue to shift until it's firmly established in that area." He estimated Middle Eastern reserves to be as much as 300 billion barrels of oil (they have since proven to be more than twice that). "Those numbers made clear," said Painter, "that the Middle East would be a region of tremendous geopolitical importance."

Ibn Saud set off on the two-day journey to the Great Bitter Lake of the Suez Canal from his home in Jiddah on February 12. Having never before left his country, the king did so at great risk: in his tribal culture, his throne could be easily usurped if he wasn't there to defend it. But the opportunity this meeting represented was too good to pass up. It was the first time a leader of the West would meet with a leader of the nation known simply as "the kingdom."

As recently as a few years earlier, the high priority granted this meeting would have been unimaginable. For all but the last decade of its history, the Arabian Peninsula had been one of the world's poorest regions. Its past had been marred by conflict: desert tribes warred with one another throughout the Middle Ages, and then faced the onslaught of the Ottoman Empire when it laid claim to the peninsula in 1514. From the eighteenth century on, a fierce tribal resistance tied to a conservative Islamic movement known as Wahhabiya grew in opposition to foreign rule.

King Ibn Saud was a descendant of one of the first Wahhabis. He was born in the Arabian desert, but as a boy he was exiled with his family to Kuwait by a rival group known as the Rashidis. In 1901, then only in his twenties, Ibn Saud led a bold and ultimately successful assault on the Rashidis. A tall, stately man, known as a magnetic leader, Ibn Saud came to be revered for his successes in battle. By 1925, after nearly continuous fighting, he had laid claim to much of the peninsula, including the religious sites of Mecca and Medina. By 1932, the kingdom of Saudi Arabia was born, with Ibn Saud as its leader.

Ibn Saud had high hopes for his new nation, despite its barren soil and forbidding climate. Around the time Saudi Arabia was formed, a vast oil field had been discovered in neighboring Bahrain, which had a strikingly similar geological composition. Saud's years of exile in Kuwait had given him a clear eye for geopolitics: at the time, Turkey, Britain, Russia, and Germany had various interests in and negotiations with Kuwait over trade routes, warm-water ports, and railroads. But Saud was wary of opening his country up to oil prospecting. American representatives of Socal (the Standard Oil Company of California, a spin-off of Rockefeller's empire) met with the king's representatives—who were "hard-headed, smart, patient, tenacious, wary . . . bargainers worthy of anyone's steel," as author Wallace Stegner later described them. A deal was reached in 1933, after extensive negotiations, granting Socal exclusive oil rights.

But early wildcatting turned up nothing. After years of sinking exploratory well after exploratory well in vain, Socal finally hit pay dirt in 1938. Following the discovery, other U.S. companies—Texaco, Mobil, and Exxon—poured into the region to join Socal, and collectively became known as the Arabian American Oil Company (Aramco). The first oil tanker left Saudi shores in 1939, within months of the war's outbreak. By 1945, the nation was producing 21 million barrels per year, and there was abundant proof that the kingdom—which today is known to hold fully a fifth of the world's petroleum reserves—was richly endowed with oil.

ARABIAN KNIGHT

Ibn Saud arrived for his voyage to the Suez meeting in full regal splendor—flowing robes and headdress, a retinue of Bedouin bodyguards, a private physician, the royal astrologer, ceremonial coffee servers, cooks, porters, slaves, and a flock of forty sheep. Arrangements had been made for the king to sail north aboard the battleship USS *Murphy*. His initial entourage numbered more than 140 people, but the *Murphy* had room for only 47.

When American soldiers offered the king the commodore's cabin, he refused. Accustomed to the vast Arabian desert, the king cringed at the idea of sleeping in a cramped, four-walled cabin. So Persian rugs were spread across the ship's deck, a royal tent was pitched, and the king's favorite armchair was placed at the bow next to a jury-rigged sheep pen. Saud had declined to eat the American meats aboard the ship since they had been preserved in a refrigerator—he didn't trust the technology and thought it unsanitary to eat animals that had been dead for more than twenty-four hours. (The king did, however, try an apple pie à la mode and was so taken with it that he had American apple trees planted in Saudi soil upon his return.) He insisted on feeding a royal supper of freshly killed mutton to everybody aboard the ship, including the lowest-ranking American soldiers. At prayer times, the ship's navigator would give the king the exact location of Mecca by compass; once the king had confirmed this reading with his astrologer, he would bow toward the holy city and lead his retinue in prayer.

When USS *Murphy* finally approached USS *Quincy*, with the dramatic spread of rugs, tents, and robed guests visible on its bow, Roosevelt kept marveling to his staff, "This is fascinating. Absolutely fascinating!"

The president, who was a chain-smoker, quickly stabbed out a cigarette before welcoming the king onboard, as traditional Wahhabi principles prohibit tobacco smoking.

The meeting took place on Valentine's Day. Wearing his customary cloak over his shoulders, the American president brought no entourage of his own, with the exception of a translator. Despite the clash of cultures, the two leaders became fast friends. King Saud offered Roosevelt complete Arabian wardrobes for his family. Roosevelt, whose legs had been paralyzed in his late thirties by a bout with polio, candidly discussed his ailment with the king, a similarly larger-than-life man who had a lame leg among other injuries from his earlier years in battle. At the end of the visit, FDR gave Saud a spontaneous gift—a state-of-the-art wheelchair, Western technology that had not yet made its way east. The president also offered his Arab ally a Douglas two-engine airplane and a supply of penicillin—more newfangled innovations that impressed the king.

William Eddy, a U.S. Marine Corps colonel who spoke Arabic and served as a translator during the meeting, described the rapport between the two leaders in his memoir: "The King and the President got along famously together. Among many passages of pleasant conversation I shall choose the King's statement to the President that the two of them really were twins: (1) they were both of the same age . . . ; (2) they were both heads of states with grave responsibilities to defend, protect and feed their people; (3) they were both at heart farmers . . . (4) they both bore in their bodies grave physical infirmities." In the aftermath of the gift-giving, the king said, "This [wheel]chair is my most precious possession. It is the gift of my great and good friend, President Roosevelt, on whom Allah has had mercy."

It was during the *Quincy* rendezvous, amid exotic meals, astrology readings, and high-tech gifts, that the leaders of the West and East established the foundations of a potent and high-stakes relationship: America would provide military support to keep the royal dynasty in power. King Ibn Saud would, in turn, continue to offer Americans privileged access to his kingdom's oil.

The issue of oil was allegedly not the centerpiece of the discussion between Saud and Roosevelt. In the absence of a transcript, aides have recalled that Roosevelt primarily stressed the urgent matter—following on the Holocaust—of resettling European Jews in Palestine. He asked for the king's support in this effort, but Saud retorted that Arabs should not have to redress the sins of Adolf Hitler, and made a remark that has since proven ominously prescient: "Arabs would choose to die rather than yield their land to Jews."

Even though the two leaders could not find agreement on the question of Zionism, Ibn Saud made a formal request for FDR's friendship and support. This was the kind of simple and direct question often used to seal alliances between tribal leaders. Saud said that he especially prized America's friendship because the United States was the only global power that had never made any attempt to colonize or enslave another country—it was, instead, a champion of freedom. The king said he valued nothing more than Saudi Arabia's independence. He had cited a similar reasoning when, just before the war, the negotiations that granted Socal exclusive oil rights had concluded. Though Socal paid a stiff price, Saud claimed he could have gotten far more from British and Japanese prospectors yearning for the bid. "Gentlemen, the Japanese offered me twice as much for one-third of what you now obtain!" he'd told the American entrepreneurs, adding that no amount of money was more important to him than the assurance that his American partners would not interfere in his domestic affairs.

FDR accepted Ibn Saud's request for friendship, and vowed that under his leadership, the United States "would never do anything which might prove hostile to the Arabs." This was one of Roosevelt's final acts as president—he died of a brain hemorrhage six weeks after the encounter.

Roosevelt's promise would be reaffirmed again and again to the Saudi royal family by subsequent American presidents. In 1950, President Harry Truman wrote to Ibn Saud: "I wish to renew to Your Majesty the assurances which have been made to you several times in the past, that the United States is interested in the preservation of the independence and territorial integrity of Saudi Arabia. No threat to your Kingdom could occur which would not be a matter of immediate concern to the United States."

No other scene in modern history intrigues me more than the meeting between Roosevelt and Ibn Saud. All the presidential administrations that have followed Roosevelt's—including the Truman, Eisenhower, Nixon, Carter, and Reagan administrations—sought to build on this precedent and further cement U.S. relations with the Saudis. They worked to ensure that the oil so crucial to our domestic prosperity and military victories would flow freely from the region's wells.

The George H. W. Bush administration cited the Roosevelt–Ibn Saud meeting as part of its rationale for launching Operation Desert Storm. "We do, of course, have historic ties to the governments in the region," said Dick Cheney, then defense secretary—ties that "hark back with respect to Saudi Arabia to 1945, when President Franklin Delano Roosevelt met with King Abdul Aziz on the USS *Quincy* . . . and affirmed at that time that the United States had a lasting and a continuing interest in the security of the Kingdom." The security of the kingdom, after all, had become tantamount to the security of the increasingly oil-dependent United States.

Twelve years later, Cheney would again voice his intention to protect Saudi security as he made his argument for invading Iraq and removing Saddam Hussein from power. At a Veterans of Foreign Wars convention in August 2002, Cheney said that with access to weapons of mass destruction, Saddam "could then be expected to seek domination of the entire Middle East, take control of a great portion of the world's energy supplies, [and] directly threaten America's friends throughout the region."

LIFELINE

The urgent memo sent by Major General Richard Zilmer from the fields of Fallujah in 2006 was not, in fact, the first document addressing concerns inside the Pentagon about the military's dependence on fossil fuels. I was surprised when a quick online search led me to a 130-page report prepared by the Defense Science Board (DSB), the military's most prestigious technical advisory committee. This 2001 document, titled *More Capable Warfighting Through Reduced Fuel Burden*, made detailed recommendations for ramping up the fuel efficiency of military equipment, facilities, and overall strategy. The report exposed Department of Defense (DoD) negligence on this issue: "Although significant warfighting, logistics and cost benefits occur when weapons systems are made more fuel-efficient, these benefits are not valued or emphasized in the DoD requirements and acquisition processes." The Pentagon, it stated, had also failed to evaluate the full cost of the fuel it was purchasing for its military operations: "The DoD currently prices fuel based on the wholesale refinery price and does not include the cost of delivery. . . ." That cost of delivery, it later added, could be more than 100 times greater than the cost of the fuel itself.

One little-known agency called the Defense Energy Support Center (DESC) coordinates all of the DoD's energy purchases, ensuring that the fuel sustaining the U.S. military is available at all times, wherever needed—including in fields of combat. I traveled to the DESC with two goals in mind: to better understand the equation of fuel with military power (an equation that had both motivated and been sealed by that first meeting aboard the *Quincy*) and to better grasp the hidden costs of our continued operations in the Middle East.

Tucked inside the bucolic Fort Belvoir army base in Fairfax County, Virginia, the DESC headquarters is located in the new Andrew T. McNamara Headquarters Complex. The sprawling six-story complex looked like an upscale suburban strip mall, with a freshly paved parking lot the size of an eighteen-hole golf course, immaculate landscaping, a brick-and-steel façade, and expansive tinted windows. The center also houses the Defense Threat Reduction Agency, the Defense Logistics Agency, and several other divisions whose exact functions were hard to discern, but all, I could see, meant business.

I reflexively stiffened my back as I entered the building, with its walls of brushed steel and blond wood, black lacquered granite floors, and a rigorous security detail. The man who was screened before me, in his mid-forties, wearing a crisp black suit and mirrored aviators, surrendered a thick black leather suitcase that looked borrowed from the set of *Mission: Impossible* along with two firearms, one from an ankle holster.

I was led silently down a long corridor to the office of Colonel Shawn Walsh, the DESC official who oversees bulk fuel contracts for the war on terror. A direct and serious man in his mid-forties, Walsh was sporting camouflage fatigues and a crew cut. He came to the DESC after serving in 2003 in Iraq, where he led the 1,200 soldiers of the 240th Quartermaster Battalion in missions that included securing fuel convoys. His office was spartan and unadorned, with the exception of a framed collage given to him as a parting gift from his battalion. Featured there were snapshots of armored fuel convoys snaking like anacondas along dusty roads under cover of night, wide sunrise vistas of Iraqi deserts, and grinning compatriots at a Hawaiian-themed party on the eve of Walsh's departure from Iraq.

Walsh began our conversation with a basic primer on the Pentagon's fuel supply chain in the Middle East: "We use just over a million and a half gallons of fuel a day," he said. I tried to imagine a 1.7-million-gallon body of liquid—that's enough fuel to fill about four Olympic-sized swimming pools. "It adds up to roughly 12 million gallons a week, 50 million gallons a month, and—with the supply trucks carrying about 10,000 gallons of fuel apiece—a heck of a lot of truckloads."

Surprisingly, none of this fuel Colonel Walsh delivered actually came from Iraq's copious oil fields, nor could it be transported by pipeline. "The Iraqi infrastructure is not to a point where we can do that," Walsh explained. Despite the tremendous volume of oil lying beneath Iraqi soil, the country's drilling equipment, refining facilities, and pipelines have been devastated—first by the Iran-Iraq war, then by United Nations sanctions imposed in 1990 after Iraq's invasion of Kuwait (sanctions which proscribed open-market fuel purchases from Saddam Hussein's regime), and finally by the combat and chaos sparked by the U.S.-led invasion in 2003. Intelligence officials had predicted prior to the invasion that Iraqi fuel production would increase nearly 30 percent "within several months of the end of hostilities," and that this would then aid in fueling and funding the occupation, but in fact just the opposite has happened: Iraqi oil production fell by more than a third after the 2003 invasion (a supply plunge that contributed to the concurrent surge in global prices). By 2008, Iraq could barely meet its own internal demand for refined fuels, let alone serve the needs of the U.S. military.

So Walsh and his logistics officials sourced bulk purchases of fuel from refineries in four neighboring countries: Kuwait, Pakistan, Turkey, and Jordan. The fuel comes in several different varieties for aviation, ground vehicles, and diesel generators. To facilitate fuel distribution, Walsh's battalion spent the first six months of 2003 laying a tactical pipeline from the supply countries into Iraq, but eventually abandoned the effort: "It was an aboveground pipeline and very vulnerable to attack, to sabotage," Walsh explained. His soldiers had to guard the pipeline twenty-four hours a day (wearing infrared goggles at night) to fend off thieves who would try to puncture the pipeline and siphon off fuel or, worse, bomb it. "The pipeline also proved to be an obstacle in the desert because we couldn't maneuver vehicles over it. So we scrapped it for both security and strategic reasons and moved to a system in which all fuel distribution into Iraq is made by trucks."

The distribution is organized via a hub and spoke system. The DESC contracts with truck drivers to transport fuel from several major refineries in the neighboring countries to central hubs inside Iraq, known as "bag farms" for the collapsible fabric bags in which the fuel is stored. These bags, which look like gigantic pillows, can hold anywhere from 10,000 to more than 210,000 gallons of fuel each, and can be easily disguised to prevent looting and avoid aerial attacks. Local units then help transport the fuel from the bag farms to their base camps—the "spokes" of the distribution system.

Some seven hundred supply trucks circulate on the roads of Iraq each day carrying everything from Lucky Charms and personal mail to bulletproof vests, fuel, and ammunition. Civilian truck drivers—not soldiers—drive the vehicles that transport fuel between the refineries and hubs, while military personnel man the choppers and Humvees that surround the convoys for protection.

KBR, the biggest employer of civilian truck drivers in Iraq, won multibillion-dollar contracts to repair Iraq's infrastructure and provide fuel delivery in the early stages of the war. A Defense Department investigation later uncovered evidence that the company had significantly overcharged taxpayers by as much as $61 million for oil distribution to troops; KBR contested the findings, but nevertheless, the Pentagon reportedly terminated the company's fuel deliveries in Iraq.

The job of driving supply trucks is undoubtedly dangerous, but the money has drawn many to sign on. KBR drivers who made roughly $30,000 a year in the U.S. working in truck fleets for Home Depot, SYSCO, and Walmart could get paid between $80,000 and $120,000 a year for shepherding fuel convoys in Iraq, and were eligible for a handsome tax break if they stayed in the field more than 330 days. Houston resident Stephen Heering took the assignment in 2004 of driving fuel trucks in Iraq for Halliburton (then KBR's parent company) to help dig his family out of debt and build a nest egg for his young son's college education. Thirty-three years old at the time, he described himself as being tired of living paycheck to paycheck; convoy driving promised financial security. But four months into his job on the roads of Iraq, Heering quit after rebels ransacked his truck and threatened him at gunpoint. "[KBR] said it would get better, but people started getting hurt bad," Heering said in a *Time* interview. "They'll find new meat. I guess that's the way it is in the money world. If it makes 'em money, they don't care if it costs them a life."

In 2005, a Halliburton convoy of four fuel trucks suffered an ambush in which three drivers were killed. Preston Wheeler, of Mena, Arkansas, captured the attack on a video that later became a YouTube sensation: "Truck 5 cannot move, please help me, I am taking fire," Wheeler pleads in the background. "I am fixin' to get killed, goddammit . . . I have no gun. I am by myself." Wheeler was shot and lost some mobility in his right arm. Soon thereafter, he lost his job at KBR. "They don't no more care about me," Wheeler later said in a television appearance, "than they care about a dog walking on the road."

In addition to the threat of ambushes, convoy drivers face frequent breakdowns caused by the merciless heat, engine-clogging sand, and the unforgiving terrain of pocked dirt roads. Civilian truckers wear body armor but are not permitted to carry firearms. The convoys are required to run their routes even in the worst conditions of weather and unrest—after all, the more challenging the conditions get, the more supplies the soldiers need to fend off attacks, stay cool, and keep the hospitals running. Not surprisingly, the military has struggled with a shortage of drivers. "We got trucks," Army sergeant Frank Vallejo told the *Baltimore Sun*; "what we don't have is drivers, and we are scraping below the bottom of the barrel."

When I questioned Colonel Walsh about the long-term viability of the fuel supply chain for the war in Iraq, he stressed the "unwavering commitment" within the military to keeping it running smoothly. "It's as important to the survival of our soldiers as food itself—more so, even." He told the famous World War II story of General Patton's Third Army: when it was running low on fuel at the Meuse River during its race across France into Germany, having outrun its fuel supplies, the general ordered his men to drive until they ran out of gas and "then get out and walk." "My men can eat their waist belts," Patton told his friend and commanding general, Dwight D. Eisenhower, "but my tanks need gas." Tragically, Patton couldn't get new fuel supplies fast enough to penetrate Berlin. (This gave the Germans time to regroup, and it was the Russians who later took the city.) The general called this failure the "unforgiving minute"—the window of time in which he would have made his place in history had he been allocated the fuel.

Having recalled this story, Walsh summed up his team's work securing fuel convoys in Iraq as "mission critical." There's no getting around the fact that "we are a hydrocarbon military," he added. "Everything runs on oil."

EMBARGOED

In the decades following World War II, in the hope of maintaining access to cheap and abundant oil supplies, the United States began offering ever bigger favors to its allies in the Middle East. It began to furnish the Saudis (and at times Iran and Iraq as well) with military aid, including state-of-the-art weaponry and combat planes, private security forces, and military training to protect American-friendly regimes. Despite these efforts, the United States eventually got a bitter taste of what it feels like to be oil-starved.

I have no memory of the Arab oil embargo; I was born in 1974, the year after it began. But when I asked my neighbor how it affected her family, she recalled bicycling to work so that there was enough gas in the car to get her kids to school. She also vividly remembered the shock of seeing lines for the first time at the gas pump. As the embargo crippled U.S. oil supplies, stations began limiting drivers to 10 gallons of fuel at each fill-up and shut off their pumps on Sundays. "To many Americans," commented one journalist, "it was impossible to understand how their standard of living was now being held hostage to obscure border clashes in strange parts of the world."

In 1973, Saudi Arabia and some of its partners in the Organization of Petroleum Exporting Countries (OPEC) ordered a total embargo on oil exports to the United States. The reason for the embargo was precisely the conflict of interest that had been outlined decades earlier on the *Quincy*: the United States could not reconcile its strong alliances with both the Jewish and Muslim worlds. Israel's longtime foes Egypt and Syria had launched a surprise attack on two fronts in Israel, the latest in an ongoing series of conflicts since Israel proclaimed statehood in 1948. It was the afternoon of October 6, 1973—Yom Kippur, the most sacred Jewish holiday. Israel was badly in need of aid, and the United States was faced with the hard decision of choosing between its allies. Nixon launched Operation Nickel Grass to secretly airlift supplies to Israel. "Send everything that can fly," Nixon told his secretary of state Henry

Kissinger. The supplies—more than 22,000 tons—were supposed to be airlifted under the cover of night, but a glitch caused some planes to arrive during the day for all to see. The Muslim world was infuriated by what it perceived as a blatant show of U.S. support for its enemy. King Faisal of Saudi Arabia (a son of Ibn Saud, as have been all the kingdom's rulers) had previously warned the United States that "America's complete support for Zionism and against the Arabs makes it extremely difficult for us to continue to supply the United States with oil, or even to remain friends with the United States." As a punishment, OPEC's Middle Eastern members enacted the embargo on October 17.

The economic fallout that ensued transformed OPEC into "a household word—not just an obscure acronym" throughout the United States, noted one article, "and for the first time turned Western attention to the distant lands they unknowingly relied on." Within months, the price of oil had jumped from under $3 a barrel to almost $12 a barrel. In turn, consumer prices soared, along with unemployment and inflation. Nixon's hands were tied—there was the option of forcibly seizing Middle Eastern oil fields, but it would have been all but impossible to generate popular support for such tactics in the wake of the Vietnam war.

The government reluctantly made a necessary—but in ways distinctly *un-American*—move: it required consumers to curb their energy consumption. Though fuel conservation had been viewed as a patriotic sacrifice during World War II, Nixon's action was a radical measure given the energy-lavish habits of the postwar boom decades. The president instructed home owners to dial back their thermostats, and companies shortened their working hours. Though the embargo was ended in March 1974, its ripple effects on world oil prices and inflation would continue to be felt throughout the decade.

During the late 1970s, the Carter administration set forth a doctrine that very clearly harked back to FDR's original promise to King Saud. President Carter vowed that the United States would from that time on take the lead in defending the Gulf. Access to Persian Gulf oil is a vital national interest, he acknowledged, and to protect that interest the United States would be prepared to use "any means necessary, including military force." This liberal president was spelling out in no uncertain terms the intimate ties between U.S. foreign policy and petroleum.

But Carter also tried to tackle the oil problem from a different, demand-side angle, offering a prophetic address to the American public calling for a substantial reduction in U.S. oil demands:

> Tonight I want to have an unpleasant talk with you about a problem unprecedented in our history. With the exception of preventing war, this is the greatest challenge our country will face during our lifetimes. The energy crisis . . . is a problem we will not solve in the next few years, and it is likely to get progressively worse through the rest of this century. . . . By acting now, we can control our future instead of letting the future control us.
>
> Two days from now, I will present my energy proposals to the Congress. . . . Many of these proposals will be unpopular. Some will cause you to put up with inconveniences and to make sacrifices. The most important thing about these proposals is that the alternative may be a national catastrophe. Further delay can affect our strength and our power as a nation. . . . This difficult effort will be the "moral equivalent of war"—except that we will be uniting our efforts to build and not destroy.

Carter was right about many things, including the fact that his proposals were unpopular from the get-go. He developed efficiency standards for appliances and buildings. He created a well-funded R & D program to promote the development of solar power, wind turbines, and electric cars—programs that led to the first commercial applications of these technologies. He installed solar panels on the roof of the White House. He kept the thermostat low and gave his presidential addresses in a trademark wool cardigan that came to represent his persistent pleas to the American public to voluntarily curb energy use.

During the Carter era, America managed to reduce its daily petroleum consumption by 18 percent, thanks in part to strict new efficiency standards for cars. This achievement was soon to be reversed.

By 1981, when President Reagan took office, the embargo was a fading memory; oil supplies were bountiful, and prices were sinking back to levels not seen for a decade. Reagan ordered the solar panels removed from the White House roof. Celebrating the restoration of cheap oil, Reagan promptly cut the funding for Carter's alternative energy program and eased fuel economy standards. In doing so, he encouraged a trend of upward-climbing energy demand in America that deepened our dependence on the oil-rich Middle East. It also intensified U.S. military vigilance in the region, as any kind of volatility there could spell another oil shock.

STORMY DESERT

Trouble bubbled up again in the Middle East in the 1980s as combative neighbors Iran and Iraq fought out a long-running territorial skirmish in a costly and inconclusive war. At the war's end, the regime of Saddam Hussein was heavily in debt and the country's petroleum infrastructure was in disrepair. Saddam, who had assumed the presidency in 1979, saw an opportunity to erase his debt by seizing the thriving oil fields of neighboring Kuwait. Holding nearly 10 percent of global oil reserves, tiny Kuwait was on a per capita basis one of the wealthiest nations in the world. On August 2, 1990, Saddam Hussein's forces invaded and rapidly succeeded in occupying Kuwait.

The United States responded with what was soon dubbed Operation Desert Storm—the first Gulf War. In January 1991, the George H. W. Bush administration declared its intention of removing Iraq from Kuwait—in part to protect the Gulf oil fields. When Dick Cheney, who was than the secretary of defense, described the motivation for war, he cited the Carter doctrine: "that basic fundmental doctrine I think is still in effect today." Cheney listed among the United States' "major concerns . . . the very real possibility that should Saddam Hussein . . . be allowed to keep Kuwait, that he would be in a position to directly control over 20 percent of the world's proven oil reserves." Cheney envisioned a snowball effect in which Saddam would eventually command all the oil in the Middle East:

> Altogether, the Persian Gulf region contains over 70 percent of the world's proven oil reserves. And the prospects that a man like Saddam Hussein, with his enormous military machine, would be allowed to sit astride that resource without any countervailing force, would be allowed to control the flow of oil to the world's economy, and . . . use the enormous wealth that would be generated for nefarious purposes is a prospect that I think most of the world's civilized nations find abhorrent.

It was a remarkable comment: U.S. leaders had justified *war* with explicit reference to defending foreign oil reserves vital to America's interests.

Having repaired its alliance with Saudi Arabia after the embargo, the United States assembled half a million troops in the region, poised to protect the 475-mile Saudi border and to expel Iraqi troops from Kuwait. The Desert Storm air campaign began on January 17, 1991, followed by a ground offensive on February 24. Just four days later, Iraqi forces were driven from Kuwait, and President Bush terminated combat operations. But Saddam didn't leave without revenge: he ordered his military to set fire to nearly six hundred Kuwaiti oil wells. These wells burned about 5 million barrels of oil per day for months on end, with flames reaching hundreds of feet high, until the massive blazes were extinguished by U.S. firefighters. A *Washington Post* correspondent reported that smoke over the region was "so thick . . . that military officers are reading maps by flashlight at noon."

Rather than advance to Baghdad and eliminate the Saddam Hussein regime entirely, the George H. W. Bush administration had opted for a "policy of containment"—one that required the U.S. military to maintain a heavy presence within the borders of Saudi Arabia in order to keep watch on neighboring Iraq—monitoring no-fly zones, for instance—and the region. This proved diplomatically disastrous for U.S.-Arab relations. As King Ibn Saud had expressed to FDR and his early American oil part-

ners, Saudi Arabia valued its independence above all, and would cooperate with the United States only so long as it did not interfere in Saudi
Arabia's domestic affairs or engage in imperialist activities. Though King
Fahd (the fourth of Ibn Saud's sons to rule) had welcomed American
troops on Saudi soil to help maintain regional security, to many Saudis—
most prominently Osama bin Laden—the U.S. military presence in
Saudi Arabia smacked of imperialism.

Bin Laden, born to an elite Saudi family that has investments in
oil, complained in a 1998 interview that Americans "have stolen $36
trillion from Muslims" by procuring oil from Persian Gulf countries at
low prices. (Oil at the time was $11 a barrel, and bin Laden claimed the
fair price would be $144 a barrel.) He accused the Saudi royal family of
kowtowing to American interests, and in 1998 along with other Islamic
leaders issued a fatwa that galvanized his growing body of disciples:

> For over seven years the United States has been occupying the lands
> of Islam in the holiest of places, the Arabian Peninsula, plundering
> its riches, dictating to its rulers, humiliating its people, terrorizing
> its neighbors . . . to kill the Americans and their allies—civilians
> and military—is an individual duty for every Muslim . . . in order for
> their armies to move out of all the lands of Islam.

Bin Laden and his men began to stage acts of terrorism on American
military facilities and embassies in the region. Their aim was to once
and for all destroy the alliance between Washington and Saudi Arabia
that had been established more than half a century earlier. Bin Laden
also spelled out his intention to sabotage the fuel supply that America
so desperately sought to protect. "By striking the oil tanker in Yemen
with explosives," bin Laden allegedly said after a 2002 explosion on a
French tanker, "the attackers struck at the umbilical cord of the Christians, reminding the enemy of the bloody price they have to pay for their
continued aggression on our nation and robbing our riches."

On September 11, 2001, bin Laden orchestrated the single largest attack on American soil since Pearl Harbor. His weapon was oil: the gas tanks of three commercial airplanes filled with combustible jet fuel. The coordinated assaults by nineteen Al Qaeda terrorists on the World Trade Center, the Pentagon, and United Airlines flight 93 claimed nearly three thousand lives. Oil had evolved from a fuel for war machines to a catalyst for war to a lethal weapon.

THE FEW, THE PROUD, THE GREEN

In the early autumn light of 2007, the trees that fringed the Pentagon were just beginning to turn—a sea of green etched so subtly with red and gold that they looked like an unfinished painting. Despite the imposing neoclassical façade I'd seen in so many photographs, the building, as I approached, looked surprisingly humble, unadorned, and low-slung. No sign of the tragic events of six years before remained on the building's exterior. But inside, a string of police line tape marked "Do Not Cross" still demarcated a section of the structure's destroyed west side.

I had gone there to discuss the military's fuel consumption with Dan Nolan, who oversaw energy projects for the Defense Department's Rapid Equipping Force. (He recently retired.) Nolan, who graduated from West Point and has an engineering degree from the University of Southern California, procured in-field equipment ranging from tents to tanks for the Pentagon. He dealt with the nitty-gritty practical challenges of upgrading military combat machinery—meaning he's the one charged with answering General Zilmer's request for sustainable power stations. Just what exactly would it take, I wondered, to transform a hydrocarbon military into a petroleum-free enterprise?

I met with Nolan in a windowless, soundproof room with cinder-block walls and a two-way mirror in the basement of the Pentagon, where interviews with the media are often scheduled. Bald and muscular, with a bejeweled U.S. Army signet ring, Nolan offered me a metal folding chair directly across from his, interrogation-style. Though the setting was

austere, the conversation rolled amicably. Nolan was eager and passionate about the military's green future, and painted a buoyant picture: "I
can see a future," he said, "where we have base-camp generators powered by garbage, surveillance aircraft powered by the sun, hybrid-engine
tanks many times more fuel efficient, soldiers' clothing that harvests solar
energy to charge their electronic field gear . . . footwear that converts the
kinetic energy from movement into stored energy, buildings and facilities
operating entirely on renewable energy . . . it's all in the works."

Thus far, Nolan's most successful energy-efficiency programs had
been comparatively low-tech. He devised a superinsulating spray foam
that could be applied to the outside of soldiers' tents in Iraq to save on
air-conditioning demands: after the DoD spent $95 million on insulat-
ing foam for base camps in Iraq, the agency earned that back in energy
savings in just sixty days. The security benefits are perhaps more impressive: DoD data show that if all U.S. military base camp tents in Iraq
were spray-foamed, the number of fuel convoy trucks needed would be
reduced by thirteen per day.

Nolan was additionally collaborating with a start-up called SkyBuilt
Power to meet the demand for renewable in-the-field power stations. Sky-
Built had developed a mobile power station that fits into a standard shipping container and uses a mix of solar, wind, and hydro power to augment
diesel generators. The hitch was cost: this contraption is priced at roughly
$100,000, compared with just $7,500 to $10,000 for a basic diesel gen-
erator. For that reason, Nolan had been able to deploy only two of the renewable power systems in combat zones since General Zilmer issued his
request.

The range of green innovations the Pentagon is working on is
impressive—extending beyond specialized military applications to products with potentially vast commercial potential: a combined-cycle jet
engine with 40 percent greater efficiency, jet fuels derived from algae,
low-cost lightweight titanium applications to replace heavy steel, ultraefficient batteries. The Pentagon has partnered with companies including Boeing, General Motors, and General Electric to try to bring some
of these products to market, but the time frame is vague at best. "Hard

to say," Nolan replied when I asked him how soon some of these products will be commercially viable. Another reality check is the Pentagon's annual budget for developing efficient and alternative technologies: $300 million. That's less than one half of 1 percent of its total R & D budget— indicating that fossil fuel reduction is not exactly a top priority.

The energy consumption rate of our forces in Iraq and Afghanistan is many times what it was in World War II and substantially higher than that of operations Desert Shield and Desert Storm. Today, an F-15 fighter jet burns about 1,580 gallons of fuel in one hour—more than the amount of gasoline an average American household consumes over the course of a year. A KC-10 (an aerial refueling tanker) burns even more—about 2,650 gallons per hour. The B-52 bomber burns a staggering 3,266 gallons per hour. It's worth repeating: that's more than the quantity of fuel needed to sustain an American household for two years burned inside of *one hour*. As for ground vehicles, an Abrams tank, which runs on technology designed in the 1960s, travels less than 2 miles per gallon of fuel; a Bradley fighting vehicle, which entered service in 1981, also gets fewer than 2 miles per gallon. An armored Humvee gets 4 miles per gallon. Much of this equipment has gotten more energy-intensive over the years because of additional armor, which adds weight and drag.

While the army had a program back in 2004 that aimed to replace the engine of the Abrams tank with a more efficient diesel-burning engine, it was canceled due to budget constraints. Instead, engineers are fitting the tank with a battery pack to save fuel while the vehicle idles. Similarly, efforts to develop hybrid engines for tanks and Humvees have been tabled due to funding shortages.

TRUE COST

What would it take to get the funding and political capital necessary to significantly ratchet down the military's energy footprint? To find out, I went to speak with Al Shaffer, the executive director of the Pentagon's Energy Task Force. Shaffer has a big-picture handle on all the moving parts of the Pentagon—not just the army, navy, and air force, but also the dozens of other divisions that handle logistics, long-term strategy, and, most important, budget.

Shaffer's office was cheery and bright, with a map of the world spanning an entire wall, mementos from his travels decorating the windowsills and bookshelves, and, at the center of his conference table, a bowl of pretzels, a "swear jar" charging himself and his colleagues one dollar for every bad word uttered, and a menagerie of state-of-the-art engine parts.

Affable, lanky, and energetic, Shaffer looked more like a TV anchor than a career soldier, with tousled Anderson Cooper–style silver hair and an oxford shirt with the sleeves rolled up. He joined the air force at the age of twenty-one while a student at the University of Vermont, and "quickly realized it was kinda neat to play GI Joe and have a mission and a purpose in my life." Shaffer trained as a meteorologist to supply weather information to frontline combat units. He spent twenty-four years in the Armed Forces before retiring and joining the Department of Defense. He formed the Pentagon's Energy Security Task Force in 2006 in response to the spike in fuel prices after Hurricane Katrina. "All of a sudden we realized we had a problem at the Defense Department," Shaffer told me, "because a $10 increase in the price of crude resulted in a $1.4 billion upsurge in our operating costs for the next year." These cost surges arrived like an exclamation point just weeks after Zilmer's urgent memo, heightening the focus within the DoD on a shift toward leaner, greener energy strategies.

Shaffer emphasized the progress the military has made installing renewable energy on its bases. "Did you know that the world's largest photovoltaic farm is on an air force base?" he asked, adding that the DoD currently derives 12 percent of the electricity for its facilities from renewable sources—making it one of the world's largest consumers of green energy. The agency has vowed to increase that to 25 percent by 2025, and reduce the energy usage of its facilities 30 percent by 2015. Shaffer noted the growth in the military's green R & D efforts to push the development of efficient and renewable technologies. "This is my favorite," he said, snatching up one of the exhibited engine parts. It was a standard piston with a low-friction nickel-boron coating that eliminates the need for lubricating oil and improves efficiency, power, and torque by 25 percent, he said, "for a total cost of about $250. It's not ready yet for the mass market, but it's close."

All of this sounded promising, but I asked Shaffer how the U.S. military can talk about a secure energy future when its own B-52 bomber uses up to 45,000 gallons of fuel in a single mission. And as the Defense Science Board's 2001 *More Capable Warfighting Through Reduced Fuel Burden* report stated clearly, fuel usage is not something the military can actually restrict: "Because DOD's consumption of oil represents the highest priority of all uses, there will be no fundamental limits to DOD's fuel supply for many, many decades."

Shaffer nodded slowly, indicating he understood this problem all too well. "Energy security is critically important—and it has become dramatically more so in recent years because of the increase in cost of oil. This is scary." He paused. "Our cost for energy went up just shy of $3 billion from fiscal year 2005 to fiscal year 2006, even though we reduced our overall usage of energy by about 5 percent during that same time period." Shaffer stressed how sudden this spike in fuel prices had been: from early 2005 to their peak in 2008, oil prices almost quadrupled, from roughly $35 a barrel to nearly $140 a barrel. Despite recent price fluctuations, oil prices in the long run will almost certainly rise, Shaffer acknowledged. "This is an enduring challenge—it's not going away."

The shift to a greener and more efficient military, said Shaffer, will accelerate as the Pentagon adjusts to volatile oil prices and rethinks the way its fuel costs are calculated. The Defense Science Board report pointed out that the Pentagon calculates the cost of the fuel it uses according to wholesale refinery price—roughly the price we pay at the pump—and does not factor in the cost of delivery in the field of combat. DSB analysis showed that the total cost of fuel when delivered to army bases over short distances is roughly $10 per gallon. That number quadruples over long distances to "at least $40–$50 per gallon" and rockets up to "more than $400 per gallon" when fuel is delivered by aerial tankers to aircraft in flight. "This produces a sub-optimal allocation of resources," the report concluded.

These kinds of added costs are particularly pronounced in the Iraq operation given that a pipeline infrastructure can't be used, and given how dispersed the combat activity is—concentrated attacks are cheaper and more efficient. "The war on terrorism is a lot like guerrilla warfare," explained Shaffer. "When we send out a convoy, there are vehicles in front of the fuel trucks, there are vehicles behind the fuel trucks, and there are aircraft flying overhead. It's very, very complex. Takes a lot of kids, a lot of our young troops. Every time they go out and do this type of run, it puts their lives in danger."

Once the Pentagon starts factoring in the true cost of fuel, it makes the new renewable energy alternatives look cheaper by comparison. "We don't do things to be green," Shaffer told me. "We do things for operational efficiency, for improving our capability to perform our mission, whatever that mission may be. And it just so happens that being sustainable is now a smart thing to do." Historically, the Pentagon was built to be effective, not efficient. So long as oil was cheap and easy to distribute, an energy-guzzling military was just as effective as an energy-efficient military. As Shaffer pointed out, not since World War II—when the Germans were bombing American oil tankers and interrupting the Allies' fuel supply routes—has the Pentagon had to worry about having affordable and abundant energy to buoy its military operations. Now the picture is changing radically.

"Suddenly efficiency presents us with multiple benefits," Shaffer told me. "We save money; we simplify our logistics supply line, which makes us a more effective fighting force; we free ourselves from dependence on oil controlled by our adversaries; and above all we save lives. With better energy conservation, we simply would not have as many combat casualties."

In March 2008, the Defense Science Board publicly released

a report titled *More Fight—Less Fuel*. This report reiterated many of the requests it had made in its 2001 document and berated the Pentagon for failing to implement these recommendations in a meaningful way. Though it praised initiatives like those of Nolan and Shaffer, it exposed the underfunding that plagued such programs. (Although the Pentagon's budget for alternative-energy development rose from $400 million in 2006 to $1.2 billion in 2009, that's still a tiny fraction of its R & D spending.) This sequel report quoted one general's emphatic plea to "unleash us from the tether of fuel."

THE PATRIOT

James Woolsey, the former chief of the Central Intelligence Agency under the Clinton administration, lives in a rambling farmhouse in rural Maryland. I arranged to meet with Woolsey after my Pentagon visit in the hope of getting a wider perspective on the marriage of war and oil. Woolsey is in many ways a consummate Washington insider—someone who served as one of the voices in the room when key decisions were made about America's foreign policy in the Middle East. But since he left the CIA he has become a twenty-first-century Paul Revere, warning the public of the national security threats posed by America's dependence on foreign fuel.

A dry-witted, sharp-featured brainiac who was a Rhodes scholar at Oxford and got his law degree at Yale, Woolsey, now in his sixties, spearheads an increasingly influential group of conservative and liberal experts alike inside Washington who argue that weaning America off foreign oil is the most ethical, effective, and affordable way to win the war on terror. This group is known as "cheap hawks."

"Americans need to know that their oil demand is fueling Al Qaeda," he told me as we ambled through a bare cornfield near his house. "Nearly $100 billion has been spent on spreading anti-U.S. propaganda around the Middle East, and almost all of that is oil money."

To reduce the flow of those "billions for hate," as Woolsey calls it, he and his fellow cheap hawks have advocated a demand-side reduction plan similar to the one President Carter promoted in the 1970s. Woolsey and other cheap hawks see alternative energy technologies as a tactical necessity—not just in the field of combat, but as a means of moving America at large off of oil. They have been pushing for federal policies promoting the widespread adoption of hybrid cars, renewable power, and fuels such as ethanol and biodiesel that are made from corncobs, garbage, manure, and switchgrass. A far cry from green idealists, cheap hawks come on strong as pragmatists.

War, reasons Woolsey, is a grossly expensive and inefficient way to defend America's energy interests, and is becoming ever more so. "Our military energy burdens will only grow heavier as the war on terror demands more fine-tuned and complex tactics, including an ever more agile and dispersed force," he told me, "and more extended deployment in the field." Woolsey also cautioned that terrorism in the Middle East is much more likely to drive oil prices up than is peak oil. "The biggest threat to our economy in terms of oil dependence is the possibility of a terrorist strike against the oil and pipeline infrastructure in the Middle East. Overnight, we could see oil prices double or triple to $200 or $300 a barrel."

But for all his caution and concern, Woolsey is an optimist who fervently believes in the power of technology to end America's oil dependence and reduce the need for its military involvement in the Middle East. In fact, he has fashioned his own home into a kind of living laboratory of the types of practical solutions that he believes will dig the United States out of its energy crisis. He took me on a tour: the roof was rigged with solar panels, the walls and windows were double-insulated, the fireplaces were equipped with "heatilators" to ventilate warmth from the chimney, and the kitchen appliances were Energy Star–qualified—all of it serving to make the house electrically self-sufficient "in case of shortages or terrorist attack," he said.

By far the most important and underutilized weapon in the war on terrorism, according to Woolsey, was the gleaming white plug-in hybrid Toyota Prius in his garage, which sported the bumper sticker "Bin Laden Hates This Car." "The most immediate and important technology challenge for America right now is reforming our car fleet—that's the biggest deadweight dragging on our oil demands," he told me. "It's not just our environmental duty to drive efficient cars, it's our patriotic duty."

These fellas are fast, proud, fearless
go-getters with rebel hearts—that about
sums up the American spirit, don't it?

—NASCAR fan

3

Road Hogs

WHY A HUNDRED YEARS OF JOYRIDING
HAS US RUNNING ON EMPTY

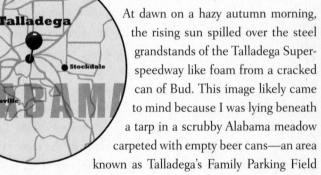

At dawn on a hazy autumn morning, the rising sun spilled over the steel grandstands of the Talladega Superspeedway like foam from a cracked can of Bud. This image likely came to mind because I was lying beneath a tarp in a scrubby Alabama meadow carpeted with empty beer cans—an area known as Talladega's Family Parking Field C. The 2.66-mile Talladega racetrack, located about 50 miles east of Birmingham, is the world's second-largest car-racing venue, with a mile-long grandstand built to accommodate more than 140,000 fans. Around my A-frame L.L. Bean tent were some 40,000 parked vehicles, most of them flatbeds, SUVs, Winnebagos, and camper vans filled with groggy pilgrims rising to greet a day that would bring them the nation's biggest semiannual NASCAR racing event.

The National Association for Stock Car Auto Racing holds "17 of the top 20 most-attended U.S. sporting events," according to its Web site, and its races are widely referred to as the nation's most popular spectator sport. This was my first visit to a NASCAR event, and I had come to see what may rank among the world's most lavish displays of fuel consumption: forty hot rods, each getting about 5 miles per gallon, hurtling around a strip of asphalt in an infinite loop. I admit I came to the event with a certain lack of regard for its premise: burning huge amounts of fuel and rubber for the sole purpose of driving around in circles. The ritual seemed careless to me at a time of war in the Middle East, unchecked global warming, and soaring energy prices. But hours later I would leave Talladega with a less skeptical take on the NASCAR phenomenon and a better understanding not just of carburetors and checkered flags, but of who we are as a nation—a thrill-seeking, speed-loving, self-propelled, forward-charging culture.

Talladega is NASCAR's XXL, Big Gulp–sized speedway—the most treacherous and most exciting. Its long straightaways and unusually wide track allow for cars to build up to and sustain speeds of more than 200 mph and to run three or four abreast. Racers don't brake for turns at Talladega the way they do at smaller tracks; instead they mash their gas pedals to the floor. These conditions raise fans' expectations for the "big one"—a massive, harrowing multicar wreck.

Field C, which a week earlier had housed only wildflowers and Alabama Longleaf pines, was now a sprawling tribal village with makeshift neighborhoods and orderly avenues webbed throughout the settlement. Families had been dwelling there for days before the race, many erecting well-appointed encampments with awnings trimmed in Christmas lights, lawn chairs, picnic tables, movie projectors, outdoor grills, and coolers stocked with cold American beer. Hoisted above the camps were Confederate flags and tributes to the denizens' favorite racers, above all Dale Earnhardt Jr., #8. "Junior" (who has since changed to #88) is the son of the legendary NASCAR champion who lost his life in the last turn of the 2001 Daytona 500. The drivers were competing on this fall day in a 500-mile race that was part of "the Chase" (the ten-race playoffs) for the Sprint Cup, the top prize in racing.

I had awoken to the ambient stench of beer-soaked crabgrass, cigarette butts, fire pits, and the charbroiled remains of the previous night's cookouts. I groped for soap and toothpaste and made my way to a public trailer marked "$5 Showers." En route, I caught sight of my neighbor shuffling out of his tent wearing nothing but his briefs. He nodded hello, and as he leaned over a propane stove to flip his pancakes, I saw the numeral 8 shaven expertly into his thicket of back hair. This tribute to Junior was a single-digit poem about America's devotion to speed—a display of fan loyalty so brash, intimate, and wholehearted that I stopped in my tracks, feeling awed and strangely jejune. I'd never been a sports fan of even mild convictions, let alone known loyalty so absolute. And though I've done my fair share of gas guzzling—including driving ATVs (four-wheel off-road motorcycles) in the backwoods of Maine throughout my childhood—I can't say I've been a real aficionado of horsepower, either.

The world of stock cars was new territory for me. But not for a large percentage of our population. NASCAR claims 75 million fans. During the nine months of the racing season, it's second only to football as the most-viewed professional sport on TV. It broadcasts its races in more than 100 countries, and has speedways in Mexico and Canada. But by no means does it have an international ethos—NASCAR is as much an American export as are blue jeans, Coca-Cola, and cherry pie. The sport grew out of the 1930s Prohibition era in America's Deep South, when rural bootleggers rigged standard-looking cars with high-powered engines to outrun the law. The forefathers of NASCAR, wrote historian Neal Thompson, were "a bunch of motherless, dirt-poor southern teens driving with the devil in jacked-up Fords full of corn whiskey—the best means of escape a southern boy could wish for."

Fans still cling to this unabashedly roughneck image. NASCAR's feisty southern outlaw spirit was on full display at Talladega. Many in the overwhelmingly white crowd wore T-shirts and baseball caps featuring Confederate flags. "Redneck by Choice, Southern by the Grace of God" went the refrain on one such garment. "If You Don't Like My Flag, You Can Kiss My Rebel Ass" went another. The sport's macho undercurrent was also supersized at Talladega: while there were plenty of women in attendance, most of them seemed unfazed by the hand-scrawled signs reading "Show Us Your Tits." (For that matter, one twentysomething in my vicinity was happy to comply.)

The atmosphere was festive and—setting aside considerations of gender and racial demographics—even good-natured. Talladega had the carnival air you might find at a county fair, only a thousand times bigger. Concession booths with all-American delicacies were nearly as prevalent as the fans themselves—you couldn't spit without hitting a purveyor of beer, corn dogs, fried chicken, or funnel cakes. (The speedway's Web site notes that "12,000 pounds of Ballpark Franks are sold by concessions during a race weekend at Talladega, which when laid end-to-end, would circle the entire 2.66-mile track 1.14 times.") After my $5 shower, pumped from a water tank by a purring diesel generator, I grabbed a double cheeseburger and followed the jostling crowd toward the track.

PIT STOP

At 1:00 p.m.—just after the national anthem blared over the loudspeakers and a squadron of B-1 bombers buzzed overhead—the green flag dropped. In seconds the chorus of twelve-cylinder combustion engines was echoing through the grandstands with a collective shriek as though the universe was being torn in two. Speed rumbled through the ground and into my bones, and my heart knocked against my rib cage. The air filled with the acrid odor of burnt rubber, hot asphalt, and spilled fuel.

Hoping for an up-close, under-the-hood look at the action, I made my way into the pit—the restricted area in the center of the track where the cars are fueled and tuned between laps. The race car teams were assembled side by side along the track's interior, each congregating in its own two-story open-air tent with VIP guests arrayed up top and the mechanics working below. A distinctive group of women who were predominantly blonde and trim—the wives of owners and drivers—sat perched in the VIP lounges wearing stiletto heels, shades, and cocktail gowns, sipping frosty drinks as they watched the action on the track.

Each of the drivers has a pit crew of over a dozen mechanics responsible for gassing the cars, changing the tires, cooling the engines, and assessing track and vehicle conditions throughout the race. The mechanics were outfitted in helmets and matching Crayola-colored jumpsuits—cherry red, royal blue, canary yellow. Their polished metal tools—wrenches, jacks, pressurized gas pumps shaped like giant baby bottles—glinted in the sunlight. Covered in branded sponsor decals, the race cars themselves looked like a flotilla of Times Square billboards rocketing across the black pavement, flashing their chrome and rainbow colors against the sky.

Roughly every forty laps, or when a hitch such as a flat tire occurred, the cars would come screeching in for pit stops. "Races are won and lost in the pit," one fan explained to me. In an instant, the mechanics of each team would assemble into a collective organism around their car like an army of ants, everyone performing his own role within the whole in miraculous harmony. These technicians are exhaustively trained for their particular functions—cooling the radiator, jacking up the car, changing the tires, clearing the windshield, even hydrating the driver with a squeeze bottle. NASCAR mechanics, like most athletes, tend to be young. They practice their roles with Olympian rigor so they can perform with the utmost speed—trying to shave off fractions of a second to get their cars back in the race and ahead of the pack. I watched one hapless wheel changer who was small and wiry (these members of the team are almost always small and wiry, the better to crouch low and wrestle with lug nuts) lose control of the tire he was rolling toward a car. It slipped out of his reach and cost his team several seconds—the equivalent of perhaps a 50-yard advantage. That's a big enough mistake to jeopardize his job and the six-figure salary that goes with it.

So rhythmic and stylized was the event that the whole experience—even the smoldering car wrecks—felt choreographed, scripted, cinematic, like those set pieces from the classics of American cinema and TV in which motor vehicles have figured in the plots like main characters—*The French Connection, The Great Escape, Bonnie and Clyde,*

Grease, American Graffiti, Easy Rider, The Dukes of Hazzard, Thelma and Louise. Only it takes a lot more fuel to keep this high-octane pageant in motion. How much, exactly?

Between pit stops, as mechanics lounged on spare tires and casually dragged on cigarettes, I pressed them for some answers. The cars get anywhere from 4 to 7 miles per gallon, which means that in a 500-mile race such as this one, averaging 5 mpg, each car would consume roughly 100 gallons of fuel. Multiply that by forty-three cars per race, and each event as a whole consumes approximately 4,375 gallons of gasoline (assuming all cars finish). With about ninety-six U.S. NASCAR races per year spread out across several divisions, that totals over 1 million gallons (factoring practice rounds and adjusting for some shorter races).

You also have to factor in the tires for every race. Several gallons of oil go into the production of a synthetic rubber tire. One car competing in a NASCAR event burns through forty to eighty tires per race. The fuel demands keep growing when you consider that there are more than a thousand other NASCAR-sanctioned races each year at affiliated tracks throughout the country. Additionally, each team has a convoy of eighteen-wheelers that hauls its race cars across the country from track to track, cumulatively traveling hundreds of thousands of miles per year. Fully loaded, these trucks get around 4.5 miles per gallon, which means that millions of gallons are consumed in just getting the cars to the races.

These numbers are small when compared to the volume of fuel that goes into America's military endeavors or our daily commutes, let alone our total oil demand. What's fascinating about this particular form of fuel consumption is that its purpose is sheer entertainment. This is gas consumption as an art form. "Fans come here for speed," as one mechanic put it to me, gesturing at the audience. "I don't see many eco-greens in that crowd worryin' about gas mileage."

As NASCAR fans enjoy the spectacle, they also participate in one of the most successful marketing platforms in the history of commerce. Drivers can pull in more than $30 million in annual commercial sponsorships. On one car alone I counted the insignia of more than twenty sponsors, including Mountain Dew, Chevrolet, 3M, Amp Energy, CompUSA,

Budweiser, Sunoco, the American Automobile Association, and the National Guard, and there were over a dozen more—including Wrangler, Sprint, and Goodyear—affixed to the driver's uniform. In total, four hundred companies spent more than $1.5 billion in 2008 to sponsor races, cars, and drivers.

NASCAR drivers have been rolling platforms for advertisers since the earliest days of the sport. This marketing strategy gained traction when Detroit automakers noticed, as racing grew in popularity over the 1950s and '60s, that American consumers would go in droves to their local car dealers the day after a race to buy the model of the winning vehicle. NASCAR entries are limited to certain approved "stock" cars— models that can be purchased at any standard dealership: the Chevrolet Monte Carlo, Ford Thunderbird, Pontiac Grand Prix, Dodge Charger, and Toyota Camry (only recently added to the otherwise all-American lineup). A driver's car choice can be a deciding factor for fans. One Talladega audience member, for instance, told me that he abandoned his loyalty to Tony Stewart when Stewart's racing team switched its affiliation from General Motors' Chevrolet to the foreign-owned Toyota.

The use of stock cars has enabled NASCAR fans not just to empathize with particular drivers—and root for their success—but also to identify with and own the very products that carried them to victory. "Win on Sunday, sell on Monday" went the sales mantra first introduced in the 1950s. The marketing strategy was so successful that it soon lured other advertisers into the mix. If a car can sell Detroit's brands, the logic went, why not make the vehicles platforms for other products?

This, in no small part, is what makes the sport a multibillion-dollar business. NASCAR devotees are known among sponsors to be ardent brand-loyalists—more so than the fans of any other sport. Little wonder, then, that NASCAR itself has become a brand: practically any wearable item, from hats, shoes, and T-shirts to boxer shorts and bikinis, can be purchased bearing NASCAR car numbers, driver mug shots, and sponsor logos. Also available are branded clocks, jewelry, bedding, tableware, refrigerators, wallpaper, and upholstered leather sofas. The merchandise garners sales of an estimated $2 billion a year.

REBEL HEARTS

With 145 laps completed and 43 to go, the #17 car shuddered and began to skid out. A collective groan went through the stands. The engine had blown out, not an uncommon occurrence in racing given the strains put on car engines. I didn't have long to wonder whether this was the "big one." As #17 began to fishtail, it clipped the nose of another car riding its fender and set off a domino effect. At 200 miles an hour, even the slightest disturbance can force air under a vehicle's chassis, popping the car up like a Frisbee and flipping it on its back. The result was an eleven-car pileup that brought the race to a stop and the entire audience to its feet.

Looking up, I was struck by the appearance of the crowd. For all the wealth of competing logos and gear available to them, by far the stand-out choice among the Talladega fans was patriotic garb: the grandstands looked like a pointillist painting in red, white, and blue. I approached one bystander, a sixty-three-year-old account manager at a North Carolina carpet company who had been coming to NASCAR races since they were held on dirt tracks in the 1950s, and asked him about this apparent connection between stock car racing and patriotism. "Those fellas are fast, proud, fearless go-getters with rebel hearts," he said, nodding toward the track. "That about sums up the American spirit, don't it?"

I'd take it a bit further to say that no consumer product more wholly embodies the American ethos than the automobile—"the heartbeat of America," as Chevrolet famously dubbed it. The word derives from the Greek root *auto,* "self," and the Latin *mobile*, "moving"—words that could be said to define the American dream: we each propel ourselves toward the life and destiny of our own choosing. In these individual pursuits, we also consume on average 1.5 gallons of gasoline per person per day. This fuel consumption—roughly quadruple that of the average European—is due in part to the great distances traveled in our largely suburban, auto-dependent lifestyles, but also to the fact that we have some of the lowest fuel economy standards of any industrial nation—lower even than those of our up-and-coming rival China. All of which contributes to a habit of domestic consumption that far exceeds our ability to produce domestic oil.

There were moments when I felt like the whole NASCAR enterprise should be illegal—just as racing was prohibited during World War II in order to save fuel for the troops, and just as gas was rationed during the Arab oil embargo. Why shouldn't Americans be asked to give up activities such as NASCAR as we grapple with war, dwindling supplies, *and* a growing environmental crisis?

But while I half expected to feel some indignation, I didn't expect to enjoy the whole experience so much. The body-rumbling speed, the roar of the crowd, the open-throttle sense of freedom, the cars shifting and moving in a V-shaped flock formation like mechanical birds in flight—the race had a thrill factor and a beauty that seemed, if only for moments, to justify all the fuel required to bring it to life. I came to understand that these questions were more gnarled and complicated than I'd thought. That NASCAR was more universally American—more *me*—than I'd ever realized.

What I couldn't get my mind around was how, exactly, it had come to this. I wanted to know the story behind America's long-running love affair with cars. How did we develop a fetish for cars so consuming that we'd spend money and time watching them drive around in circles? How did Detroit automakers, and American consumers, become so reckless about our fuel consumption? How did so much of our lifestyles, our very identities—our neighborhoods, shopping centers, transportation networks—come to revolve around the combustion engine?

AMERICA'S TOP MODEL

Henry Ford, the father of the American automobile industry, was born just outside of Dearborn, Michigan, in 1863. The descendent of Belgian and Irish immigrants, Ford attended a one-room schoolhouse and worked on his family's farm until the age of sixteen when he left for

Detroit. By day he worked in a machine shop and by night he repaired clocks and watches. Eventually he landed a job as chief engineer with the Edison Illuminating Company. He dabbled in inventions in his spare time—building a horseless carriage and a steam engine tractor—until in 1884 he read a magazine article about the four-stroke internal combustion engine newly invented by German engineer Nicholas Otto. Ford spent the better part of a decade tinkering in a small brick shed in his garden, eventually constructing the "Quadricycle," a two-cylinder motor affixed to bicycle wheels with no body or means of reversing.

By August 1899, Ford had raised enough money to leave his job at Edison Illuminating with the hope of building "a car for the great multitude." Whereas the cars of Europe were affordable only to the elite, Ford's founding vision for the motor company he chartered in 1903 was to give America—its rich and poor alike—the gift of motion. One of the most comprehensive books on American car history is James J. Flink's *The Automobile Age*, which documents the meteoric rise of Ford's empire. In 1908, Ford released his Model T at a cost of $850 ($19,300 in today's dollars), the first motorcar affordable to the middle class, and one that could be replicated quickly enough to meet large-scale demand. The Model T sold hundreds of units in the first year, but Ford kept pushing his innovation—refining his assembly process and dropping his price.

By 1912, the Model T was selling for $575 ($12,200 today) and the assembly of the chassis took one-sixth the amount of time it had taken just a year earlier. By 1916, the car was priced at $345, and produced at a record-high annual rate of 738,811 units—roughly half the total number of cars then on U.S. roads. And by 1926, "assembling an automobile took only ninety minutes," wrote Flink, "and cars rolled off [the] four final assembly lines every thirty seconds." Ford sold more than 15 million cars between 1908 and 1927, at which point the price had been reduced to a record low of $290 ($3,400 today)—a car even bootleggers could afford.

In fact, as mentioned earlier, the roots of stock car racing were formed during the Prohibition era when bootleggers ferried illegal booze throughout the South. Ford became a folk hero to these renegades—

no small irony given that he was a puritanical, compulsively disciplined man so opposed to drinking that he made all his employees swear off alcohol consumption. (Ford declined even a taste from the jar of home-made moonshine one fan offered him during a trip to Asheville, North Carolina.)

For all his success, Ford lived in a relatively modest home. His fire-place bore the inscription "Chop your own wood and it will warm you twice." He openly questioned the limits of his wealth: "Money means nothing to me," he said in a 1923 interview. "There is nothing I want that I cannot have. But I do not want the things money can buy." He wanted, among other things, political power—at one point he contemplated a bid for president—but his reputation as a teetotaler, a corporate autocrat, a religious eccentric (he reportedly believed in reincarnation), and a bigot (he ran anti-Semitic comments in a newspaper he owned) limited his voter appeal.

In truth, Henry Ford probably had more influence on America from outside of Washington than he ever could have had from inside the capital. By far his biggest cultural and economic impact came in the form of the assembly line, which became the basic building block of large-scale manufacturing. "The way to make automobiles is to make one automobile just like another automobile, to make them all alike, to make them come through the factory alike," said Ford, "just like one pin is like another pin when it comes from the pin factory, or one match is like another match when it comes from the match factory."

When coupled with a seemingly unlimited domestic supply of coal and oil, Ford's assembly-line innovations helped propel America's ascent. In fact, it was the success of the automobile that kept the oil industry alive and thriving. Ford's cars debuted just as Edison's invention of the lightbulb was killing the demand for lighting oil. In the nick of time, Ford had unwittingly created a new market for oil—transportation.

In the decades after World War II, America became known as the world's industrial behemoth, thanks to a combination of abundant energy reserves and the mass production techniques first established and re-fined at Ford's Highland Park, Michigan, headquarters. "She sits bestride the world like a Colossus," the British historian Robert Payne wrote of the United States in 1949; "no other power at any time in the world's history has possessed so varied or so great an influence on other nations. . . . Half the wealth of the world, more than half of the productivity, nearly two thirds of the production of the world's machines are concentrated in American hands; the rest of the world lies in the shadow of American industry, and with every day the shadow looms larger, more portentous, more dangerous."

CHROME SWEET CHROME

There was another vision of the automobile evolving in America, one that ultimately came to dominate the industry. The Ford Motor Company was all about functionality—making cars reliable, available, and affordable to the masses. It focused on technology and manufacturing challenges, with the goal of putting a standard car model into widespread use. "The customer can have any color he wants," Ford famously said, "so long as it's black."

Ford's chief competitor, General Motors, run by Alfred Sloan, was all about style. The GM focus was on selling consumers cars not for their mechanical performance but for the social status and sense of identity they conferred—car designs in great varieties, gussied up with grilles, fins, medallions, chrome detailing, stereo systems, leather interiors, and fanciful names like Le Sabre and El Dorado. The focus wasn't under the hood but on the packaging. Sloan didn't just want to put cars into use; he wanted to put them into the hearts and minds of Americans.

The 1950s, you could say, was the dawn of car fetishism, an era that forged an emotional relationship between consumers and cars that is still very much alive today. Alfred Sloan was an unlikely catalyst for such a trend. A sharp, austere, by-the-books executive, he had studied electrical engineering at the Massachusetts Institute of Technology and by the age of twenty-four had become president of Hyatt Roller Bearing, a company that made ball bearings for machinery. Sloan rarely strayed from his monochrome business uniform and scarcely looked the part of a legendary tastemaker. He was concerned about one thing only—growing General Motors into the world's most powerful company. "The primary object of the corporation," Sloan wrote, "was to make money, not just to make motor cars."

Sloan's emphasis on design was driven by two beliefs. First, he viewed cars as technologically mature, arguing that little more engineering was needed to improve vehicle performance (an attitude that pervaded Detroit until very recently). Second, Sloan felt that the market for cars was already saturated. In 1950 there were just over 150 million citizens in the United States and almost 50 million cars on the road—more than one for every household. To Sloan, this meant that consumers needed a new incentive to buy cars—functionality alone wasn't enough. He wanted to offer consumers variety, a goal he trumpeted in his motto "A car for every purse and purpose."

Sloan devised a strategy to make car owners quickly tire of their car models and desire upgrades. He believed the American car market could be broken down into status categories, wherein car models were equated with personal achievements and income brackets. In his book *The Fifties*, David Halberstam summed up GM's automotive class structure:

> The Chevy was for blue-collar people with solid jobs and young couples just starting out who had to be careful with money; the Pontiac was for more successful people who were confident about their economic futures and wanted a sportier car—one thinks of the young man just out of law school; the Olds was . . . for the white-collar bureaucrat or old-fashioned manager; the Buick was for the town's doctor, the young lawyer who was about to be made partner, or the elite of the managerial class; the Cadillac was for the top executive or owner of the local factory.

Sloan wanted consumers to aspire to new, more expensive car purchases just as they aspired to new stations in life. He wanted the car to be seen not as a long-term possession but as "an economic benchmark on life's journey to the top," as Halberstam put it. If you got a new wife, a new kid, or a job promotion, the thinking went, it was time to trade up your car model, even if your current one was perfectly functional. Sloan called his philosophy "the constant upgrading of product." Easing the financial burden of frequent new car purchases was GM's consumer credit program—the first of its kind, which allowed customers to pay for their vehicles in installments over time.

The General Motors business model was directly at odds with Henry Ford's philosophy of offering consumers one basic, static car model at an ever-shrinking unit price. The strategy of rapidly outmoding products has become all the more pronounced today, when merchandise from blue jeans to iPods is often viewed as obsolete in mere months. Sloan's emphasis on both style and product variety became so expensive from a production standpoint that it drove many smaller auto manufacturers out of Detroit, paving the way for the reign of the Big Three.

Sloan left the razzle-dazzle part of his strategy up to his wingman, Harley Earl. Also known as "the Cellini of chrome," Earl did look the part of a style maven. He wore boldly colored suits, flashy silk ties, and blue suede shoes, and had a cavernous closet in his office filled with freshly pressed clothes in case he found himself wrinkled before a meeting. Over six feet five inches tall, with broad shoulders, slicked-back hair, and a high forehead, Earl reportedly never hired anyone who exceeded his height, in order to maintain his dominant stature.

Sloan imported Earl from Hollywood, where he had made his name designing customized cars for the rich and famous—including a car body with a saddle on its hood for "the king of cowboys," Western movie star Tom Mix. (In The Beverly Hillbillies, Jed Clampett moves to Hollywood so that he can live next to his hero Mix.) Earl charged top dollar for this work: Fatty Arbuckle reportedly paid him $28,000 (the equivalent of $350,000 today) to build a custom car with specialized lights, body design, and leather detailing.

Earl was the first to use modeling clay to render car designs in 3-D, a method he had seen the movie studios use in designing their props. He drew inspiration from offbeat sources, most famously World War II's

Lockheed P-38 fighter plane. With their curved windshields and pointed hoods, Earl's auto designs echoed the plane's stealthy, aerodynamic lines and gave the illusion of motion even when the cars were parked. "True innovation is like beauty," he often said, "hard to find, but easy to recognize when you see it." Sharks were another one of Earl's eccentric inspirations, making their way into his designs in the form of fins. Jutting above the taillights, these fins quickly became the top design icon of the 1950s. Such decorative details, said Earl, gave customers an "extra receipt for their money in the form of a visible prestige marking for an expensive car." They symbolized an age of "too-muchness," of "no-expense-spared largesse," wrote historian Karal Ann Marling, a "doctrine of luxury for all."

The name of the game under the Sloan-Earl regime at GM was non-functional styling, and perhaps least functional of all their cars' style elements was their size. "Bigger is better" was the logic behind most things built in the 1950s, and cars were no exception. To Sloan, bigger meant pricier: increases in the size of his car models every year enabled him to ratchet up the unit prices on the same basic designs, yielding sizable growth in profit margins. As for Earl, bigger meant a broader canvas on which to hang his glitzy design details and more room for luxury add-ons. According to James Flink, the average wholesale price of a car jumped nearly 50 percent between 1950 and 1960, from $1,270 at the beginning of the decade to $1,822 at the end. Far from discouraging consumption, these price increases accompanied a massive surge in sales: in that same period, the number of automobiles on American roads also increased nearly 50 percent, from roughly 50 million to about 74 million.

But Detroit's emphasis on impractical design would come to haunt U.S. automakers. As Detroit fixated on frivolous details, its foreign competitors were engineering safer, more reliable, and technologically superior cars with better-performing engines, brakes, suspension, steering, and fuel economy. U.S. consumers began to purchase more automotive products from abroad than they did from homegrown companies, and Japanese and European automakers including Toyota, Honda, BMW, and Mercedes-Benz rapidly gained market share. "Complacency carried a high cost," noted Flink. "The United States shifted as early as 1967 from being a net exporter to a net importer of automotive products."

Gas mileage on American-built cars plummeted, a trend that continued into the mid-1970s. Whereas a large Caddy got 20 miles per gallon in 1949, by 1973 American cars averaged 13.5 miles per gallon due to their continued growth in size and weight, and added creature comforts. Fuel economy was hardly a concern during the fifties, when oil prices were at record lows: a vast new infrastructure of pipelines and refineries had been built during World War II to provide a reliable oil supply for the military. The year 1943 saw the birth of Big Inch, a pipeline roughly 1,200 miles long that funneled petroleum from Longview, Texas, to Linden, New Jersey. Little Big Inch, built the following year, was an even longer conduit, extending almost 1,500 miles, transporting oil pumped from Beaumont, Texas (home of Spindletop), to the Northeast. These pipelines revolutionized oil delivery. Less than 5 percent of U.S. oil traveled by pipeline in 1942; just three years later, the volume was more than 40 percent. This expedited distribution system flooded U.S. consumer markets with cheap and abundant petroleum after the war.

In the meantime, Americans were coming to depend on oil in ways that went far beyond the business of getting from point A to point B. Automobiles were quickly and irrevocably shaping not only our preferred modes of transport but also our destinations. The proliferation of cars among mainstream American consumers gave rise to the first drive-in fast-food joint (1921), the first shopping mall (1922), the first motel (1925), and the first drive-in movie theater (1933).

The mainstreaming of cars also influenced rituals of romance. The automobile helped bring about an era of sexual liberation, enabling teenagers to escape the surveillance of their parents and journey to the fabled "lookout point." Ever the killjoy, Henry Ford allegedly designed the backseat of a Model T to discourage sexual frolicking. "But determined couples found ways to thwart Ford's intentions," wrote Flink, "and the automakers came to facilitate lovemaking in cars with such innovations as heaters, air-conditioning, and the tilt steering wheel." Flink cited a 1967 survey in the magazine *Motor Trend* showing that of 1,100 marriages, some 40 percent of grooms had popped the question in an automobile. The American landscape in which these young couples would be raising families was, meanwhile, increasingly becoming one of suburbs linked to cities and other suburbs by a sprawling network of highways.

ROAD SCHOLAR

As auto sales boomed throughout the 1950s, American roadways and bridges were in a state of disrepair. The growth in car purchases had far outpaced the growth in infrastructure. Along came Dwight D. Eisenhower.

Born in Denison, Texas, the third of seven boys, Eisenhower was admired throughout his career for coupling a small-town folksy charm with the imposing authority of a five-star general. As a boy, Eisenhower worked the night shift at the local creamery to help put his brothers through school. He aspired to become a professional baseball player until he was accepted into West Point. Serving stateside in World War I, he quickly rose to become a leader of his battalion. He ascended through the ranks during peacetime training military staff, and after the 1941 attack on Pearl Harbor was tapped by Army Chief of Staff General George C. Marshall to become a leading architect of U.S. military strategy. By 1944 he had become the supreme commander of the Allied forces in Europe, and supervised the legendary D-Day invasion of Normandy.

Eisenhower, who had trained in tank warfare as a young recruit, was profoundly impressed by the German autobahns, a network of superhighways constructed during the 1930s and later used to facilitate the military maneuvering of Germany's jeeps, trucks, and tanks in this first fully mechanized war. An autobahn was distinguished by its elevated overpasses, the strength of its bridges, and the breadth of its tunnels. When he returned to the United States after the Allied victory, Eisenhower found American roads in "shocking condition," and envisioned a network of highways that would rival the autobahns' high-caliber engineering, but on a vastly greater scale. "After seeing the autobahns of modern Germany and knowing the asset those highways were to Germans, I decided, as President, to put an emphasis on this kind of road building," Eisenhower later reflected. "This was one of the things I felt deeply about, and I made a personal and absolute decision to see that the nation would benefit by it."

Eisenhower's interest in highways was also rooted in an earlier experience. As a nineteen-year-old army recruit, he had joined America's first cross-country truck convoy, an expedition designed to rally support for the U.S. military. Interstate roads at the time were virtually nonexistent, and Eisenhower recorded a long list of tire punctures, engine blowouts, busted axles, and broken fan belts throughout his trip across the nation's rutted dirt roads. "Some days when we had counted on sixty or seventy or a hundred miles," he wrote, "we would do three or four." His convoy had departed Washington, D.C., on July 7, 1919, and arrived in San Francisco a laborious sixty-one days and hundreds of breakdowns later. It was enough to convince him that a sophisticated road network would be crucial to America's long-term growth and security.

Before Eisenhower became president in 1953, road building had been largely overseen by the states. America's highways were patchy and disjointed: only half of the nation's 3 million miles of roads were paved when Eisenhower took office, and of these just a few thousand miles had been properly maintained. Eisenhower championed Detroit's great boom in auto sales, arguing that more cars meant "greater convenience, greater happiness and greater standards of living." But he warned that the level of congestion and the state of disrepair on American roads was dangerous, and could stymie economic growth. The president's personal conviction would ultimately grow into the biggest interstate highway network in the world, at 40,000-plus miles, and the biggest and costliest public works program in history.

As conservative as Eisenhower's politics were—he favored states' rights and small government—he knew that an interstate highway system had to be built with centralized federal oversight. Rallying the political will for such a project would be a huge challenge, particularly given the astronomical cost and accelerated time frame Eisenhower had in mind: $50 billion ($6.6 trillion in today's dollars) invested over ten years.

Eisenhower intended to stage a dramatic unveiling of his plan at a governors' conference in 1954. It was held in the wood-paneled rooms of the Sagamore Inn on Lake George in New York's Adirondack Mountains. The president was unexpectedly detained by a death in the family, so his vice president, Richard Nixon, had to make the appeal. "Our highway net is inadequate locally and obsolete as a national system," Nixon told the governors, adding that it was a "haphazard" and "arbitrary" road system "designed for local movement in an age of transcontinental travel." The health risks posed by the system were "comparable to the casualties of a bloody war," Nixon stated, with some forty thousand Americans dying in car accidents a year.

Nixon next made his plea for financial backing, arousing a din of chatter from the assembled governors: the project would pay for itself if states would approve a gasoline tax and a system of highway tolls—a controversial request coming from a conservative administration. Nixon stressed, above all, Eisenhower's national security argument: "the appalling inadequacies to meet the demands of catastrophe or defense, should an atomic war come." Seventy million city dwellers throughout the country would have to be evacuated to safety if a nuclear bomb struck American soil, said Nixon, and the quickest route would be wide, paved superhighways that didn't yet exist.

Critics of the plan abounded. Urban planners predicted that the highway program would signal the decay of cities, accelerating suburban sprawl and stymieing the development of mass transit—concerns that have only escalated today.

But the Eisenhower administration wasn't about to give up. The highway network, argued the president, would benefit average Americans not just by making them safer but by creating scores of jobs, making the growing American economy more robust and efficient. Billions of dollars, he said, had already been squandered because of perilous and congested road transport. Moreover, the highway network would further unify a nation that had recently drawn together to emerge successfully from world war: "Our unity as a nation is sustained by free communication of thought and by easy transportation of people and goods . . . over a vast system of interconnected highways," Eisenhower wrote in a 1955 appeal to Congress. "Together the united forces of our communication and transportation systems are dynamic elements in the very name we bear—United States. Without them, we would be a mere alliance of many separate parts."

While politicians bickered over the particulars of financing the highway plan—should tolls, for instance, be placed on the travelers who would use the roadways or taxes levied on the industries that would profit from them?—business leaders did their best to speed the project along. Automobile makers, truckers, car dealers, oil companies, rubber and cement companies, trade unions, and real estate developers collaborated in an alliance known as the "highway-motor lobby." This secretive and amply funded group, which had been working for decades to advocate road building, helped wage a political blitzkrieg to win congressional votes in favor of the national highway network.

Lobbyists argued that new roads would stimulate the growth of their varied industries. Real estate development would spring up alongside the thousands of miles of freshly paved highways. Greater car use would increase tire manufacturing and demand for oil and gas. Mail order and product delivery services would enjoy a boom thanks to more rapid and reliable transportation, as would cement and asphalt production for the roads themselves and the production of aluminum and paint used for highway signage.

"Obviously we have a selfish interest in this program, because our products are no good except on the road," James J. Nance, president of the Automobile Manufacturers Association and Studebaker Packard, testified before the House Public Works Committee in 1955. "Unless we know that there is going to be an expansion of the roads in this country . . . it is very difficult for us to plan over the next ten years as to what our expansion is going to have to be." Nance was describing an outgrowth of the stunning chain of events set off by Ford's discovery of mass production—a vast network of enterprises that all connected to and abetted one another, all expanded and enhanced the American dream, and all dangled from the same invisible thread of cheap oil.

LITTLE BOXES

Among the many beneficiaries of Eisenhower's interstate highway system, real estate developer William Levitt arguably had the biggest impact on the American lifestyle. On July 3, 1950, Levitt's portrait appeared on the cover of *Time* magazine against a backdrop of freshly bulldozed land and rows of multicolored boxlike homes as neatly aligned as squares on a checkerboard. "House Builder Levitt: For Sale: A New Way of Life," read the caption. Levitt was the architect of America's first mass-produced suburb, Levittown, which would serve as a leading model for suburbs throughout the country.

This planned community of some 17,400 homes sprang up just outside New York City in the potato fields of Long Island between 1947 and 1951. It was built for a target consumer—the more than 10 million American soldiers who had recently returned from war to their families. There was at the time a tremendous housing shortage, and thousands of former soldiers were living in makeshift dwellings including army Quonset huts and even converted trolley cars. These young men were entitled to major home investment subsidies via the GI Bill, and they were eager to settle down and start families, eventually fathering the baby boom.

William Levitt was an indifferent student who dropped out of New York University in his junior year because he "got itchy. I wanted to make a lot of money," he told *Time* in 1950. "I wanted a big car and a lot of clothes." The article described him as almost foolishly overconfident: "At 43, the leader of the U.S. housing revolution is a cocky, rambunctious hustler with brown hair, cow-sad eyes, a hoarse voice (from smoking three packs of cigarettes a day), and a liking for hyperbole that causes him to describe his height (5 ft. 8 in.) as 'nearly six feet' and his company as the 'General Motors of the housing industry.'"

While the GM comparison drew ridicule at the time, Levitt does, in retrospect, deserve comparison to another auto giant—Henry Ford. Levitt brought Ford's concept of mass production to the housing market. Just as Ford introduced low-cost automobiles to the American mainstream, Levitt used the assembly line to bring affordable homes to the burgeoning American middle class. And just as Ford saw himself as an instrument of capitalism and purveyor of market freedoms, Levitt had a similar raison d'être: "No man who owns his own house and lot can be a Communist," Levitt once said. "He has too much to do."

The homes of the first Levittown had uniform floor plans—two bedrooms, one bathroom—and were located on grassy plots of identical shape and size with trees planted identical distances apart. Levitt divided the construction process of his homes into two dozen steps—among them digging the foundation, setting pipe, pouring the concrete, framing the house, installing the floor, setting the tile, mounting appliances, painting the walls—with specialized crew members assigned to each task. The teams were so specialized that, for instance, those that applied red paint were distinct from those that applied white. Levitt prebuilt as many parts as possible, trucking them to the sites where they could then be snapped together like Tinker Toys. New power tools such as electric saws and paint sprayers hastened the assembly process.

Levitt next built an identical 17,000-unit development in Pennsylvania. His company produced its homes at rates unthinkable even by today's standards: "a new one was finished every 15 minutes," *Time* reported. Levitt offered homes in four varieties—the Rancher, the Levittowner, the Jubilee, and the Country Clubber—that ranged in price from $8,000 to $16,000 (roughly $68,000 to $135,000 today). Prospective buyers only needed a $90 deposit to clinch a home owner's contract, and made small monthly payments in financing plans much like the consumer credit programs GM was offering to car buyers.

Drawing still further from the lessons of Detroit, Levitt adopted GM's strategy of yearly models. "Houses were updated annually with slight style changes. Instead of tail fins and chrome, here were washers and stoves," wrote historian Dan McNichol. Like Sloan's cars, each home came with accessories Levitt dubbed "built-ins"—the 1950 model came with the standard refrigerator and washing machine, plus an 8-inch television. The following year's model came with a 12-inch television.

Levitt's initial units presold before construction had been completed. Shopping centers, playgrounds, libraries, movie theaters, churches, and restaurants sprang up to accommodate Levittown residents. What Levitt built wasn't just a product, it was a lifestyle—namely, for America's youth. "Levittown has very few old people," reported *Time* magazine. "Few of its . . . residents are past 35; of some 8,000 children, scarcely 900 are more than seven years old." The article described this "entirely new kind of community" as an environment with cultlike restrictions:

> The community has an almost antiseptic air. Levittown streets, which have such fanciful names as Satellite, Horizon, Haymaker, are bare and flat as hospital corridors. Like a hospital, Levittown has rules all its own. Fences are not allowed. . . . The plot of grass around each house must be cut at least once a week; if not, Bill Levitt's men mow the grass and send the bill. Wash cannot be hung out to dry on an ordinary clothesline; it must be arranged on rotary, removable drying racks and then not on weekends or holidays.

There were no African Americans among Levittown's residents. Levitt prohibited nonwhites from buying in to his communities for more than twenty years—well after civil rights laws had made such a policy illegal. "I have come to know that if we sell one house to a Negro family, then 90 or 95 percent of our white customers will not buy into the community," he claimed. "As a company, our position is simply this: We can solve a housing problem, or we can try to solve a racial problem. But we cannot combine the two." The discrimination didn't end there—Jewish and Hispanic residents were not welcome, nor were gay, single, or cohabiting residents.

America's political leaders were nevertheless enamored of Levitt's suburban experiment. When the Soviet leader Nikita Khrushchev visited the United States in 1959, President Eisenhower put Levittown at the top of his list of sightseeing attractions—a showcase of the fruits of American mass production that the president boasted was "universally and exclusively inhabited by workmen." This was a striking irony given Levitt's outspoken aversion to Communism.

American social critics lamented the rise of postwar suburbia. In his 1956 book *The Crack in the Picture Window*, John C. Keats described suburbs as "developments conceived in error, nurtured by greed, corroding everything they touch." But the trend was unrelenting: in the three decades after 1950, no fewer than 60 million Americans flocked to the suburbs. In 1970, the number of suburban residents in the United States had outstripped the number of city-dwellers.

DERAILED

There was, for a time at least, an alternative to gas-guzzling cars in the evolving American landscape: public transportation. Historians still debate the factors that led to the triumph of automobiles over railroads, trolleys, and streetcars: did consumers decide that they preferred the convenience and comfort of the automobile, or was there a deliberate corporate effort to sabotage the intricate network of railways that once webbed throughout the nation? The answer is probably a little of both.

Critics argue that Detroit saw public transportation as its biggest competitor. At its height, the nationwide intercity rail network reached more than eighty-five thousand U.S. towns and cities, and almost every major city had a mature rail system within its borders. Detroit saw an opportunity to expand its market by moving rail passengers into motor vehicles. In the early 1930s, General Motors began to purchase trolley and streetcar lines from the railroads and electric utility companies that owned them, and then convert them to combustion-engine bus lines. "Streetcar companies were bought up," wrote James Flink, "then resold after being motorized."

In 1925, GM acquired Yellow Coach, the nation's top manufacturer of buses, and roughly a decade later the company "abruptly curtailed its production of electric-powered trolley buses," explained Flink, "in favor of new diesel-powered buses." GM also became the largest stockholder in Greyhound bus lines. In 1936, GM became part of a holding company known as National City Lines (NCL); other NCL participants included Standard Oil of California, Phillips Petroleum, Firestone Tire and Rubber, and the trucking company Mack. This consortium had ownership in or control of rail systems in forty-five cities in more than a dozen states. By the 1960s, most of these railways had been shut down and their cars scrapped or sold overseas.

In 1974, the Senate held antitrust hearings that addressed GM's role in the depletion of the nation's rail lines. GM maintained that public streetcars and interurban transit trains had already been in decline when it began bus conversion. But critics took strong exception. "I cannot accept the argument that rapid transit systems broke down because of their complete inefficiency to serve the public," San Francisco mayor Joseph Alioto told the panel. "If it is true that the streetcar companies were breaking down of their own weight, why was it necessary for General Motors to join with Standard Oil and the tire companies to go in and buy the systems and tear up the tracks?"

Congress never amassed enough evidence to charge GM with coordinated conspiracy, but critics still bristle over this issue, claiming that public transport systems languished without adequate federal assistance or intervention while the auto and petroleum industries enjoyed windfalls of profits and government support.

The shift from electric trains to fuel-powered motorized transport is one of many factors that led to the rapid rise in American oil usage in the following decades. Historian Daniel Yergin explained that in the twenty years between 1950 and 1970, America's consumption of oil nearly tripled from roughly 6 billion barrels a year to more than 16 billion. At the beginning of that time span, coal accounted for more than 65 percent of global energy use; by 1970, the roles of the two leading fossil fuels had been reversed—it was oil that comprised 65 percent of energy use.

Consumers, of course, barely noticed the shift. It was only in the mid-1970s, when America's cheap and invisible oil lifeline came under

threat during the Arab oil embargo, that the hazards of our oil dependence became suddenly, eerily apparent. People began to carpool and
ride their bikes, and moved nearer to public transportation hubs. Detroit
increased the average fuel economy of its fleet by nearly 50 percent in
the span of seven years. Even NASCAR reduced its race distances by 10
percent to conserve fuel.

IT'S A SPRAWL WORLD

Today, the American lifestyle is heavily suburban and car-reliant. According to a 2007 analysis by the U.S. Census Bureau, nine out of ten Americans drive to work, and 77 percent of these commuters drive alone rather
than carpooling. The number of carpoolers increased a mere 0.1 percent
over 2000 levels, despite the rocketing fuel costs of the time period. So
zealous a culture of commuters are we that in recent years the muffler
company Midas actually gave an award to the American with the longest
commute—a man who boasted a seven-hour daily commute of over 372
miles between his home in rural California and his San Jose office.

Our penchant for long-distance driving is not surprising in a geo-
graphically expansive country that now has nearly 4 million miles of
heavily subsidized, well-maintained roadways, low gas taxes, and a hob-
bled rail system—a country in which driving has become, on the whole,
significantly more convenient than public transit.

A rural existence may be even more auto-dependent than a suburban
one: when gas prices shot up in 2008, numerous rural school districts
from New Jersey to Minnesota responded by cutting back on school bus
routes and even school days because the cost of transporting students
the substantial distances from home to school was simply too high.

Long-distance joyrides are more than a convenience in America,
they are a passion. "Whither goest thou, America, in thy shiny car in
the night?" Jack Kerouac asked in his novel *On the Road*, in which road
trips signify the searching, restless soul of the nation. Cars are by far
the preferred mode of recreational transport in the United States. Even

in the summer of 2008, when gas prices hit record highs, some three-quarters of Americans vacationed in cars. According to the Department of Transportation, the average American driver travels between 30 and 40 miles per day or nearly 14,000 miles a year—the distance around the equator every 1.8 years.

European drivers travel considerably shorter distances by comparison. Part of the discrepancy stems from commuting patterns: while 20 percent of Europeans walk or ride their bikes to work, less than 3 percent of Americans travel to the office motor-free.

America's heavy fuel consumption is as much a function of the great distances we travel as it is of our fuel-hungry vehicles. The love affair with cars in the United States has evolved from the Sloan era of the 1950s into an obsession that is increasingly impractical—a reality that's as apparent inside the NASCAR subculture as it is in the parking lot of any American supermarket. "How much sense does it make for a 113-pound housewife to get into 4,000 pounds of machinery and drive 2 blocks for a 13-ounce loaf of bread?" the writer Harry Crews once noted. We buy big gas guzzlers and slick luxurymobiles not so much for practical reasons as because we like the way they make us feel.

Our national car fleet gets an average fuel economy of just under 23 miles per gallon—that's roughly half the efficiency of cars in Japan and the European Union, which have average fuel economies equivalent to nearly 43 mpg. "Even the godless communists in China have higher fuel economy standards than the leaders of the free world," quipped Dan Becker, an environmentalist who worked on fuel economy legislation at the Sierra Club for over thirty years. "Their laws were about 5 mpg tighter than ours in 2002, and they're about 10 mpg tighter today. The same goes for all our trade partners, including Australia, South Korea—we're coming in last in the race on fuel economy among developed nations."

Mandating stricter mileage standards has until very recently been widely perceived by a majority of U.S. voters as un-American and has therefore been wildly unpopular in Washington. Congress did pass fuel economy legislation when gas prices soared in 2007, but before that it had not voted to significantly improve fuel economy standards in nearly thirty years—since 1980, just after the Arab oil embargo. The tides have begun to change on this issue now as federal officials take on global warming and energy independence, but consumer purchasing habits have remained fickle. While SUV and light truck sales plunged 40 percent between 2003 and 2008, as gas prices more than doubled, they bounced back up in 2009 as soon as gas prices again declined. Moreover, the total number of vehicle miles traveled in the U.S. barely responded to these price fluctuations, dipping less than 4 percent when gasoline reached $4 a gallon.

This driving behavior is hardly rational, but it stems from the very same philosophy Alfred Sloan put in place half a century ago. My father remembers his first car purchase at the age of seventeen, a secondhand Alfa Romeo he bought for $800 from a neighbor, as "the niftiest thing I ever laid eyes on. It was my first real taste of freedom, my passage into adulthood." We all still love cars not just as a means to get from point A to point B but for the sense of identity they give us. Cars and highways have shaped who we are, for better and for worse—the contours of our neighborhoods, the conveniences of our drive-throughs and shopping malls, the freedom to travel where we want, when we want, stopping when we please.

ROAD WORRIERS

Packing up my sagging tent in Field C at Talladega, I struck up a conversation with an amiable family from Missouri camped out nearby. The four boys, aged twelve to nineteen, and their parents had driven 600 miles from their home in Lebanon in an RV they'd named "Bigfoot." It's a voyage they make every year because, as one of the kids told me. "Getting here is half the fun."

Clearly a large portion of the other Talladega fans were in agreement. I conducted an unofficial survey of the license plates in my area of Field C and found that while about half came from neighboring states in the southeastern United States, at least a quarter came from places as far away as Maine, Iowa, New Mexico, and South Dakota. It would have been futile to try to accurately assess the total amount of fuel burned in the collective pilgrimage to this event (not to mention the many other NASCAR races per season), but it can be estimated conservatively in the tens of millions of gallons.

I asked my Missouri neighbors how much their fuel bills—in a Winnebago that gets 8 mpg—had been affected by rising gas prices. The father, a tall, bearded man in his fifties cooking a hot dog on a fork over his smoldering fire pit, answered, "It'll cost you. But we adapt—cutting back on the restaurant stops, maybe going direct instead of taking the scenic route." But as oil prices swing toward cost extremes, he conceded that eventually Bigfoot may not be able to make the journey from Lebanon to Talladega.

As I looked out over the sea of campers and Winnebagos in Field C, I wondered what would happen to this scene if oil stopped flowing tomorrow. The answer, simply, is that NASCAR would go with it, along with a piece of the American identity and a slice of the American dream.

The quick-change artistry of plastic is
absolute: it can become buckets as
well as jewels.

—Roland Barthes, philosopher

4

Plastic Explosive

FROM BAGGIES TO BOOB JOBS—OUR
LOVE AFFAIR WITH SYNTHETICS

A snub-nosed, two-seater Smart car whizzed past a GMC Yukon in my neighborhood recently like a mouse outrunning a buffalo. The miniature car is cartoonish in appearance, looking more like a product of Pixar than of the makers of Mercedes-Benzes. Measuring just 8 feet in length from bumper to bumper—the width of a standard parking space—it can be parked crosswise in a parallel spot, nose-first against the curb. Notwithstanding its toylike appearance, this vehicle means business: the German-based Daimler AG has sold more than a million Smart cars worldwide since their 1998 debut. Smart cars have also been racking up waiting lists in the United States, where they were initially expected to flop given the American consumer's penchant for extra-large automobiles. Part of the model's cult appeal—even seven-foot-one basketball legend Shaquille O'Neal has had one custom-made—is that it's pennywise and planet friendly. Just over 1,500 pounds, a Smart car is one-third the weight of a large SUV and gets about 40 miles per gallon on the highway—almost as much as the Toyota Prius, but at half the sticker price. The secret to success in this case isn't break-through engine design but something far simpler: plastics.

Daimler engineered fuel savings not just by shrinking the size of the car but also by lightening its load through the use of featherweight ma-terials. The company swapped out portions of the standard heavy steel-and-glass exterior of the car with two kinds of plastics—polypropylene, a synthetic that can be found in everything from carpets to car batter-ies, and polycarbonate, the same superstrong material used to make bulletproof glass. Polypropylene is molded into the panels that make up the Smart car's exterior shell, and the polycarbonate is used for its transparent roof. The exterior panels are infused with color while still in liquid form, cutting the need for a paint job (a costly and energy-intensive process). They can be easily popped off the car frame, replaced, and entirely recycled. While critics have argued that the downsizing and down-weighting of vehicles sacrifices safety, Daimler contends that the Smart car's crash test performance rivals that of similar all-steel vehicles. Add to that the car's environmental benefits: the use of plastics in place of glass and steel results in annual fuel economy savings equivalent to millions of tons of CO_2 emissions fleetwide when compared to a conven-tional car fleet of the same size.

By all accounts, plastic car shells are becoming ever more prevalent in the auto industry. "You're going to see steel phasing out—everything is going lightweight," one Detroit materials engineer told me. "It's the quickest and cheapest way to improve gas mileage."

The hitch here is that the very same resource these lightweight materials are intended to conserve—oil—is also one of their main ingredients.

Polycarbonate and polypropylene are, like the vast majority of plas-tics, derived from petroleum and natural gas. Using heat and other agents of chemical transformation, manufacturers refine fossil fuels into by-products including gasoline, motor oil, and petrochemicals, which serve as the raw materials for plastics. Petrochemicals are then subjected to another blast of heat (sometimes thousands of degrees Fahrenheit), and "cracked" into molecular compounds known as "hydrocarbons" that can be blended with various additives and transformed into nearly eighty thousand different types of commercial plastics.

Look around you—petrochemical by-products are everywhere. I had this stunning realization as I was leaving Talladega's Field C. The thousands of gallons of fuel I had just witnessed propelling the NASCAR fleet were only part of the problem. Trapped in the slow-moving exodus of fans, I surveyed the vast array of plastic creature comforts lying discarded and forgotten in that Alabama meadow. I counted scattered Styrofoam burger boxes and coffee cups; plastic soda bottles and water jugs; polyethylene shopping bags; empty candy wrappers; deflated sacks of corn chips, Funyuns, and fried pork rinds; crumpled wads of Saran Wrap; sandwich baggies; Igloo thermoses and melted ice packs; abandoned diapers and pacifiers; old tires; tarps; folding beach chairs; and foam pads for sleeping bags.

The inside of my rental car, too, was fully sheathed in plastics, from the dashboard, steering wheel, gear shift, and stereo modules down to the flexible PVC floor mats and polyester seat coverings. Even my body was draped in synthetics—my jeans and T-shirt were made from cotton-Lycra blends, and my socks and undergarments were woven with elastic polymers. The plastic water bottles resting in my resin cup holders and the plastic bags holding my snacks were also part of the tally.

These fossil-fuel-based plastics helped to satisfy a very different, more *human* purpose than the oil burning in my gas tank—that of feeding, hydrating, warming, and comforting bodies. All of the discarded creature comforts in that field had been worn, held, rested on, and eaten out of by people.

Most of us come into contact with more plastic every day than we do human skin—or any other substance, for that matter. Just think about how many times a day your fingers touch plastic keyboards, containers, escalator handrails, elevator buttons, door handles, and Formica countertops; how often your mouth touches plastic forks, straws, coffee lids, bottle tops, and phone receivers; and how often your skin touches plastic fabrics—spandex, fleece, Lycra, Modal, polyester, rayon, nylon,

Ultrasuede, and leatherette. These oil by-products fulfill a role in our lives far more intimate than fuel.

But in recent years, the use of plastic products has acquired a stigma to rival that of gasoline consumption, in part because of all the energy used to produce these items. Taken together, the fossil fuels used to manufacture plastics account for roughly 5 percent of total annual U.S. energy consumption. That may not sound like much, but it translates into the energy equivalent of roughly 600 million barrels of oil.

As we look to a future of lighter-weight cars, more electronic gadgetry, and dwindling reserves of raw building materials such as trees and granite, it appears likely that plastics will only become more—not less—pervasive in our lives.

I've tried to reform my plastic-hoarding habits in recent years. I've started carrying my own canvas bags to the grocery store and bringing filtered water to work in Mason jars. I now decline the plastic utensils when I get takeout, the plastic bags on my dry cleaning, and the Styrofoam cups at my coffee shop, filling up a travel mug instead. I get refills for soaps rather than buying new dispensers, and purchase my nuts and grains from the bulk section of the grocery store to cut down on plastic packaging. But these efforts have only made me more aware of how little there is in my universe that isn't plastic—and, moreover, of how few good natural substitutes exist for the mountains of plastic products I use.

As I dove into my plastics research, one particular passage in the book *Plastic: The Making of a Synthetic Century* by Stephen Fenichell stood out in summing up this predicament. Fenichell describes a phase in his life when he was seized with a fit of "anti-plastics passion" and

rushed out to buy a pure cotton shower curtain, a varnished oak toilet seat, a wood-handled toothbrush with genuine bristles—the kind they pull out of the backs of pigs in Third World countries. But soon I regretted my reckless embrace of Mother Nature. After my next shower, the new curtain wouldn't dry. I had to buy a nylon liner to protect it from water. My beautiful bristle toothbrush also refused to dry. The idea of inserting moist animal parts in my mouth began to seem a little too au naturel. The final indignity occurred when my oak toilet seat cracked setting me back another 40 bucks. A full-fledged plastic purge is a luxury few of us can afford.

I'd go a little further to say that a full-fledged plastic purge is not just unaffordable, it's virtually impossible. Where are you supposed to get a nonplastic cell phone? A natural-fiber laptop computer or iPod? A sippy cup made of glass (what would be the point)? A car without a plastic bumper or dashboard? An electronic gadget without plastic-coated wires (without this safety measure, the devices could short circuit and catch fire)? Even some products that appear to be plastic-free—cardboard milk carton, for instance, and aluminum soda cans—are lined with invisible layers of plastic to help keep the contents fresh.

While the sheer ubiquity of synthetic materials is unnerving, plastics can have environmental benefits that outweigh their costs—like the Smart car panels that lighten the load and increase gas mileage (considerably less oil is used to make the car's plastic panels than is saved over the product's lifetime by its fuel economy benefits). Or the polyurethane foam insulation inside your refrigerator and freezer—without it, these appliances would use significantly more energy.

There's even a good argument to be made that cutting down trees for building materials and carving marble out of mountains for countertops is worse for the planet than is using synthetic substitutes. There are many problems associated with the use of plastics—like the inadequate recycling practices that lead to their filling up landfills. Moreover, the presence of toxic ingredients such as bisphenol A in certain plastics has raised serious public health concerns. These issues deserve careful scrutiny, but with all the current criticism of plastics, we tend to lose sight of how miraculously versatile—and, in many cases, irreplaceable—these materials really are.

MODERN ALCHEMY

Plastics don't get rusty, rotten, weathered, dull, or tarnished. At most, they can warp or crack, but as much as we pound our keyboards, drop our iPods, plod across our vinyl floors, and slap our phones into their receivers, the plastic generally holds up to the beating. Most plastics were engineered to do just that—to break down either incredibly slowly or not at all. Plastic bags and food containers can take decades to decompose; and many plastics dumped in landfills, including standard water bottles and foam products (like those colorful sling-back Crocs shoes, Styrofoam boxes, home insulation, and my cherished Tempur-Pedic mattress), may never—as in *never ever*—decompose. Some plastics are, as the saying goes, forever.

This presents significant and dire problems from an environmental standpoint, as we glut our landfills and sully our oceans with permanent waste. But the unparalleled durability of plastics can be a good thing when it comes to safety, for instance. You wouldn't want your synthetic rubber brake pads quickly wearing down and giving out as you do your weekly errands, or your safety helmet cracking if you fell off your motorcycle, or the lenses of your glasses shattering if you bumped into a door, or the vinyl siding on your home eroding in a downpour, or a bulletproof vest giving out because it wasn't made of Kevlar (come to think of it, before plastics we didn't *have* lightweight bulletproof armor for soldiers and law-enforcement personnel).

What this means, in an era of unstable oil prices and dwindling supplies, is that eliminating plastics from our lives would redefine our identities on a level far deeper and more emotionally immediate, even, than removing oil from our gas tanks.

This is a paradox of plastics: one of the qualities that make them such an environmental nuisance—their durability—is also what makes them so impressively utilitarian. Moreover, plastics are much easier to form into products than materials like wood, ceramics, and metal, which require a great deal more input of energy to shape. That's what makes plastics so cheap, so conducive to mass production, so disposable, and so easy to replace—and therefore both so convenient and so menacing to the environment.

For all the problems that have yet to be solved within the plastics industry concerning disposal, recycling, and in certain key instances toxicity, these concerns should not entirely discount the many virtues of synthetics.

Nowhere is the durability—indeed, the *permanence*—of plastics more valuable than in the field of medicine. Pacemakers and artificial heart valves are among the dozens of implantable medical devices made from plastic—devices that you wouldn't want decomposing or abrading inside your body.

You doubtless know someone who has benefited from plastic body components. My father-in-law had a plastic-encased pacemaker; my mother-in-law has two titanium-and-plastic hips; my grandfather had a plastic prosthetic leg; my dad and I wear contacts—disposable plastic lenses that are essentially temporary implants that enable us to see better. Athletes are among the most common recipients of plastic implants. Yankees pitcher Jimmy Key has plastic to thank for his victory in Game 6 of the 1996 World Series. Fifteen months earlier, Key (a lefty) had undergone surgery for a torn rotator cuff. Doctors inserted plastic stitch anchors into Key's left shoulder bone and sewed the ruptured tendons and muscles to the anchor with bioabsorbable polymers (plastics that disintegrate over time in the flesh), enabling Key to fully recover.

My friend Malia received a difficult diagnosis of breast cancer at the age of thirty-eight, underwent a mastectomy, and had reconstructive surgery that involved a new state-of-the-art silicone breast implant material. She marveled that her implant "came with a twenty-five-year warranty—like you bought a tire or some mechanical part. It's right out of *Star Trek*. I'm practically bionic." Far from being put off by the permanence of her implant, she found it comforting, telling me wryly, "I already had one faulty boob—I really don't need another."

As I was rolling out of Field C surveying the plastic wreckage that would soon be swept up, stuffed into bags, and hauled to landfills, I spotted a Pamela Anderson lookalike with breasts the size of twin volleyballs. They were also the gravity-defying *shape* of volleyballs, a fact that led me to the conclusion that I was looking at a product of modern medicine, not of nature. Cinched up in a Daisy Duke–style gingham blouse, these miracles of plastic surgery struck me as the very epitome of petroleum's versatility. Bionic, indeed! The same substance that fuels a B-52 jet can also be embedded into human flesh in the form of a translucent pouch. If this isn't modern alchemy, what is?

The most intimate terrain in America's energy landscape, I realized, is the human body—one of the last things I'd expected to explore when my research on petroleum began. How, I wanted to know, is oil transformed into a medical implant? Why are petroleum by-products the best materials for such devices, and what sorts of alternatives—if any—exist? What are the risks and benefits of plastics, and what does it actually take to embed these materials inside a human being?

WEIRD SCIENCE

First, a quick primer on polymers. The word *polymer*, often used to describe plastics, comes from the ancient Greek words for "many parts." It's an apt description, since plastics are essentially long chains of simple building blocks called hydrocarbons—compounds consisting of carbon and hydrogen. Most plastics have their own unique assemblies of hydrogen and carbon atoms arranged in different formations (some also contain other atoms including nitrogen, chlorine, and silicon). Hydrocarbons can be found in nearly all organic substances—coal, plants, animals—but they are most abundant in and most easily extracted from crude oil and natural gas.

It is no small miracle that American engineers found a way to transform natural gas and petroleum—the very same substances that power war, heat homes, and fuel vehicles—into products as diverse as toaster ovens, condoms, toilet seats, bulletproof vests, and even body parts. After the petrochemicals are refined from petroleum or natural gas and polymerized, they continue through the production process in the form of liquids or solid powders or pellets, winding up in manufacturing plants where they're molded, stamped, or otherwise formed into millions of different shapes and items. The energy equivalent of roughly 1 gallon of oil yields 3 pounds of plastics such as polypropylene.

While I understood that plastics come in every imaginable form—materials that are gossamer-thin, cinderblock-thick, soft as cashmere, hard as steel, bendable, squishy, bouncy, absorbent, waterproof, odorous, and scent-free—I knew almost nothing about the science and business of plastics. So I sought out a tutorial from Daniel Schmidt, an assistant professor at the University of Massachusetts, Lowell, which has the top plastics engineering program in the country.

Polyethylene (PE), he told me, is that familiar soft plastic in plastic shopping bags, milk jugs, and shampoo bottles, and is produced at a global rate of 65 million tons a year. Polyethylene terephthalate (PET) is found in soda bottles, camera film, fleece jackets, upholstery blends, and other versatile fabrics and materials. Polycarbonate, used in the shell of the Smart car, is also used to make hard hats, eyeglass lenses, and water-cooler jugs. Polystyrene (PS) comprises everything from computer and appliance casings to Scotch tape dispensers and Styrofoam. Polypropylene (PP) is used in Tupperware and is also the quick-drying insulating fiber in many types of long underwear. Then there's polyvinyl chloride (PVC), used in products such as industrial-strength sewage pipes. Acrylics are another versatile class of plastics that includes Plexiglas, latex paints, and "Superglue" (which, when drying, actually polymerizes into a superstrength plastic). Polyamides are used to make everything from nylon stockings and suitcases to skateboard wheels, bulletproof vests, and fire-resistant clothing. Finally, polyurethanes are used in automobile clearcoats, furniture foam, and home insulation.

Many medical implants and prostheses, Schmidt explained, are made of silicone, which is a flexible, lightweight, inert plastic that can be manufactured to bear a remarkable similarity to the look and feel of flesh. Silicone is made from two basic ingredients—silica, a nonorganic substance derived from quartz, and a petrochemical known as methane, commonly derived from natural gas. Processing the silica and combining it with methane requires a series of complex chemical reactions and the application of temperatures of up to 4,000 degrees Fahrenheit, which makes it expensive and energy-intensive as plastics go. (Common plastics used in everyday products are cheaper because they require simpler and fewer chemical reactions.)

Countless nonsilicone plastics are also commonly used in medicine. A majority of today's medical equipment—syringes, blood bags, surgical gloves, dressings, catheters, and IV tubes—is made of polyethylene, nylon, and flexible PVC. These disposable materials guarantee sterility, cutting down on potential infections that were far more common in the pre-plastics era. As Malia recalled her months of chemo, she told me: "Plastics were my lifeline during that time—literally. A fresh syringe was used for each injection, a new bag and tube was used for each round of chemo, I was constantly handled with surgical gloves. Honestly, I appreciated the sterility of it all—it was comforting."

Amazingly, certain forms of chemotherapy also have core petroleum-derived ingredients called nitrogen mustards. In fact, as Professor Schmidt told me, "the vast majority of pharmaceuticals come from petrochemicals." Oil-derived carboxylic acids and anhydrides are used to make painkillers such as Novocain and the acetaminophen in Tylenol, as well as sedatives, tranquilizers, decongestants, antihistamines, and antibacterial soaps. Esters and alcohols derived from fossil fuels are used in the fermentation process that produces antibiotics. Glycols and celluloses are used to coat pills and bind together the contents of tablets. The list of medical products in which petrochemicals may be found includes everything from penicillin, cough syrups, and rectal suppositories to radiological dyes and X-ray film. Petroleum by-products also find their way into most cosmetics—lipsticks, foundations, mascaras, cleansers—in the form of moisturizing agents, alcohols, binders, and aromatic chemicals. Petrochemicals are also the basic ingredients in industrial glues and adhesives, as well as in the dyes that make up the ink in these letters and the paint on the walls of your room.

This basic primer from Schmidt on petrochemicals revealed to me an impressively vast and varied galaxy of products—one that made the most common form of petroleum, gasoline, look simple and uninspired by comparison. Still, I wanted a deeper understanding of these strange and surprising fossil fuel by-products—an in-the-flesh, real-world tutorial. I wanted proof that the same substances that fight wars, fuel our commutes, and generate electricity can also be fused with the human body.

THE BODY SHOP

The operating room of Dr. Grant Stevens is located in a nondescript mixed-use office building just off the San Diego Freeway and minutes from the Los Angeles airport—easy access for his many patients who travel from afar for their procedures. I had made arrangements to observe one of the most common surgical procedures in the United States. Between 300,000 and 400,000 breast implant operations are performed annually in America. Southern California is an international mecca for plastic surgery, and Stevens is among its most sought-after practitioners. The office was hard to find, and deliberately so—no placard identifies the business, only an anonymous suite number, in order to protect patient privacy.

Once inside, I found the lobby to be an oasis of calm. It was decorated in earth tones, with linen-textured taupe wallpaper, an ecru suede couch, freshly misted orchids, and sepia-toned watercolors of old-fashioned schooners. I half expected a massage therapist to walk out and offer me a bathrobe. On the sound system, Sade crooned, "You gave me the kiss of life." Piles of photo albums were arranged on the coffee table filled with before-and-after shots of breast enhancements, nipple repair, face-lifts, dermabrasion, and wrinkle fillers. "Build a better you with Botox!" read one caption. "Get yourself a pair o' bears!" read another, referring to the latest in breast implant technology—a material called high-strength cohesive silicone that is more commonly known, in nip-and-tuck argot, as a "gummy bear" breast implant for having a texture like that of gummy candies.

The field of plastic surgery did not derive its name from the polymeric substances I've just described. Both the material and the medical discipline share their roots in the Greek word *plastikos*, meaning "to mold" or "to shape." Modern-day plastic surgery had its origins in war—more specifically, in the efforts of the New Zealand–born Dr. Harold Gillies to repair the grievous burns and other disfiguring facial injuries received by British soldiers in World War I. Though the term *plastic surgery* today is often associated with minor, voluntary cosmetic procedures, its practitioners continue to perform essential work, including reconstructive surgeries after accidents and illnesses, administering skin grafts to burn victims, and repairing cleft palates.

Stevens performs a broad range of plastic surgeries, including the insertion of breast implants that can be reconstructive (as in Malia's case) or cosmetic. His task on the day of my visit fell into the latter category—what Stevens refers to as a "technology upgrade." The thirty-two-year-old patient would be exchanging her saline breast implants for a pair of the more technologically sophisticated "gummy bears." Dr. Stevens proudly claims to have coined this term. He is one of only a handful of doctors in the United States who have received the FDA's permission to implant cohesive silicone in patients, and is the technology's most zealous advocate.

Unlike the liquid filling of conventional implants, the "gummy bear" implant is made of a pliant, semisolid silicone that can't tear or leak. The cohesive material was first designed in U.S. labs in 1992, a time when health concerns were tarnishing the reputation of silicone in fluid or gel form: people worried that if the pouch that contained it ruptured, liquid silicone could enter the bloodstream, possibly causing cancer or autoimmune diseases such as lupus and arthritis.

While "gummies" have been widely adopted throughout the rest of the world, they are still banned for common use in America, where "silicone is practically a four-letter word," says Stevens. In 1992, the FDA called for a voluntary moratorium on all silicone breast implants while it investigated whether the material could cause autoimmune or connective tissue disorders. This action was a response to a series of ongoing lawsuits by women alleging that the silicone implants they'd received had caused these health disorders. What did not make sense about the moratorium was that other types of implants made with silicone—from pacemakers to artificial testicles—had not fallen under scrutiny. (Saline implants also contain the offending plastic—the pouches themselves are made of silicone, even though their filling is simply salt water.)

In 2006, after fifteen years of research, the FDA quietly dropped the moratorium against silicone gel, having found no conclusive link between this substance and disease. (While the old liquid silicone technology now has approval, the new cohesive silicone is still making its way through the regulatory hoops.) Saline implants accounted for 90 percent of the breast implant operations performed annually in the United States up until 2007. Dr. Stevens predicts that the tide will soon turn toward cohesive silicone, which is already used in more than 90 percent of breast implant procedures performed outside the United States.

As part of a clinical trial to prove the safety of cohesive silicone to the FDA, Stevens has implanted more than nine hundred gummy bears since 2005—significantly more than has any other U.S. surgeon. I had come to witness his 917th procedure, and as I sat in the waiting room I battled the jitters—would I vomit, I worried, or faint?

Dr. Stevens, an exuberant man in his fifties with thick sandy brown hair, a chestnut tan, and a blinding white grin, came barreling through the door in Tommy Bahama–style scrubs decorated with green palm fronds and exotic birds. Stevens attended the University of Oregon with a double major in psychology and science and a minor in fine art. He paid his way through college working odd jobs at a chicken processing facility and driving eighteen-wheelers. He shifted his focus to medicine after a strange twist of fate: One day during a jog, Stevens encountered another runner who had sprained his ankle, and offered him a ride home. The man was a prominent cardiac surgeon who invited Stevens to sit in on an open-heart surgery. "It was the coolest thing I'd ever seen; I decided on the spot to become a cardiologist." He put himself through Washington University medical school in St. Louis working the graveyard shift in the Emergency Room. Stevens quickly found his niche in medicine: "I realized plastic surgery has so many more artistic things that I can do. This is way more fun than cardiac surgeries where they just sew in the vessels and put in the valves."

Stevens, who now teaches at the University of California, Los Angeles medical school and the University of Southern California alongside his private practice, radiates pride in his career, describing the gamut of plastic surgeries he performs almost as acts of public service: "I'm not naive enough to say that plastic surgery isn't sometimes frivolous," he told me. "But I see my role as giving people the emotional freedom of self-confidence—whether it's a car accident victim who has been disfigured or a flat-chested woman who feels unfeminine and ashamed."

Once I'd suited up in mint green medical scrubs, a hair net, and face mask, Dr. Stevens briefed me on the patient: "We've got a thirty-two-year-old, five foot six, 120 pounds." I pressed him on her frustrations with the previous saline implant. "The old technology is archaic—it ripples and wrinkles the skin; it feels hard, unnatural. This girl is a volleyball player, and she told me, 'I don't like the feel, I don't like the look, it sloshes'—patients say it to me all the time. When she's out there on the beach jumping around it sounds like she's got a couple of water balloons in her chest." Gummies are by far the superior technology, Stevens added—"a breakthrough, really, in synthetics."

SILICONE VALLEYS

The patient was lying on a surgical table, her head concealed behind a hanging curtain and her lower body beneath a white hospital sheet. The general anesthesia had taken effect and the woman's body was motionless and pale, her breasts covered in markings—arrows and dotted lines that indicated where the incisions should be made and the seams sewn. A "D-" was scrawled below her collar bone in red marker—an indication to Stevens that she wanted a small D-cup. A medical fellow of Dr. Stevens who would play the role of wingman stood beside him at the table in green scrubs, and a team of four nurses stood by to ready the surgical instruments and cleaning agents, while an anesthesiologist in a white lab coat monitored the patient's vital signs. The overhead lights were blindingly bright, bouncing off the polished steel counters and the lacquered floor. A heart monitor beeped in the background—a constant reminder to me that this surreal experience was happening not to a headless mannequin but to a live woman.

Despite my nervousness going in, the surgery proved to be remarkably tidy and swift. Stevens approached the procedure with the sangfroid of a man who often performs eight surgeries a day—rapid, assured, aloof. He made a 3-inch incision at the base of the breasts as his medical fellow kept the blood flow at bay with an electric wand that cauterized the cuts. Working in close concert, Stevens and his medical fellow then held open the two pockets of skin, removed the old implants with a large pair of metal tongs, and tossed them in a biohazard bag. "Put your finger in there and see the size of the cavity," he told his fellow, handing him a dummy implant known as a "sizer," this one filled with 400 milliliters of water. "Let's see how the four hundred looks—it may be a smidgen big."

Once Stevens found the right size, a nurse pulled the corresponding set of implants off the shelf, where dozens of boxed gummies were tidily stacked. Stevens and his fellows popped the new implants out of their sterile plastic pods, rinsed out the open cavities with a miniature hose, stuffed in the fresh state-of-the-art silicone fillers, and swiftly but neatly sewed the patient up with biodegradable plastic sutures. I was shocked to look up at the wall clock and see that the entire process had taken about twenty-five minutes flat. (In a patient who hasn't previously had implants, there is a more involved process of stretching open the pockets into which the implants will be inserted, but in this case they were already there.)

Stevens is very irked by the fact that the United States "is fifteen years behind the rest of the world" when it comes to the caliber of polymers used for breast implants. In his storage room after the procedure, Stevens lined up examples of different types of implants on a counter for me to feel for myself, telling me, "I've felt hundreds of thousands of breasts in my life and I can tell you they're not supposed to feel like water balloons. It's ludicrous—who should have to live with breasts that slosh and wrinkle?" Saline implants are also more susceptible to leaking, requiring additional surgical procedures for repairs.

I approached the display of rounded objects. The saline was, yes, a sloshy bag of water. The liquid silicone gel was a little more realistic, like a pouch of shampoo. The cohesive silicone, meanwhile, felt like a ripe peach, a mound of firm Jell-O, a fist-sized gumdrop—remarkably like flesh. In the future, said Stevens, this material will be molded into different crescentlike shapes for the left breast and the right breast (rather than standard circular pouches) so that the result will look more natural.

What shocked me most was how this man-made substance had been so easily accepted by a human body. Why wouldn't the patient's immune system reject it or attack it as an invasive object, as something that shouldn't be there? Miraculous as it may seem, silicone is a substance that has been proven to be biocompatible with human bodies—a fact corroborated by the FDA when it lifted the moratorium on silicone gel breast implants. Silicone is odorless, is impervious to mold and rot, has high elasticity, and can withstand extreme temperatures. (It has one of the highest melting points among plastics, which is why it is often used to make items like bakeware, oven mitts, and the handles on pots and pans.) It is, in short, an ideal substance for implantable devices—be they artificial kidneys, pacemakers, heart valves, or breasts—that need to be able to safely dwell for decades inside a warm body without corrosion or decay. It's also the ideal substance to mold into prosthetic devices for patients who have suffered the loss of ears, noses, limbs, and even eyes.

You may have heard about the emerging field of "green plastics," in which carbon is extracted from cellulosic plant crops such as corn rather than from fossil fuels, and manufactured into materials that can be entirely biodegradable. For the most part, these bioplastics are significantly more expensive to make. Research is under way to make the process cheaper and the products more sophisticated, but for now bioplastics applications are largely limited to simple products intended for short-term usage, such as food containers. Stevens told me of an effort in the mid-1990s to manufacture "natural" breast implants made of fillings derived from soy and bamboo; however, many among the hundred or so test subjects for these products found that the implants began to decay internally, giving off a rotting-garbage smell, and had to have them promptly removed. The concept has never been attempted again.

I learned at Dr. Stevens's office how deeply embedded—figuratively and otherwise—synthetic materials have become in our lives. But still the history of plastics seemed elusive. I wanted to understand how Americans came to accept this odd, mercurial, long-lasting, nonrusting, and decidedly unnatural substance—and to trust it enough to eventually adopt it as surrogate flesh.

A STAR IS BORN

The history of plastics, I would discover in the archives of a brightly lit research library at the DuPont chemical company's Wilmington, Delaware, headquarters, reads like a history of successive consumer fads—described by witnesses as "manias," so miraculous and exciting did these novel everyday products seem to American shoppers. The first wave of plastic mania came in the form of cellophane, which its manufacturer, DuPont, billed tantalizingly in 1923 as "Thin as tissue but hard to tear, like paper but not paper, transparent as glass but not glass—a non-fragile waterproof product with a wide range of uses. . . ."

Cellophane did not originally come from petroleum. Nor did other early plastics, including the celluloid used for film in early cameras and motion pictures. These plastics were actually derived, as their name suggests, from cellulose—the structural fibers found within green plants (the same plant fibers that today are used to make ethanol and other biofuels). Common feedstocks for early plastics were wood, cotton, and hemp, but these ingredients presented cost barriers. "It takes two chemical reactions to get commodity plastics from petroleum. It takes many more reactions to get these plastics from raw materials such as trees and plants," explained Rudolph Deanin, a professor emeritus and mentor of Daniel Schmidt's at the University of Massachusetts, Lowell. "A general rule is that each chemical reaction multiplies the cost of a product by two and a half times." So when early plastics engineers began to substitute petroleum for cellulose, they suddenly had a product that was not only multipurpose and durable but also cheap and affordable enough for mass consumption.

Cellophane, which DuPont first debuted at a National Confectioners Association meeting, was initially used to unit-pack individual products with a see-through wrapper that kept freshness in and dust out. The Whitman candy company readily adopted the new clear wrapping for its chocolates, and quickly saw a boom in sales. Other confections followed: "A national grocery chain store reported a 2,100 percent increase in doughnut sales in *two weeks* after wrapping its doughnuts in cellophane," wrote Stephen Fenichell in his book *Plastic*. The makers of Camel cigarettes seized on the plastic wrapping to increase the shelf life of their tobacco products, and saw such a surge in sales that Lucky Strike promptly tried to out-cellophane its competitor by adding a special easy-to-tear tab for quick removal of the glossy packaging.

"The boom was on," noted a 1949 *New Yorker* retrospective on the cellophane craze. "The country was suddenly overrun with cellophane, and soon thereafter with cellophane gags—crumpled-up cellophane, instead of ice, was dropped into people's highballs, Hollywood starlets appeared in cellophane bathing suits, and magazines . . . printed innumerable cartoons in which the [joke] depended on something being, or not being, wrapped in cellophane." By 1934, Cole Porter had canonized the substance in his song "You're the Top" for the hit Broadway musical *Anything Goes*. The song was inspired by a cruise on Germany's Rhine River during which Porter asked fellow guests what they considered to be the best things in life:

> *You're the purple light of a summer night in Spain*
> *You're the National Gall'ry*
> *You're Garbo's sal'ry*
> *You're cellophane.*

By the mid-1930s, a decade after its introduction to the American public, cellophane encased everything from the original documents of the Constitution and the Declaration of Independence to magazines, golf balls, cosmetics, and pickles. "Civilization as we know it today," the *New Yorker* remarked, "would practically disappear . . . without cellophane and its kindred products." In 1936, cultural critic H. L. Mencken's newspaper, the *American Mercury*, complained: "At the rate this peekaboo covering is being slapped on things, it will soon be a novelty to find something you can still buy in a state of nature."

The World's Fair of 1939 had its grand opening in late April in New York City. All eyes were on the future as the Great Depression gave way to a new economic momentum. The theme of the fair was "The World of Tomorrow," and attendance was overwhelming—more than 29 million people came from around the world. General Motors presented its famous "Futurama" exhibit, which carried each visitor in an armchair on a conveyor belt through an imaginary 1960 landscape featuring cars shaped like raindrops. But one display drew more popular response than any other: the display featuring nylon, the world's first purely synthetic fiber. Developed by DuPont as an affordable substitute for silk, nylon was introduced to America in the form of lustrous, featherweight ladies' stockings displayed in the process of being knitted by a machine, and as a finished product on the legs of models. "A fiber as fine as the spider's web, yet strong as steel, and more elastic than any of the common natural fibers," said a trade magazine.

American women went wild: cheap, sexy, and durable, these wardrobe staples were a radical improvement over cumbersome woolen leggings and fragile silken stockings available only to the rich. DuPont successfully promoted the nylon hose as a homespun, patriotic alternative to silk—an industry dominated by the emerging military enemy Japan. What ensued was a public sensation known as "nylon mania." Approximately 64 million pairs of nylons were sold within their first year on the market. *Fortune* magazine raved: "It is an entirely new arrangement of matter under the sun. . . . In over four thousand years, textiles have seen only three basic developments aside from mass production: mercerized cotton, synthetic dyes, and rayon. Nylon is a fourth."

The nylon mania halted abruptly in 1941 when America entered World War II—but the production of nylon continued to grow. Suddenly, the material was needed to fabricate parachutes and waterproof mold-resistant tents and uniforms for the Armed Forces. In droves, women who had just purchased the coveted nylon stockings donated their gossamer treasures to be recycled into gear for U.S. troops. Newspapers reported women "taking 'em off for Uncle Sam" and "sending their nylons off to war." What had once been a patriotic alternative to Japanese goods was now a patriotic feedstock for military equipment.

World War II presented a turning point for the plastic industry. "When the armed services faced shortages of aluminum, brass, or rubber," wrote historian Jeffrey Meikle, "they called on the plastic industry for substitutes often superior to the originals." Plastic products were used for everything from mosquito netting and rotproof shoelaces to tank tires, gears, bearings, and explosives. American companies produced roughly 130 million pounds of plastics in 1938. By 1958, annual production had skyrocketed to 4.5 billion pounds. The American plastic industry has since continued to see exponential growth, with production numbers leaping to 60 billion pounds in 1988 and then almost doubling to roughly 100 billion pounds annually in recent years—nearly a quarter of all global plastics production.

World War II proved to be as great a boon for plastics as it was for the automobile and suburbia, having provided ample evidence that plastics were valuable materials in their own right, not just as imitations of the real thing. Soldiers flocked home attesting that synthetics had protected and sustained them when their lives depended on it. "Plastics had moved from lighthearted aesthetic indulgence," wrote Meikle, "into tough functional seriousness." In 1945, *Life* magazine ran the headline "War Makes Gimcrack Industry into Sober Producer of Prime Materials." Another headline of that period trumpeted: "Plastic—The 'Stand-In' That Became a Star."

PARTY TIME

In the postwar era, even nylon mania couldn't keep pace with the rising phenomenon of Tupperware. The growth of plastics throughout the 1950s mirrors the car culture that was evolving at the same time. Much like General Motors, DuPont jettisoned the search for a single perfect model of plastic and instead vowed to create "a function for every plastic and a plastic for every function."

Earl Tupper was a farm boy from New Hampshire whose tree nursery business had been driven to bankruptcy during the Great Depression. Out of work, he took an entry-level job at the DuPont chemical factory. Earl was a self-taught engineer who kept daily diaries coaching himself on the art of invention, with entries titled "How to Invent," "How to Realize an Invention," and "The Purpose of My Life"—all revealing a tireless drive toward self-improvement not unlike the single-minded pursuits of John Rockefeller, Henry Ford, and William Levitt.

Using scraps of polyethylene borrowed from DuPont, Tupper experimented during the 1940s in his makeshift home laboratory, purifying the material and molding it to create lightweight, unbreakable containers. His novel injection-molding process could affordably produce an infinite variety of shapes and sizes. To accompany these products, Tupper designed a spillproof lid modeled after the airtight covering of a paint can. He had high hopes for his humble containers—saving leftover food, he believed, would "help dispel the discontentment of a wasteful consumer society," wrote historian Alison Clarke. Tupper's invention debuted in 1946 in hardware and department stores. His flagship product: the Tupperware Wonder Bowl.

During most of his adult life, Tupper "has described himself as 'a ham inventor and Yankee trader,'" read one 1947 profile of the inventor. "By last week, one of his inventions—an unbreakable, flexible, shape-retaining plastic which can be molded into all sorts of containers—was forcing him to temper the 'ham' and drop the 'trader' entirely." Initial sales were respectable, but Tupper's invention wasn't an instant success. The shy, balding, reclusive man could not devise a good marketing plan for his invention—until, in 1951, a perky, attractive single mom living in Detroit named Brownie Wise stumbled across Tupper's product.

Though she hadn't been to school past the eighth grade while growing up in rural Georgia, Wise was a natural-born marketing genius who was to become the first woman ever featured on the cover of *Business-Week*. Struggling on a small secretarial salary, Wise had been working as a door-to-door salesperson for Stanley Home Products to supplement her income. Immediately she saw the mass appeal of Tupper's storage dishes. She began staging "Poly-T Parties" at which she would educate her audiences on the best way to "burp" the container lid for an airtight seal. She often demonstrated the virtues of Tupperware by flinging a Wonder Bowl filled with stewed tomatoes across an immaculate living room and letting it drop to the floor without springing a leak. She began to assemble a team of similarly gutsy saleswomen throughout the region, many of whom were able to outsell a store's typical sales volume.

Tupper caught wind of Brownie's marketing prowess and hired her to head up a new sales operation: Tupperware home parties. The entrepreneur set his sights on expanding his product line to include offerings such as tea sets, turkey basters, egg poachers, apple slicers, ketchup funnels, and cake domes. Wise, meanwhile, began to train a nationwide army of Tupperware saleswomen with one "basic merchandising philosophy," as *BusinessWeek* described it: "If we build the people, the people build the business."

Sometimes appearing in her "bonnet"—an inverted Tupperware bowl perched on her head—Wise, according to a recent PBS documentary, proudly promoted her own brand of 1950s feminism: "Women who had worked in factories, or five-and-tens, or on farms, were now dressed in white gloves and hats, self-assured, able to speak publicly with confidence." By 1955, Wise's sales force had grown to more than half a million women. Referred to as the "Prophet in Plastic," she adopted cultish motivational tactics to inspire these workers. She literally buried in the ground tens of thousands of dollars' worth of merchandise—mink stoles, appliances, diamond rings, and plane tickets to Europe—and had her saleswomen dig for their treasures. She held "séances" at which hundreds of women at a time "shut their eyes, rubbed their hands on a block of polyethylene, and wished—that they might sell more Tupperware," reported *BusinessWeek*. The media was transfixed by Wise's unabashed celebration of consumerism; one Methodist preacher hailed her tactics as a "bulwark against communism."

In 1958, Tupper forced Wise out, reportedly having grown jealous of this media attention. Shortly afterward, Tupper garnered $16 million for the sale of his company to Dart Industries and retired to a private island. Today, Tupperware still maintains a stronghold in the global plastics industry, employing 2.2 million consultants worldwide.

In postwar America, Tupperware provided the bridge between a cold, soulless industrial material and the warm intimacy of the home. This was the first time plastic had been successfully promoted to average consumers as a material with its own special look and feel—not just a product wrapping but the product itself. It was advertised not as a substitute for silk, steel, or other "natural" products but as a material that could not exist in any better or more original form. Tupperware's phenomenal popularity proved to be a tipping point of sorts, signaling the mainstream acceptance of plastic in domestic settings. Tupper's products were promoted in advertisements against backdrops of comforting images—housewives pulling casseroles from ovens, white picket fences, kitchens with state-of-the-art appliances, living rooms cluttered with creature comforts. These images captured the curious combination of decency, frugality, and indulgence that defined 1950s America.

It was during this era of proliferation that physicians performed the first medical implants of synthetic organs. Dr. Willem J. Kolff, a native of Holland, developed the first artificial body part—an external dialysis machine that performed the role of a kidney—in a remote hospital during World War II. Soon thereafter a Kentucky-born surgeon performed the first medical implant: Dr. Charles Hufnagel made global headlines in 1952 when he implanted the first partial heart replacement—a prosthetic heart valve composed of silicone and Dacron (a form of polyester).

Having immigrated to the United States, Kolff then completed the first total artificial heart implantation into a dog in 1957. Heading up the University of Utah's Division of Artificial Organs and Institute for Biomedical Engineering, Kolff worked with his students to perfect the artificial heart design until, in 1982, the invention was famously implanted into the chest of sixty-one-year-old Barney Clark. It kept him alive for 112 days, and was still ticking after he died of other complications. By then, Kolff had founded the first commercial supplier of plastic internal body parts, which produced its hearts for $20,000 apiece. The motivation behind his work, Kolff said, was the simple belief that "what God can grow, Man can make."

PLASTIC GOES POP

As doctors used plastics to save lives and housewives used them to save leftovers, artists were forging an entirely different approach to synthetic materials, splashing them across canvases and molding them into sculpture. In the 1960s, plastics reached the height of their cultural "cool." The vehicle for this elevated status was Pop Art, a movement commonly associated with the irreverent, droll, unabashedly commercial work of Andy Warhol. In Pop Art, there was no distinction between "high art" and "low art"—mass-produced works were considered to be just as sacred as originals. Pop artists were engaged in a restless, relentless celebration of consumer society and advertising, as exemplified by Warhol's repetitious silkscreened ode to cans of Campbell's soup.

Warhol popularized the use of bright, shiny synthetic paints such as acrylic, latex, and liquitex as alternatives to traditional oil-based paints. Cheap, quick-drying, and available in a high-intensity color palette, acrylic paints had been ignored if not shunned by the world of fine art until Warhol came along. He was the first to sell highly valued paintings and silkscreens made out of this low-value medium. Roy Lichtenstein, known for his paintings of blown-up comic book images, painted vast landscapes on sheets of plastic that gave his work an ethereal shimmer.

The artist Christo used plastic fabrics he described as "packaging" in installation art such as his 1969 wrapping of 1.5 miles of coastline in Sydney, Australia.

Perhaps the most striking example of the 1960s love affair with plastic was Warhol's "Exploding Plastic Inevitable," a series of theatrical events held in cities nationwide—deliberately gaudy, highly stylized celebrations of modern pop featuring the music of the Velvet Underground, with plastic-clad go-go dancers and graphic multimedia movies displayed in the background. It wasn't just products and performance that were "plastic" to Warhol—people, too, were plastic, like pliable decorative objects. He openly discussed having cosmetic surgery on his nose. "I love Los Angeles. I love Hollywood. They're beautiful. Everybody's plastic, but I love plastic," he said. "I want to be plastic."

Certainly not all artists of the time felt this way. The painter Mark Rothko, known for filling his canvases with subtly shifting blocks of color, refused to show his work at top galleries that had also showcased Pop Art. "For Rothko's generation," noted one art journal, "plastic was the emblem of everything antithetical" to fine art. Whereas Rothko wanted to lure viewers into the nuanced depths of paintings, Warhol wasn't interested in depth of any kind. "Just look at the surface," Warhol said—both of himself and his art. "There's nothing behind it." He wanted the experience of his work to be clean, smooth, light, and airtight—not unlike the feel of Tupper's Wonder Bowls.

It was, in fact, Tupper himself who had inadvertently paved the way for the Pop Art movement. His polyethylene creations were selected in 1956 as part of a permanent design collection of the Museum of Modern Art. The Tupperware brand "entered middlebrow folklore—and highbrow scholarship," wrote Jeffrey Meikle, "as one of the most common symbols of the time."

Plastic symbols quickly spread beyond people's kitchens and into their living rooms and wardrobes. Throughout the 1960s, plastic furniture began to adorn the homes of trendsetters: fashionista Tatiana Lieberman, for instance, filled her living room with cheerful plastic patio furniture. *Better Homes and Gardens* celebrated the freedom that plastic gave designers to "create objects that could never be built before," objects that "uniquely fit today's way of life—innovative, flexible, and mobile." Herman Miller fabricated chairs entirely from injection-molded plastics, making furniture design more an act of sculpture than of carpentry. Transparent coffee tables made of Lucite and inflatable lounge chairs were all the rage among young consumers. "Some people in the industry are saying the surface has hardly been scratched in potential uses for plastics," reported *BusinessWeek*. "Already the product engineers have come up with an all-plastics house. They recognize no boundaries."

Even the most esteemed names in architecture and design—Le Corbusier, Mies van der Rohe, Rem Koolhaas, Alexander Calder—incorporated utilitarian plastics into their buildings, sculpture, and furniture. In the world of fashion, designers from Betsey Johnson to Pierre Cardin filled their runways with garments made of cellophane, vinyl, and pleather, topped with whimsical plastic props such as goggles and helmets. The models looked like space-age embodiments of the plastic Barbie dolls that Mattel had introduced to America in 1959. Meanwhile, fabrics from spandex to double-knit polyester began to pervade everyday fashion trends.

The 1960s was also the decade of "vinylmania"—when music was recorded and avidly collected on tens of millions of vinyl records. It was the decade when plastics became part of the language of cinema, itself a medium produced on plastic. *Barbarella*, *Planet of the Apes*, and *2001: A Space Odyssey* were set against ethereal, whimsical landscapes of plastics. In a class all by itself was 1967's *The Graduate*, which emblazoned into cultural memory the line spoken as unsolicited career advice to protagonist Ben Braddock by his father's business partner: "I want to say one word to you, Ben. Just one word. Are you listening? Plastics. There's a great future in plastics."

SUBSTANCE ABUSE

There had been inklings of a cultural pushback against plastic in earlier decades, as captured by this 1947 article in *Collier's*: "The witchery the chemist performs turns them first into something unearthly, that gives you the creeps. You feel, when you go into a chemical plant where plastics are made, that maybe man has something quite unruly by the tail."

Just as plastics inspired manias of consumer acceptance, a material at once so foreign and so omnipresent also inspired waves of passionate rejection. A countercultural revolution was bubbling up in the 1960s, giving rise to Woodstock, hippies, and the back-to-nature movement in which disillusioned youth by the thousands started eating granola and moving away from chemical-based foods and materials. The light, ephemeral, surface-driven vanity of the prevailing plastics culture seemed to have brought about a form of malaise—a longing for a more enduring, earthy rootedness. To many of America's youth, plastic came to represent all they railed against.

In *The Graduate*'s most quoted line, there is an implicit critique: dazed and disaffected, Ben responds to this unsought advice on plastics by nodding blankly, and soon afterward bolts up to his room in a state of alienation and despair. Here, plastics came to represent a certain inauthenticity blindly embraced by many baby boomers but rejected by their children. This sentiment was also conveyed by a line in the movie's theme song, "Sound of Silence" by Simon and Garfunkel: "And the people bowed and prayed / To the neon god they made."

Perhaps no author lamented this strange beast mankind had harnessed more forcefully than Norman Mailer. "We gave away our freedom long ago," Mailer wrote in his *Esquire* column in 1963, beginning a multiyear attack on plastics. "We divorced ourselves from the materials of the earth, the rock, the wood, the iron ore; we looked to new materials which were cooked in vats, long complex derivatives of urine which we called plastic. They had no odor of the living or what once had lived, their touch was alien to nature." He later compared the proliferation of "the faceless plastic surfaces of everything . . . built in America since the war" to "metastases of cancer cells."

Mailer's cancer reference outraged the plastics industry, but it soon proved prescient: reports emerged in the early 1970s showing high rates of liver cancer among workers at polyvinyl chloride (PVC) factories. Studies on the health risks of PVC proliferated, and ecologists and public health advocates began to expose the threat, putting the public on high alert. In 1973, the Food and Drug Administration banned PVC from certain food and beverage containers. American consumers remained wary in the face of other scandals tied to the plastics and chemical industries throughout the decade—most notably, the tragedy that came to be known as "Love Canal."

In the mid-1970s, local residents of the Love Canal community in western New York discovered that they were living on top of roughly 21,000 tons of toxic waste that had been dumped by the Hooker Chemicals and Plastics Corporation in the 1940s and 1950s. The EPA reported that nearly a dozen of the eighty-two chemical compounds dumped by the company were carcinogenic; the toxins were linked to an epidemic of health problems including shockingly high rates of miscarriage and birth defects. Nearly the entire community was evacuated.

Throughout the 1970s, the plastics industry also absorbed the blow of the Arab oil embargo. The soaring cost of crude pushed up the cost of plastics just as American consumers were growing wary of their dependence on oil and all its by-products. In 1973, the year of the embargo, environmentalist Barry Commoner released *The Poverty of Power*, a book that explored the perils of America's petroleum habit, including the health and ecological impacts of plastics and their petrochemical ingredients. He made the case that petrochemicals existed "not so much to serve social needs as to invent them," and argued for a return to higher-cost, natural materials. By the late-1970s, plastics industry leaders began fearing irreversible damage to their once-transcendent image: "Ecological concerns increased so steadily after the first Earth day of 1970," wrote Jeffrey Meikle in *American Plastic*, "that insiders feared the crisis might 'really end the industry.'"

But even as the reputation of plastics bottomed out, chemical plants and factories continued to churn out plastic products at a steadily climbing rate. All cultural and other resistance notwithstanding, by 1976 plastics had become the most widely used materials in the world, surpassing wood, steel, copper, and aluminum as the most common substance in all products. Today, when roughly 100 billion pounds of plastics are produced in the United States per year, the material has become so prevalent that "it's getting difficult," said Keith Christman of the American Chemistry Council, "to find products that don't have plastic in them."

BOTTLED UP

Today, the plastics mania has reached new heights. After the decades of cultural sensations surrounding cellophane, nylon, Tupperware, acrylics, and vinyl, the turn of the millennium will likely be remembered for another obsession—water bottles. In recent years, Americans have thrown away more than 25 billion plastic water bottles annually, according to beverage market analysis. The Pacific Institute, an environmental nonprofit organization, estimates that in one year manufacturers used the equivalent of 17 million barrels of oil—that's enough to fuel more than 1 million vehicles for a year—merely to produce the bottles needed to meet U.S. demand for bottled water, and still more to transport the water from remote locations like Evian, France, and the islands of Fiji and to chill it in store refrigerators.

While a bottled water backlash has been gaining momentum—high-profile restaurants have refused to serve it, big-city mayors have urged people toward the tap, and bottled water manufacturers have begun offering their product in bottles made from 30 percent less plastic—consumer purchases of bottled water have continued to increase.

Plastic bags, too, have become something of a modern epidemic. More than 500 billion plastic bags are used globally each year, 100 billion of these in the United States alone. An estimated 12 million barrels of oil annually goes into the production of the plastic bags used by American consumers. Ireland, Sweden, and Germany have imposed taxes on the use of plastic bags. Campaigns to ban nonbiodegradable plastic bags have been implemented in cities including San Francisco and Shanghai, and throughout France.

Criticisms of plastic bags and bottles go beyond the oil consumption that accompanies their use. When dumped in the ocean, plastic bags and other disposables can choke or suffocate wildlife. Environmental scientists have also raised concerns about "endocrine disrupters" in plastics, such as pthalates, that can mimic hormones when they enter the bloodstream, possibly contributing to infertility and cancer. Another endocrine disrupter, bisphenol A (BPA), was until recently widely used as an additive in hard plastics such as baby bottles, causing fears that the toxin could leach into infant formula, and potentially interrupt hormone production in infants. The toxin was also reportedly present in the big 5-gallon plastic jugs used in water coolers. The wakeup call made me nervous, so I checked in with Daniel Schmidt at UMass Lowell.

The purpose of BPA and other potentially toxic additives such as pthalates, he told me, is to give certain plastics their desired softness or rigidity. In theory, such toxins can be removed and replaced with benign alternatives (as they have been in the case of BPA in baby bottles), but it often takes years of medical research to show which chemicals present health concerns and which don't. This raises a serious concern about today's plastics industry: chemicals such as BPA have been put into commercial use before sufficient medical research has been conducted to test their safety. More sophisticated screening and governmental oversight of the industry's chemicals could go a long way toward alleviating this problem.

Another concern about America's profligate use of plastics today is that we have few viable, affordable alternatives. In theory, you could make plastics from plant matter—a hydrocarbon is a hydrocarbon, whether you get it from petroleum, corn, or elsewhere. "Plastics could be produced from trees and plants," explained the materials engineer Rudolph Deanin. "It just takes many more chemical reactions to do it."

Which means moving away from fossil-fuel-based plastics would sharply increase the cost of our products. More research could streamline the manufacturing of green plastics and make them more affordable—this, too, is an area that could benefit from more focus and funding.

Better recycling methods and facilities are also needed. In recent years, less than 7 percent of the 60 billion pounds of plastics discarded annually in the United States were recycled. And only a tiny, almost immeasurable fraction of that number was made from new green cellulosic plastics derived from plant crops. Presumably, almost all plastics can be recycled, but our current systems of consumption and disposal aren't designed to support this outcome. When you consider the tremendous value and versatility of these materials—and their full cost to the environment—wasting them in landfills begins to seem foolish, if not wildly irresponsible.

One of the biggest problems with recycling today is that there is such a vast array of different plastics (some eighty thousand types) and they can't all be lumped together in a single recycling process. Each has to be sorted and recycled separately, a monumental chore requiring costly human labor. Short of a direct tax on plastic waste, it is simpler and cheaper (in dollars, not environmental terms) for individual consumers—and even for many of our current recycling facilities—to place a large percentage of plastic waste in landfills. Moreover, the chemical processes through which plastics are recycled are highly sensitive and specific—if a product has a certain kind of dye in it, for instance, or a certain ingredient in the adhesive on the label, this can contaminate and spoil its recycling.

"In order to forge a sustainable future of plastics we need to standardize the industry," said Professor Robert Malloy, chair of the plastics engineering department at UMass Lowell and a colleague of Schmidt's. Malloy suggested that we could reduce the number of plastics that are used, and standardize the use of glues, dyes, and paints to make these materials easier to recycle, thereby lowering the fossil fuel inputs of plastics manufacturing and the recycling process itself.

In the future, Malloy told me, as petroleum feedstocks become scarcer and more expensive, we may see plastics manufacturers exploiting a different buried treasure—not the riches of oil fields but the rubble of landfills. Future prospectors could extract old plastics from the ground and find new uses for them via sophisticated sorting systems—computers that "read" the different types of plastics by laser scan and sort them mechanically. "There are billions of tons of plastics buried in landfills," said Malloy, "that could have a productive future life."

An additional problem in recycling is that plastics are often "downcycled"—meaning the recycling process tends to degrade the quality of the plastic so that it can only be used for lesser, cheaper products in the future. The green architect William McDonough, who wrote the book *Cradle to Cradle: Remaking the Way We Make Things*, is working with chemists to completely rethink the engineering of plastics so that they can either fully biodegrade or be recycled over and over again without any degradation in quality. McDonough has invented an upholstery textile from plant cellulose "that can, when composted, serve as garden mulch," reported the technology magazine *Wired*. He has invented another material that can "now yield a perfectly reconfigurable nylon fiber." The new fiber can be easily broken down into its component parts, which can then be completely reassembled into brand-new products. "Instead of being inefficiently 'down-cycled' into something like a plastic park bench, your carpet can be reincarnated every time you redecorate," *Wired* reported. "Herein lies the Big McDonough Idea: 'The materials go back to soils safely, or they go back to industry. That's it. That's the new paradigm.'"

BUCKETS AND JEWELS

In 1940, vying to maintain the momentum of its 1939 World's Fair exhibit, DuPont launched a popular stage show around the country featuring "Miss Chemistry: a girl in a test tube attired from head to toe in products of chemical research." Garbed in nylon hose, a rayon dress, Lucite shoes, and cellophane bows, Miss Chemistry was the mascot of the company's "Better living through chemistry" campaign that flourished for the next few decades.

A very different campaign was launched in 2008 by the American Chemistry Council, a consortium of leading manufacturers including DuPont. These ads touted plastics, among other products of modern chemistry, as being as "essential" to American life as H_2O: images of a farm and farmer, a cityscape, doctors, firefighters, and astronauts alternate with captions that include "essential2living," "essential2safety," "essential2health," and "essential2american heroes."

In just under seventy years, the focus of these advertisements had gone from the frivolous and fun applications of plastics and other products of modern chemistry—in an effort to reassure and convince Americans of the pleasures of choosing to adopt these new substances—to their elemental, life-saving, national-security-safeguarding functions. The ads were almost certainly designed in response to the mounting public critique of plastics—to remind us of just how woven into the fabric of our lives these fossil-fuel-derived products have become. The theme might be said to have shifted from "Better living through chemistry" to "Let's face it—chemistry is here to stay."

I have come to agree. The truth is, we can far more easily imagine a future without the combustion engine than we can a future without plastics. In fact, as we look to the sustainable future, plastics are increasingly perceived as being part of the solution, not the problem—in the form of those polypropylene shells for the Smart car and other ultralightweight and fuel-efficient vehicles, for instance, better home insulation, and wood and paper alternatives to help preserve forests that act as carbon sinks in a warming world. In the next few decades, huge amounts of attention and resources will have to be devoted to developing ever more efficient uses of plastics, ever more sophisticated and mandatory recycling programs, and ever healthier and more environmentally sound methods of manufacturing synthetic materials.

It's hard to imagine a future in which plastics are entirely made from plant crops—there's simply too much demand on cropland to feed the world's booming populations and to produce alternative fuels. As we enter an era of increasingly scarce petroleum reserves, we may have to encounter hard trade-offs, cutting unnecessary uses of plastics—water bottles, plastic bags, and packaging—while eliminating the use of fossil fuels in transportation and power plants so that there are enough available to produce essential petrochemicals, including those used to make safety helmets, medical implants, and other lifesaving synthetics.

In the 1950s, upon hearing of the successful implantation of the first artificial heart valve, the French philosopher Roland Barthes announced that the "hierarchy of substances is abolished: a single one replaces them all: the whole world *can* be plasticized, and even life itself since, we are told, they are beginning to make plastic aortas." Barthes went on in his book *Mythologies* to characterize plastics as "the stuff of alchemy . . . transforming the original crystals into more and more startling objects." He praised the substance for having a certain noble humility: "It is the first magical substance which consents to be prosaic," often masquerading as something mundane.

"The quick-change artistry of plastic is absolute: it can become buckets as well as jewels," he wrote. "Hence a perpetual amazement, the reverie of man at the sight of the proliferating forms of matter." He added that the seemingly limitless potential of plastics to take on new forms gives mankind "the euphoria of a prestigious free-wheeling through Nature." Barthes's observations reminded me of the spectacle of discarded plastic strewn across Talladega's Field C—a scene dazzling in its sheer volume and variety of matter. It was also a scene dangerous and foreboding in the careless "freewheeling" that demonstrated humanity's sense of control over nature—a sense of control that can only be delusional.

In fact, our very bodies are now captive to the freewheeling substances we have created out of fossil fuels. We are bound to these substances in ways, I was to learn in the next leg of my power trip, more intimate even than medical implants beneath our skin. An overwhelming percentage of the food we eat is grown with petrochemicals—integrating with the very molecules that compose our flesh, the amino acids that combine to form our DNA.

If you don't put your fertilizer on, you'll cut your yields by half or more. . . . Worst thing that can happen to a farmer is getting a reputation for having a low yield. It's like being a race car with a lawnmower engine or a newspaper that's always a week behind.

—Ken McCauley, farmer

5

Cooking Oil

HOW FOSSIL FUELS FEED THE WORLD
(AND ENERGY SHORTAGES COULD STARVE IT)

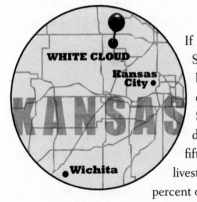

If you pinned a map of the United States to a dartboard, Kansas would be the bull's-eye. Smack dab in the center of the country, the Sunflower State is one of America's most productive agricultural hotbeds—the fifth-biggest producer of crops and livestock in the country. More than 90 percent of the state consists of farmland endowed thousands of years ago with rich glacial loam. This fertile topsoil is no longer as robust as it once was, having offered up its nutrients season after season, decade after decade, century after century to produce great bounties of wheat, corn, soybeans, sorghum, hay, and sunflowers. I could almost sense the exhaustion of the land as I drove through the back roads of northeastern Kansas one chilly November morning—past sagging wooden farmhouses silvered by age and weather, barbed wire fences with listing wooden posts, general stores and swinging-door saloons, a Native American heritage museum commemorating the Kansa tribes that once roamed and tilled these prairies, and mile after desolate mile of denuded farmland.

It wasn't that this dormant soil was incapable of producing—on the contrary, during the previous summer and fall it had yielded one of the most plentiful harvests in Kansas history, producing many times more crops than it did a century earlier, when the land was more inherently fertile. But now, like an aging bull receiving shots of testosterone, this well-worn ground reaps the benefits of modern chemistry. That late fall morning, thousands of tractors combed the Kansas countryside, priming the soil for next spring's planting with a "booster shot" of nutrients that would transform the weary earth into some of the world's highest-producing farmland. This was what I had come to see: the miraculous process by which harvested land—land that had sprung seeds, supported plants, produced its quota, and now lay depleted and resting—would be magically revived with chemical nourishment so it could bear just as much fruit again the following spring.

That chemical nourishment, also known as fertilizer, is yet another by-product of fossil fuels—one that has transformed America's economy over the last century, and expanded the global population, too, by vastly increasing the food supply. I'd encountered the basic science of fertilizers in my plastics research, learning that synthetic soil nutrients were—along with the wide universe of polymers—yet another twentieth-century innovation made from petroleum-derived chemicals. Many scientists and historians have argued, in fact, that chemical fertilizers have had a greater impact on human survival and well-being than have any other technological advances.

Potassium, phosphorous, and nitrogen are the three most common nutrients in the fertilizers applied to American farmlands; nitrogen is by far the most prevalent. The main form of nitrogen fertilizer is known as anhydrous ammonia, and natural gas is its primary feedstock. Natural gas provides hydrogen, one of the components of the ammonia fertilizer. This fossil fuel is also burned to generate the tremendous amounts of heat and pressure needed to produce the fertilizer's other main ingredient—nitrogen—and fuse it together with hydrogen and other additives to create the final product. Nitrogen fertilizers take many forms, ranging from the Miracle-Gro sold at your local Home Depot to the industrial-strength anhydrous ammonia that's used on tens of millions of acres of U.S. corn and wheat crops.

Each year, American farmers apply 6.2 billion pounds of fossil-fuel-based fertilizers to their croplands.

Fertilizers set industrialized countries—the farming megaproducers— apart from many developing countries in which low-yield farms can't afford the costly input. (Grains are produced so abundantly in the United States, and so cheaply relative to the rest of the world—thanks to chemical additives, high-tech seeds, and government subsidies—that we export vast quantities of our agricultural products to the world's poorest populations.) Rosemary O'Brien of the Agriculture Energy Alliance, an agribusiness lobby group, told me that commercial fertilizers are "directly responsible for about 40 percent of our nation's crop production." That makes them a very valuable commodity in a U.S. industry that produces over $110 billion worth of crops each year.

Put another way: if chemical fertilizers were suddenly eliminated from American farms, said O'Brien, the productivity of our farms would plummet, and food prices would soar. A Happy Meal could cost as much as a pound of caviar. What would Americans do without their 99¢ hamburgers? Our cheap-eats lifestyle today is hooked on fertilizers just as much as our lifestyle depends on the fuel that powers our transportation and the plastics that encase our creature comforts.

Environmentalists maintain that America's farmlands can flourish without the use of chemical fertilizers and pesticides. Tom Philpott, an organic farmer from North Carolina who advocates and writes about sustainable farming policies for the environmental Web site Grist.org, sees a future system of agriculture in America that's "100 percent organic." This system would replace the large-scale farms that currently dominate U.S. agribusiness, some spanning 10,000 acres with a single crop planted throughout. Instead, a vast array of mixed-use farms would grow America's food, each farm spanning hundreds of acres or less, with diverse and regularly rotated crops. Soil nutrients would be replenished by natural fertilizers such as manure and composted food. This system would have tremendous benefits, said Philpott, to public health, our land, and our waterways. But, he acknowledged, it might also spell the end of America's cherished cheap-eats lifestyle: consumers would have to adapt to food prices "that could double or more."

Cutting out fertilizer and pesticide use would not alone eliminate fossil fuels from farming. While fertilizers account for roughly a quarter of the energy that sustains American agriculture, huge volumes of petroleum and electricity are also used in other aspects of agriculture, organic and conventional alike: operating farm machinery, powering irrigation systems, processing crops into products, and packaging, refrigerating, and transporting them to stores, restaurants, and kitchens. All told, America's food system, according to a University of Michigan study, accounts for roughly 10 percent of annual U.S. energy demands.

Another study from the University of Chicago concluded that Americans actually emit more greenhouse gases from eating than we do from driving—when you combine the carbon dioxide emissions from the fossil fuels that go into our farming practices with the methane and other planet-warming pollutants routinely released from farms and feedlots.

It took me a while to digest these facts and numbers, but eventually they led me to one hair-raising realization about our current dependence on fossil fuels: Americans eat oil.

LOST IN A MAIZE

I could readily come to terms with the fossil fuel inputs in war, my car, my frequent-flyer lifestyle, my highway-centric hometown of Nashville, and the plastic paraphernalia that clutters my life. I could make sense of the fact that complex machinery, urban landscapes, and even industrial substances with names like polyethylene were part of the petroleum landscape. But somehow it was harder for me to associate our energy addiction with something as primitive and elemental as soil—and, for that matter, as unassuming as a tortilla chip or a slice of bread.

Surprisingly little attention is paid to the fossil fuel inputs in American crop production. I could find few industry and U.S. governmental studies on this subject—most of the data mentioned above is an assembly of independent research from academics, nonprofit organizations, and journalists. The most notable writing in this area is that of food journalist Michael Pollan, author of *The Omnivore's Dilemma* and a contributor to the *New York Times*. Pollan takes aim at corn in particular as the most fertilizer-hungry of all the major American crops, receiving roughly 140 pounds of nitrogen per acre. (Other crops such as conventional potatoes, tomatoes, and grapes may get fertilized even more heavily but are produced at much lower volumes.) David Pimentel, a professor emeritus at Cornell University's College of Agriculture and Life Sciences, estimates that corn requires the equivalent of 92 gallons of oil to grow per acre when you add together the energy that goes into planting, growing, and harvesting the crop.

The plant known to most of us as a simple buttered cob is in fact as versatile as plastic. According to Pollan, "There are some forty-five thousand items in the average American supermarket and more than a quarter of them now contain corn." This vast galaxy of corn products encompasses not just canned and frozen corn kernels, corn chips, corn muffins, corn oil, and corn syrup, but also lesser-known and widely used ingredients including glucose, food coloring, leavenings, preservatives such as citric acid and maltodextrin, and emulsifiers such as lecithin and monoglyceride. You're even eating corn when you bite into a conventional apple—it's in the gloss that's often applied to fruit and vegetables to make them gleam.

Even these grocery store items represent a small portion of corn applications. Much of the corn grown on American farms is inedible, Pollan explained, having been bred for qualities that are optimal for processing into products far removed from the produce aisle. Roughly a quarter of all the corn grown in America goes to make the gasoline additive ethanol, and more than half goes to animal feed. Corn is the main ingredient in cheap, high-calorie food for the cows, pigs, and chickens that, in turn, go into hamburgers, bacon, and McNuggets. The milk, eggs, cheese, yogurt, and ice cream that come from these corn-fed animals and their by-products are also part of the petroleum food chain. Fish, too, consume corn: farm-raised salmon, tilapia, and catfish have been bred to digest this grain, which would otherwise not be sustenance for aquatic life.

All of which brings a certain gravitas to the energy crisis that I hadn't considered before. I had seen the end of the petroleum era as a challenge that would transform our lifestyles, our comforts, our mobility— our identities as Americans on a deep level. But I hadn't fully grasped the role of fossil fuels as a form of *sustenance*. With the exception of certain medical innovations such as pacemakers and chemotherapy, no petroleum by-products that I'd considered before agriculture were fundamental to human health and survival.

We could all certainly survive—in fact, we would arguably thrive— without high-fructose corn syrup in our soft drinks. Some experts, for that matter, believe that we should eliminate many corn-derived products from our own diets and those of our farm animals—corn-based animal feed can have damaging affects on the digestive tracts of cows, for instance. But not all the crops grown with fossil fuels play such an optional role in the global food system. "The use of fertilizers undoubtedly has life-and-death implications," United Nations special advisor Jeffrey Sachs told me. The 2008 surge in worldwide food prices—caused in part by a sharp spike in petroleum prices—was blamed for increased rates of malnutrition and death among the poorest populations in Africa who could no longer afford to buy basic grains.

It is true that high-yield agriculture pumped with fossil fuels has contributed to the obesity epidemic in America (by offering relatively cheap domestic food loaded with sugars). It's also true that the government-subsidized megafarms that dominate U.S. agriculture generate food at such great volumes and such low prices that they've run smaller-scale family farmers out of business at home and overseas. And fertilizer use and runoff have decisively been proven to have serious environmental consequences. But it is also a certainty that petroleum-dependent crops currently keep billions of people alive. Transitioning to a post-petroleum food system in the United States would require enormous changes— changes we can't begin to contemplate without first understanding the current role of fossil fuels on American farms.

The more I dug into the details, the more I realized how little conclusive research has been done on this subject. David Pimentel's numbers are hotly contested by many in the farming industry, but I couldn't find much evidence to prove them wrong. I wondered: Can we ratchet down our use of fossil fuels on American farms, or eliminate them altogether? Why are chemical fertilizers so essential in the first place, and what are their actual consequences? How feasible is Tom Philpott's dream of going 100 percent organic?

In search of answers, I wanted to observe the fertilizing process for myself. I called up Ken McCauley, the former president of the National Corn Growers Association and a fifth-generation farmer. He and his thirty-year-old son Brad harvest 4,000 acres of cropland in White Cloud, Kansas, half of which they own; the other half they lease from other landowners with whom they share profits. McCauley agreed to show me his farm. These rolling Kansas acres were more accessible than the Gulf of Mexico, less intimidating than the Pentagon, more tranquil than Talladega, and considerably less gruesome than a Los Angeles operating room. But they were no less dramatic in the stories they told about the challenges of our dependence on fossil fuels.

FERTILITY TREATMENTS

I drove toward the town of White Cloud on U.S. Route 36, a narrow two-lane highway dotted with pro-life billboards ("Abortion is American holocaust: 1,400,000 deaths a year") and church slogans ("This blood's for you!"). The sky was a dull gray that mirrored the color of the bare earth. I drove for nearly an hour without spotting another car, the yawning landscape marked only by occasional fence posts, silos, and tractors puttering in the distance. Every 20 miles or so, I'd pass an old farm with an American flag fluttering in the front yard or a weathered porch decorated with red, white, and blue bunting.

I was passing through Ken's land long before I realized it—rolling stretches of bone-colored soil that, like so much of the land I had driven through that day, can only be compared to the ocean for its sheer reach and expanse. Sporting a plaid shirt and a thick mustache, Ken drove up to meet me at the end of his driveway in a white Ford pickup with license plates that read "CORNFED." That's also the name of his boat, a Cobalt, which he keeps on a lake at his cabin in the Ozarks. Ken attended a local junior college specializing in agriculture, focusing his studies on farm mechanization and corn genetics. Brad majored in economics at Kansas State.

Ken's home office is decorated with corn-themed bric-a-brac—a clock with a corn stalk on its second hand and kernels for every minute, antique corncob pottery, old tins of corn oil, a model of the 1906 Henry Ford prototype car designed to run on ethanol, and a tomahawk Ken found on his land "showing the roots of the people before us," the Kansa tribes, "who were the original corn farmers." To Ken, the diversity of corn's applications ensures its long-term profit potential as a crop: "The point is, there's corn in almost everything. It's what you want to be growing." That's as true on grocery store shelves, he says, as it is in his own life. "Corn feeds my family in every way—our bodies, our salary, our faith, even our car" (which runs on corn ethanol).

There's no better soil in the world for growing corn than America's heartland soil, he explained, due to its moisture, clay and mineral content, and the region's freezing and thawing seasons—all optimal conditions. That soil is a key reason why America is responsible for more than 40 percent of the total global production of corn. Nearly 86 million acres of farmland in the United States produce corn—surpassing the acreage involved in growing wheat (63 million acres) and soybeans (76 million acres). Corn production expanded steadily throughout the twentieth century in the United States as American farmers found ever-broader uses for the grain, including food additives, animal feeds, and now ethanol. McCauley's own production is 80 percent corn, 20 percent soy (the latter, he explains, is a useful rotation crop since it doesn't require nitrogen additives). All of his corn goes to food production, and the leftovers that aren't used—cobs and husks—go to ethanol production.

Ken's great-grandfather David McCauley, who bought the family's first parcel of Kansan land in 1910, was a corn farmer, too. The McCauley family records show that his yield in the early 1900s was roughly 20 bushels per acre. Now Ken gets nearly nine times that—about 175 bushels per acre—thanks to huge leaps in the technologies of genetically modified seeds, farming equipment, and, of course, fertilizers. Ken prides himself on using the best available farming technology, which he says enables him to use no more than a bare minimum of fertilizer to achieve maximum yield. He gave me a firsthand tutorial on his state-of-the-art equipment soon after I arrived.

To distribute his fertilizer, Ken hitched a 2.2-ton canister of nitrogen to the back of his apple-green John Deere tractor. A series of tubes and wires connected the tractor and fertilizer tank to a mechanism that looked like a giant rake spanning eight rows of corn. The dozens of prongs at the end of the rake were tipped with knifelike cutters that would pierce into the soil, opening it up so that hoses embedded within the blades could blast the chemical nutrients 6 inches into the ground. (If the nitrogen were simply sprayed on top of the soil, it would evaporate and blow away.) The liquid fertilizer freezes into golf ball–sized lumps in the wintertime that then thaw and release into the soil in the spring. It's best to inject the fertilizer in the late fall or early winter, Ken explained, so that the soil doesn't have to be opened up in the spring, which would release precious moisture.

I climbed up into a plush passenger seat in the tractor cab next to Ken's foreman, Nick James. Though the seats were mounted on shock absorbers, they still bounced and pitched as we trundled over the rough, hilly ground. I grabbed the dash to steady myself as we began to move slowly down the field. A mist of ammonia wafted in, burning my nostrils and lungs with the smell of airport bathroom cleanser, only a thousand times stronger. "That smell will take your breath away at first," Nick said, noting my gag reflex, "but you get used to it." I grew ever warier when he warned me not to touch any of the liquid nitrogen dripping from the tank because it could give me second-degree burns. It was hard for me to believe that this putrid stuff could possibly have life-giving properties. But I learned that a plant with no nitrogen is like a computer with no cord—without nitrogen the plant cannot "turn on," so to speak. Nitrogen is a primary element in chlorophyll—the pigment that gives plants their green color and enables them to channel energy from sunlight into the creation of the fibers and sugars that make them a source of nourishment.

Maneuvering a tractor through a cornfield is a little like steering a ship through waves—it's hard to keep the vessel in a straight line on the sloping, bumpy earth and then to repeat that straight line exactly as you traverse the rest of the field, without overlapping any areas on which you've already sprayed nitrogen. Conventional tractors routinely overlap on fertilizer application, wasting precious resources. Ken is able to overcome this costly human error because his tractor drives itself. "See the GPS system?" Nick asked, pointing to a small round blinking device on the dashboard. That device was feeding signals to a satellite monitoring the position of the tractor on the field. The satellite was then automatically feeding those location coordinates into an autopilot system that steers the tractor on a precise course, never double-applying fertilizer to the same patch of soil.

The process of dispensing fertilizers is also automated on Ken's tractor. Whereas conventional tractors indiscriminately dump the same amount of fertilizers throughout the field, new computer software enables Ken to vary his distribution of nutrients according to soil quality. A flat-screen monitor inside the cab showed a multicolored map of the field we were traversing with splotches of green, yellow, and red indicating which areas of the field have been most and least productive in past harvests. The tractor automatically dispenses the most fertilizer on the low-yield (red) areas, less on the moderate-yield (yellow) areas, and still less on the green areas, which have higher inherent fertility. Ken estimates that about 10 to 15 percent of the fertilizers applied on U.S.

farms actually go to waste because they're blindly doused on areas of soil that in fact have sufficient levels of nitrogen. Other agriculture experts I interviewed put that number even higher, saying that up to 35 percent of the nitrogen typically sprayed on farmland goes to waste, draining out of the soil and polluting nearby bodies of water.

Precision fertilizer application technologies are not yet in widespread use—only about 5 to 10 percent of American farmers have fully adopted them—but they hold promise. The software has shaved down Ken's fertilizer applications by nearly 15 percent, and he expects the savings to increase as the technology evolves and he adapts to using it. Ken also plants genetically modified corn seeds known as No. 6169, engineered by Monsanto to have built-in characteristics that enable the plant to repel insects, for instance, and withstand herbicides (so that products such as Roundup can be sprayed directly on the crops, "killing everything but the corn," as he explained it). The seeds will soon be engineered to utilize nitrogen fertilizers more efficiently than conventional seeds—an innovation that will further trim fertilizer use. "We are using 10 percent less fertilizer today than we were a decade ago," Ken said, "and in that time our yields have nearly doubled."

Those are crucial gains, he told me, at a time of unstable energy prices. As natural gas and oil prices surged in recent years, the costs of fertilizers nearly quadrupled. In 2005, when natural gas prices were low, a 2.2-ton tank of anhydrous ammonia cost under $400. When gas prices shot up in 2008, that same tank of fertilizer cost nearly $2,000. To fertilize Ken's 4,000 acres, that added up to an expense of roughly $500,000 a year—about 40 percent of his total operating costs of $1.2 million a year. Even with soaring costs, he explained, "fertilizer is the most economical thing we do because it gives you your production on the top end." In other words, while Ken spent nearly half a million dollars on fertilizers in 2008, these additives still created significantly more value in enhanced crop production.

What would happen if Ken cut out chemical fertilizers altogether? "If you don't put your fertilizer on," he told me, "you'll cut your yields by half or more. No farmer is going to stop using nitrogen altogether.

Look at the poor countries—when you travel to places that don't use the fertilizer you'll see they're raising a third of the yield." He boiled the conundrum down to six words: "Nitrogen is yield. Yield is nitrogen." And yield, he added, is everything. "Worst thing that can happen to a farmer is getting a reputation for having a low yield. It's like being a race car with a lawnmower engine or a newspaper that's always a week behind."

Organic farmers contest this, saying that farmland can be naturally replenished through sustainable farming practices including the application of animal manure, which is high in nitrogen. Michael Pollan described a virtuous cycle of nutrient recycling between crops and animals: "Sunlight nourishes the grasses and grains, the plants nourish the animals, the animals then nourish the soil, which in turn nourishes the next season's grasses and grains. Animals on pasture can also harvest their own feed and dispose of their own waste—all without help of fossil fuel."

From Ken's perspective, the drawback of these organic methods is that they require more labor and time, and in turn generate lower profits. "It's not a way to maximize production," he told me. Moreover, he argued, organic crops have to be tended, weeded, tilled, and rotated on a more frequent basis than do those on chemically treated farms, and this extra work requires hauling out fuel-guzzling farm equipment with greater frequency and pouring more fuel into the machinery, even if none goes into the crops themselves.

But Ken is also quick to agree that the current fertilizer-intensive methods of American farming may not be sustainable in a world of climbing fossil fuel prices—even with his high-efficiency technology. Seventy percent of Ken's budget is fully vulnerable to rising energy costs: while fertilizers account for half of his operating expenses, the fuel that powers his tractors and transports his products to wholesale markets accounts for another 20 percent. Ken says the annual operating costs of a typical American farm have been growing by leaps and bounds along with the cost of energy inputs. In 2005, corn farming cost roughly $100 per acre; in 2008, it was $400 per acre. If prices were to continue to climb without relief, says Ken, he and other farmers would have trouble staying afloat.

How and why, I wondered, did our food system become so thoroughly hooked on fossil fuels in the first place? What are the consequences of this dependence, and how will we begin to dig our way out?

POPULATION BOMB

One name is unavoidable in any attempt to examine the costs and limits of food supply systems: Thomas Malthus. Born in 1766 in Surrey, England, Malthus—a well-liked country parson and professor—was an unlikely prophet of worldwide famine, gloom, and doom. But the theory he set forth in his treatise *An Essay on the Principle of Population* has had profound effects upon the way we think about demography, economics, and farming practices to the present day. Malthus's theory can be boiled down to this: while population grows geometrically (2, 4, 8, 16, etc.), the food supply grows arithmetically (1, 2, 3, 4), and so at some point disaster will inevitably ensue as the populace outstrips its sustenance.

"The power of population," he grimly noted, "is so superior to the power in the earth to produce subsistence for man, that . . . premature death must in some shape or other visit the human race." He envisioned a future in which "gigantic inevitable famine stalks in the rear, and at one mighty blow levels the population." Malthus did offer one glimmer of hope: mankind's capacity for invention. "The main peculiarity which distinguishes man from other animals, in the means of his support is the power which he possesses of very greatly increasing these means."

Malthus's theory contradicted the more optimistic tenets of the Age of Enlightenment and provoked great controversy at the time. But half a century later, in the mid-1800s, his writings were echoing strongly with Europeans. England had faced a disastrous famine in the 1840s, with its population swelling and its acreage extremely limited compared to most other parts of the world.

Investigating their food problem, European scientists recognized that the single biggest constraint on food production was the amount of nitrogen in the soil. Natural deposits of pure nitrate (a nitrogen-rich salt) were rare and could only be found in small supply in South America. A pressing question loomed—where to get supplemental nitrogen?

By the turn of the twentieth century, as European food supplies were growing ever tighter, a young and headstrong German chemist named Fritz Haber came up with an answer. It was as obvious as it was revolutionary: air. The very air we breathe is a mixture of about 20 percent oxygen and 80 percent nitrogen. If harnessed, this floating sea of nitrogen could generate an almost limitless supply of nourishment. While studying at the University of Karlsruhe, Haber and his colleague Carl Bosch began to experiment with the process of splicing nitrogen out of the atmosphere. It was a daunting task: atmospheric nitrogen, N_2, is an extremely stable molecule composed of two atoms of nitrogen that are tightly bound and difficult to separate. But Haber discovered that if subjected to enough pressure and heat, those cozy molecules could be zapped apart to form pure nitrogen, which could be combined with hydrogen to form a substance capable of reviving anemic soil.

The problem that existed then with the so-called Haber-Bosch process still persists: the zapping process requires an enormous amount of energy. Today, it takes nearly the same amount of energy to power about three hundred American households for one year as it does to generate enough heat and pressure to produce the nitrogen that fertilizes 4,000 acres of corn for one harvest. In a century of cheap and abundant energy supplies, this didn't pose a problem. But today's looming energy and environmental constraints could limit the production of fertilizer—and perhaps the global food supply.

Haber's invention won him the Nobel Prize in Chemistry in 1918 and proved to have multiple—and conflicting—applications, all of them breathtaking in scope. Many historians have credited Haber with the single most significant invention of the last century. "Airplanes, nuclear energy, space flight, television and computers . . . none of these inventions has been as fundamentally important as the industrial synthesis of ammonia from its elements," wrote Vaclav Smil, a scholar of agriculture and energy at the University of Winnipeg. "The single most important change affecting the world's population—its expansion from 1.6 billion people in 1900 to today's 6 billion—would not have been possible without the synthesis of ammonia." In other words, over the course of a single century, the world population nearly quadrupled thanks to fossil-fuel-based fertilizers.

The Haber-Bosch process is still widely used today in fertilizer factories, helping to sustain populations worldwide—not just Americans enjoying cheap fast food, but populations throughout the developing world whose core crops depend on nitrogen. There is, however, a sinister twist in the story of Haber's invention: nitrogen is also the main ingredient in materials such as TNT and nitroglycerine, which are used to make dynamite and other explosives. These had been invented before Haber came along, but his process provided the ready means for their manufacture in 1914 when Germany's supply of nitrates ran low, and thereby helped to prolong World War I.

Born to Jewish parents, Haber converted to Christianity and developed deep ties with Germany's nationalist leadership, which eventually brought him into the fold as a party chemist in the years leading up to World War I. In this position, Haber came to be known as the pioneer of chemical warfare. He discovered that nitrogen could be an ingredient in both explosives and poison gas, which he suggested could offer Germany a strategic advantage in routing out enemy trenches.

"Haber actually insisted on this," his biographer Margrit Szöllösi-Janze told National Public Radio. "He said, 'If you want to win the war, then please, wage chemical warfare with conviction.'" The chemist traveled to the front to personally supervise the first successful launching of chemical weapons at the Second Battle of Ypres in Belgium on April 22, 1915. There is perhaps no more memorable description of the effects of chemical weapons than that of the British poet and soldier Wilfred Owen, who wrote of his experiences on the Western front in "Dulce et Decorum Est":

> Gas! Gas! Quick, boys!—An ecstasy of fumbling,
> Fitting the clumsy helmets just in time;
> But someone still was yelling out and stumbling,

And flound'ring like a man in fire or lime . . .
Dim, through the misty panes and thick green light,
As under a green sea, I saw him drowning.
In all my dreams, before my helpless sight,
He plunges at me, guttering, choking, drowning.

Bald and mustachioed, Haber wore a pince-nez and was always impeccably dressed, an extravagant, showboating inventor who loved an audience. At the 1918 Nobel Prize ceremony, Haber grabbed the headlines even though he was honored alongside many celebrity scientists including Max Planck and Johannes Stark. It wasn't Haber's charisma or appearance that captured the media's attention, it was the controversy he drew: protesters condemned Haber for his part in the German war machine, calling for his prize to be revoked, and two French scientists nominated for prizes boycotted the event.

Haber was unfazed—and the ceremony, held in Stockholm, Sweden, went forward as planned. Gerhard Ekstrand, the president of the Royal Swedish Academy of Sciences, declared Haber's nitrogen fixation process "an exceedingly important means of improving the standards of agriculture and the well-being of mankind," and then hung a heavy gold medal around the chemist's neck. In his acceptance speech, Haber never mentioned war, instead stating that "improved nitrogen fertilization of the soil brings new nutritive riches to mankind and [enables] the chemical industry [to come] to the aid of the farmer who, in the good earth, changes stones into bread."

In his 2005 biography of Haber, *Master Mind*, Daniel Charles makes the compelling point that nitrogen is not inherently bad or good; it's the way we choose to apply it that has moral consequences.

Nitrogen fixation teaches a broader lesson: Technology accomplishes nothing by itself. It has no will or moral purpose, any more than the law of gravity does. Human societies create new tools in their own image, and deploy them in the service of their eternal passions. Our machines do not change civilization; like giant mirrors, they reflect it. In the stream of nitrogen, we see humanity's genius, its pursuit of the good life, its inequity, carelessness, and selfishness.

Precisely the same case can be made for fossil fuels. There is nothing inherently bad or good about petroleum or coal or natural gas; it's the ways we've used them that have both benefited the economy and hurt the environment. Petroleum can help feed the masses, regulate heartbeats, grow industries, create wealth, warm homes, clothe our families, and get us where we need to go. By the same token, it can also provoke and power war, bankrupt industries, and pollute and heat the planet. It can just as easily embolden as it can corrupt.

SEEDS OF CHANGE

Though nitrogen fertilizers significantly expanded agricultural yield and helped alleviate hunger in Europe at the turn of the twentieth century, by midcentury the Malthusian scenario of mass starvation again appeared imminent—this time on a global scale. In the early 1960s, something that world leaders had been speculating about for centuries finally came to pass: the planet's supply of farmable land ran out. Population was exploding as advances in medicine were leading to longer life expectancies and higher birthrates. The world's habitable territories had been settled and "expansion hit its limits," as Richard Manning wrote in *Harper's* magazine. "There was nothing left to plow." The pace of population growth seemed finally to have outstripped the global supply of food. A report from President Lyndon Johnson's Science Advisory Committee summed up the issue: "The scale, severity and duration of the world food problem are so great that a massive, long-range, innovative effort unprecedented in human history will be required to master it."

Throughout the 1960s, the developing world—South America and Asia in particular—was haunted by hunger and drought. But then something wholly unexpected happened: global grain production tripled. These were the fruits of what came to be known as the "green revolution," a series of concurrent technology breakthroughs that included increasingly potent seeds, newly mass-produced fertilizers and pesticides, and advances in machinery and irrigation systems.

The cornerstone of the green revolution was the engineering of "miracle seeds" for the high-yield production of the three dominant grains: wheat, rice, and corn. Norman Borlaug, an American agronomist of Norwegian descent, spearheaded these breakthroughs throughout the 1950s at the International Maize and Wheat Improvement Center, an agricultural research station in El Batan, Mexico. Now in his midnineties, Borlaug has conducted countless thousands of experiments on grain seeds and he's still working feverishly to promote more productive farming practices in developing nations. ("He probably sleeps four hours per day," his daughter said in a recent interview.)

The agronomist is as obsessively humble as he is hardworking. When his wife got the news that he had been awarded the Nobel Prize in 1970, Borlaug was collecting seeds in the wheat fields of Atizapan, outside Mexico City. She arranged to be driven there to tell him of the honor and $78,000 prize. "No. No. That can't be, Margaret. Someone's pulling your leg," was his reply, and he went back to his field work. Borlaug was later awarded the Congressional Gold Medal and the Presidential Medal of Freedom, standing among only four other figures—Nelson Mandela, Mother Teresa, Elie Wiesel, and Martin Luther King Jr.—who received all three prestigious awards.

Borlaug was born in 1914 on his grandparents' farm near Cresco, Iowa, a few miles from the Minnesota border. He attended a single-room, one-teacher schoolhouse through the eighth grade, and thereafter made a daily four-hour round-trip commute to the nearest high school. Borlaug's boyhood dreams—of tending his family farm and playing second base for the Chicago Cubs—were a far cry from the globetrotting life he would lead. His grandfather pushed him to pursue a college education, saying, "it's better to fill your head now if you want to fill your belly later." Borlaug flunked his college entrance exam but eventually made his way into the University of Minnesota. It was during the height of the Great Depression, and the streets of Minneapolis were filled with people who had lost their homes and were begging for food. To pay his tuition, Borlaug worked multiple jobs, one alongside people who were starving. "I saw how food changed them," he later remembered. "All of this left scars on me and caused me to want to do something to help."

After earning his Ph.D., Borlaug worked for three years as a well-paid biochemist at DuPont. He resigned from that position in 1944 to accept a job in researching and teaching improved farming techniques in Mexico offered jointly by the Mexican government and the Rockefeller Foundation. Borlaug labored for years without seeing any real progress. "[O]ften sick with diarrhea and unable to communicate . . . I was certain I had made a dreadful mistake in resigning," he later recounted. But he kept at it, working both in the lab and in the dirt alongside farmers in some of Mexico's poorest regions until "the fog of gloom and despair began to lift." The miracle seeds Borlaug eventually pioneered there came to be known as high-yielding varieties (HYVs). These seeds, he emphasized, were engineered in response to the age of fossil-fuel-derived fertilizer: "If the high-yielding [grain] varieties are the catalysts that have ignited the green revolution, then chemical fertilizer is the fuel that has powered its forward thrust," he stated in his Nobel Prize acceptance speech.

As the seed breakthroughs proved successful in the United States and Mexico, U.S. foundations began spreading them to the world's poorest nations—literally airlifting bundled packages of seeds, fertilizers, and pesticides to these hungry populations. In one particularly dramatic incident, Borlaug traveled to the Indian subcontinent in 1965, just as war was breaking out between India and Pakistan, in order to work—sometimes at the edge of bloody battles—beside local scientists to plant high-yield wheat. His effort was ultimately successful in allaying the region's famine.

Designed for an era of cheap fertilizers, HYV seeds were bred to be more responsive to chemical nutrients: Borlaug and others in his field had unlocked a gene that allowed the grains to utilize more nitrogen and, in turn, to bear more abundant fruit. Grains were also bred to have shorter, stiffer stalks—making more energy available for growing the edible kernels than the inedible shafts, and supporting the weight of the kernels better (without bending) to facilitate the harvesting process. Scientists additionally scrambled to develop seed varieties that could hasten and expand the growing season—seeds that would mature more quickly and grow at any time of year, enabling farmers to reap more crops from the same amount of land. The new seed varieties could also perform another miraculous trick—they could soak up more pesticides to resist the scourges of bugs and disease, which intensify under high-yield farming conditions.

Former USAID director William Gaud coined the phrase "Green Revolution" in 1968 when he applauded the widespread success of the breakthrough technologies: "These and other developments in the field of agriculture contain the makings of a new revolution. It is not a violent Red Revolution like that of the Soviets, nor is it a White Revolution like that of the Shah of Iran. I call it the Green Revolution."

In retrospect, the term *green* smacks of irony, given the high influx of fossil fuels required to expand the boundaries of agriculture to lands that previously couldn't be farmed. In his *Harper's* magazine feature article, Richard Manning summed up the view held by many environmentalists that "the green revolution is the worst thing that has ever happened to the planet." Scientists have serious concerns about, among other issues, the pollution caused by fertilizers (pollution that includes both the chemicals released into the ground and greenhouse gases released into the atmosphere); the erosion of topsoil, a complex living system that builds up over centuries rich with fungi, bacteria, and protozoa (organisms that can be decimated by chemical additives); the diminished biodiversity of farms wrought by our increasing dependence on a few major grain varieties; the growing patented corporate ownership of important seed stocks; and the potential long-term effects of genetically modified crops on ecosystems and human health.

In geopolitical terms, the motives of the Ford and Rockefeller foundations, which funded Borlaug's work and much of the Green Revolution's underpinnings, have been criticized by some as a move to bring control and social stability to countries vulnerable to the spread of communism—essentially, as a bloodless way to police the world. But from a technology and human health standpoint, the results have been nothing short of miraculous: total crop production in Asia doubled between 1970 and 1995, yet the total land area cultivated for those crops increased by a mere 4 percent.

The Green Revolution is now credited with boosting world grain production in the 1960s and '70s by 250 percent. Its greatest impact can been seen in population growth: in the forty-year period from 1960 to 2000, the world's population doubled—thanks in large part to the exploding food supply. "My good friend Norman Borlaug," wrote former president Jimmy Carter, who has collaborated closely with the scientist, "has accomplished more than any other one individual in history in the battle to end world hunger . . . [his] achievements have saved hundreds of millions of lives."

From the standpoint of energy, one dire—Malthusian—question about the green revolution persists. It is a question Ken McCauley and thousands of other farmers grapple with on a daily basis: will the end of the petroleum era also spell the end of high-yield farming methods— and the populations that depend on them?

BIG BUTZ

The sprawling landscape I encountered in Kansas—and the millions of acres of farms that extend across the heartland, from western Illinois to eastern Colorado—had been shaped in fundamental ways not only by these fossil-fuel-powered innovations but also by government policies determining how such innovations would be used. From 1971 to 1976, in his capacity as secretary of agriculture under the Nixon and Ford administrations, Earl "Rusty" Butz set U.S. agricultural policy with the hubris, ruthlessness, and single-minded intensity of a czar. In this role, he was loved by prosperous farmers and loathed by environmentalists, labor leaders, and social justice advocates. To this day, critics blame Butz for killing off America's once-vast network of family farms through an aggressive expansion of consolidated large-scale agribusiness.

Butz grew up on a small farm in Indiana, but with his severe black suits, groomed gray hair, and thick-framed glasses he looked more like a mortgage broker than a sodbuster. Nevertheless, he went to great lengths to ingratiate himself with the farming community: "When he is among farmers, he drops his 'g's' and talks about plantin' and plowin'," reported the *New York Times* in 1976, "and he tells them that he can still feel the spot on his back where the plow straps once dug into his skin."

Butz's policies were based on "two principal new tenets for American farmers," according to the *Times*: "(1) produce more, (2) sell abroad." By these measures, Butz was a great success: under his five-year tenure, farm incomes shot up by nearly 60 percent and exports almost tripled.

Butz transformed agricultural trade, making America newly reliant upon global food markets and expanding its role as a "breadbasket to the world." In turn, this expanded production stoked the agricultural industry's growing reliance on fossil fuels.

When he first came to office, Butz orchestrated a $1.1 billion grain sale to the Soviet Union—the mark of a new era of large-scale global trade. The move was widely applauded at the outset, but soon resulted in scandal. The Soviet purchase was so huge that it created an imbalance in supply and demand, leading to a grain shortage in the United States that pushed prices up. Just weeks after farmers committed to sell their crop to the Soviets at about $1.35 a bushel in the early summer of 1972, grain prices shot up to more than $2 a bushel. Farmers felt cheated because Butz had priced their crop so low. The grain shortage spiraled, ramping up prices for livestock, poultry, dairy products, and bread. Consumers began protesting at grocery stores. To ease the pain of food prices, Butz tried to freeze the price of U.S. meat, which in turn led to job losses and angered Americans even further. "Beef plants were shutting down in increasing numbers," read a 1973 news story. "Layoffs were spreading. Commodity speculators were planning to take deliveries of cattle and hold them off the market till the freeze lifts. Others were shipping cattle to Canada in order to bring the beef back as imports and thus circumvent controls."

To quell the public outrage, remedy the price crisis, and open the

gates to a steadier flow of food exports, Butz plotted a soup-to-nuts overhaul of the American farming industry with twin goals in mind: slashing prices and expanding crop production. He pursued these goals with
an almost manic zeal, publicly goading farmers to plant "fencerow to
fencerow" for maximum yield, and even voicing unsubtle threats: "Get
big or get out," he famously demanded, reviving a maxim he had coined
earlier in his career: "Adapt or die, resist and perish."

Before Butz came along, the Department of Agriculture controlled
how much land could be farmed and imposed limits on the pricing of
crops. The aim was to manage fluctuations in supply and demand and
to keep food costs stable for consumers. Butz argued that these constraints were crippling the industry, and persuaded President Nixon to
yank them altogether. "Last month President Nixon signed into law a
new policy that eliminates acreage controls," *Time* reported in September 1973, "and permits a farmer to sell his crop for whatever the market
will bring; if his price falls below specified target levels, the Government
will send him a check for the difference."

This encouraged a no-holds-barred approach to agribusiness. American farmers aggressively embraced the use of newly introduced high-
yield seeds and motorized farm equipment, and ramped up fertilizer and
pesticide applications. The impact on developing-world farmers was also
significant: Without similar crop subsidies, they couldn't possibly compete with U.S. production levels and prices, a disadvantage that has contributed to keeping them largely dependent on U.S. imports.

The high costs of fertilizers, fuel, high-yield seeds, farm equipment,
and pesticides put a squeeze on small family farms. Wealthier, bigger
farmers who could benefit from economies of scale in their purchasing
and use of these goods got the upper hand over mom-and-pop farms on
shoestring budgets. (This also benefited Butz himself, who had strong
ties to three large agribusiness companies.) The policies of the Butz
era set in motion a trend of rapid consolidation, and as farm sizes have
grown, farm numbers have dwindled: the number of farms in the United
States has fallen from a historic high of 6.3 million in 1930 to roughly
2.2 million in 2008.

Butz defended his policies as a boon not only to American farmers but also to the American economy at large: "Twenty years ago, the average American consumer paid $23 for food out of every $100 take-home pay. In 1971 he spent $16 from $100 of take-home pay for food, and he will pay out even less this year," he wrote in a 1972 op-ed. "Nowhere else in the world does food take up a lower percentage of the consumer budget."

In 2007, at the age of ninety-eight (a year before he died), Butz was interviewed in the documentary *King Corn* and defended just this point. The "best-kept secret" behind the prosperity of the U.S. economy, he said, is that Americans today spend a tiny fraction of their disposable income on food—about 10 percent. "That is marvelous," said Butz. "It's the basis of our affluence now." (The nations of the European Union spend on average about 18 percent, and the developing world spends close to 50 percent.) Butz's open celebration of America's easy, cheap-eats lifestyle echoed a controversial comment made by James Bostic Jr., the former deputy assistant secretary of the Department of Agriculture: "Just stop for a minute and think about what it means to live in a land where 95 percent of the people can be freed from the drudgery of preparing their own food."

Paradoxically, the toil of food production was something Butz often praised as "the last bastion of patriotism and hard work" during his tenure as agriculture secretary. He assured farmers that the food they grew would be the key to world peace during the next quarter century: "You are the peacemakers! You are the most productive part of America!" he once proclaimed at an American Farm Bureau symposium. Although he loved to chum it up with farmers, his legacy to them is a complicated one.

Initially, his policies seemed to be working. During the 1970s, food costs began to stabilize, farmers were more productive than ever, and exports were making them rich. But in the following decades, as the policies initiated by Butz took ever-greater hold, American farmers began to produce such a mountain of grain that they could barely sell it all. Grain prices eventually plummeted and farm income began to steadily decline, forcing millions of farmers deeper into debt and toward a steady trend of bankruptcy. The collective debt of American farmers nearly quadrupled from 1970 to 1985, soaring from roughly $48 billion to $175 billion, as product and land values bottomed out. What ensued was an epidemic of bankruptcies and foreclosures: the United States lost nearly 200,000 farms between 1981 and 1986—numbers that rivaled the losses of the Great Depression.

In Brown and Doniphan counties, the lush terrain around White Cloud, Kansas, banks foreclosed on dozens of midsized farms in the 1980s—many of them belonging to old friends of Ken McCauley's family. Government measures were passed for farm relief in subsequent years, measures that have helped keep people such as McCauley in business—but the landscape around me in Kansas had been profoundly and perhaps irreversibly shaped by an era of policies that favored ever-bigger farming operations. McCauley worries that now, as energy costs trend upward, the dying breed of American family farmers may be facing another era of crippling debt.

ZONED OUT

The agricultural practices of the Green Revolution have profoundly altered the contours not just of the heartland but also of regions far removed from agribusiness. Every spring in the Gulf of Mexico, as water temperatures rise, an area known as the "dead zone" emerges. In the summer of 2008, this area reached a record size of 8,800 square miles—roughly the size of the state of New Jersey. This dead zone and thousands more like it across the globe are the result of algae blooms caused by fertilizer runoff.

"Anywhere from 50 to 70 percent of the nitrogen that's spread on a cornfield is actually absorbed by the crop," explained Fred Below, a professor of crop physiology at the University of Illinois. "Around 20 percent of the amount applied stays in the soil and the rest—around 20 percent—is lost as runoff or disappears into the air as nitrogen gas."

An algae bloom occurs when this runaway nitrogen fertilizes the plankton that lives in rivers, lakes, and oceans. The plankton then rapidly multiplies and soaks up the water's supply of oxygen, in turn suffocating aquatic animals. In Florida, for example, nitrogen runoff from farms has choked off many of the marshland inhabitants of the Everglades. The Mississippi River, which houses on its banks a large concentration of fertilizer plants, channels heavily fertilized effluent from these factories down into the Gulf of Mexico. The dead zone in the Gulf consists of a massive expanse of overgrown algae that looks like a thick layer of red scum hovering above a graveyard of clams, crabs, starfish, and marine worms—the slow-moving creatures that aren't able to swim beyond its reach.

The airborne impacts of evaporated nitrogen are also grave. As fertilizers build up in the soil, bacteria rapidly convert it into nitrous oxide. Better known as "laughing gas," nitrous oxide in this free state is a potent greenhouse gas and a threat to the ozone layer. While it isn't as abundant as the CO_2 emitted from tailpipes and power plant smokestacks, nitrous oxide is many times more concentrated—posing a significant long-term hazard in a warming world.

Environmentalists worry that fertilizer use will escalate in the coming decades, asserting that year after year, the flow of chemicals applied to American farms erodes the topsoil's biodiversity and nutrient levels a little further. California-based farmer and sustainability advocate Jason McKenney wrote that the "use of synthetic fertilizers to create artificial fertility has had a cascade of adverse effects on natural soil fertility. . . . Soils are less efficient at storing water and air. With less available oxygen the growth of soil microbiology slows, and the intricate ecosystem of biological exchanges breaks down." As the soil quality erodes over time, McKenney contends, ever more synthetic nutrients must be added to keep the farmland producing. The trend will continue until eventually—perhaps a century or more down the road—the dying fields may not respond to any amount of nitrogen. Tom Philpott believes that long before we get to that point, "the global warming impacts of nitrous oxide will pose a far more devastating problem."

Ken McCauley dismisses these concerns, noting that he has increased his yield in the past decade while decreasing his fertilizer use. With the help of genetically engineered seeds, no-till farming practices (which protects the soil quality), and intelligent, software-enabled machinery for fertilizer dispersal, he has managed to buck the trend. "Everybody wants to tell us we aren't sustainable," said Ken. "If our agriculture wasn't sustainable, we wouldn't be increasing our yields the way we are." Organic farmers counter that technology improvements will eventually run out.

Certainly advances in precision fertilizer application technology can slow the growth of fertilizer use in America, curtailing both the wasteful consumption of petrochemicals and the environmental impacts of algae blooms and the release of greenhouse gases. But farmers in China, India, and other developing nations with growing economies will likely continue to push global demand for chemical fertilizers ever higher. The United Nations has predicted an increase in fertilizer use worldwide of roughly 35 percent by 2030.

Jeffrey Sachs, the United Nations special advisor who wrote *The End of Poverty*, told me plainly that fertilizers will be necessary to human survival for the foreseeable future: "We will not feed 6.7 billion people on the planet without chemical fertilizers." On weathered tropical soils like those of farmlands in large portions of Africa, says Sachs, fertilizers will play a key role. "In all the world but Africa, farmers are using around 100 kilograms per hectare on average of fertilizer. In Africa it's essentially zero, which is one of the real reasons for the massive hunger there."

Add to that the growing market for ethanol, which is widely used as an additive in gasoline. You probably don't realize it, but your car is already driving on a mixture of ethanol and gasoline: more than 6.9 billion gallons of ethanol flow into the U.S. gasoline supply every year, as a means of shaving down the price of gas and the demand for imports. Since 2005, the United States has been converting about a quarter of its domestic corn harvest to ethanol, and the U.S. Department of Agriculture projects that this figure will rise nearly a third by 2015. That climbing demand for ethanol in turn creates ever more pressure to maximize corn production, increase yield per acre, and keep the nitrogen fertilizer freely flowing. The rising demand for green plastics (those made from cellulosic crops such as corn) will also put ever-greater pressure on American croplands.

All of these factors contribute to a growing demand for a limited supply of both corn and fertilizer, which helps to explain why crop prices have trended upward in recent years. The United States has also forfeited its position as the world's largest supplier of fertilizers, given our dwindling supply of available natural gas. Today we import more than half of our nitrogen fertilizers from other countries, including Canada, Russia, and the nations of the Middle East, where natural gas supplies are abundant. In the long term, this means that the well-being of our farming industry is—just like that of our oil-dependent manufacturing and transportation sectors—tied to these regions and vulnerable to their potential shortages and political conflict.

FULL PLATE

How should we approach a future in which the demand for energy inputs for our crops will likely increase, while the energy sources themselves are on the decline?

Sustainable farming advocates such as Pollan and Philpott put forth a sweeping vision: total reform of the food system, beginning with the dismantling of large single-crop farms. Simply removing fertilizers and other petrochemical additives from industrial farming is not the answer, argues Pollan, since "only a fifth of the total energy used to feed us is consumed on the farm; the rest is spent processing the food and moving it around." For that reason, he added, the large-scale organic farms that produce most of the organic products in your grocery store are, just like conventional megafarms, "floating on a sinking sea of petroleum."

Reformers want to see a network of small and midsized organic farms emerge that is organized into regional cooperatives. These aggregates would enable small farms to serve local markets but think like big farms, working together to make bulk purchases of equipment and aggregate distribution systems. They want to see crops and animals re-integrated into the same farms, naturally feeding and fertilizing one an-other, correcting the current system in which cattle, chickens, and pigs are concentrated on huge feedlots, producing an oversupply of nitrogen-rich manure far removed from croplands. They want to see municipal composting programs in which all food waste from homes, offices, res-taurants, and schools is recycled into crop nutrients. They want to see farmers proliferate in number from roughly 2 million today to perhaps 10 million or more. They want to see organic farming become a robust green-jobs initiative that attracts new young talent as the current genera-tion of farmers faces retirement (the median age of farmers in America today is sixty-three). They hope to see consumers volunteer and partici-pate in their regional farms—investing sweat equity in their sustenance rather than letting machines and petrochemicals do all the work. They hope to see Americans start eating seasonally and mostly locally, with global food trade shrinking to include only certain key items that aren't produced domestically, such as coffee and sugar.

What organic farming advocates are proposing is as much a change to American patterns of thought as it is to our food system—a change whereby people would see preparing their own food not as a form of drudgery but instead as a privilege. They are proposing to begin an era in which consumers buy food not because it's cheap, convenient, pre-washed, and ready-made, but because it is wholesome, nutritious, grown locally, and petrochemical-free.

An increasing number of Americans are supporting this vision. Demand has soared in recent years for organic food, as has support for local farming cooperatives and farmers' markets. But local and organic food still represents less than 5 percent of the total market share. This vision is assuming a lot—that American consumers will be willing and able to find substantial reserves of extra time and money to make it a reality.

Most agronomists will tell you that we can't rapidly shift to growing food without chemical fertilizer and fossil-fuel-powered machinery. The challenge today is not so much to go cold turkey on modern farming methods as it is to significantly improve methods for getting the greatest amount of food production using the least amount of fuel and fertilizer. America can begin pursuing key components of Philpott's vision even, and especially, in the midst of the current recession—engaging a new generation of young farmers, encouraging the rebirth of midsized farms, developing municipal compost programs, and supporting more research in and development of natural methods for optimizing crop yields, all of which would have the added benefit of creating jobs. Meanwhile, lawmakers can also provide incentives to accelerate the widespread adoption and continued improvement of precision fertilizer application technologies like those Ken McCauley showed me.

It would be another tremendous leap forward if we used renewable energy sources such as solar and wind to power the energy-intensive Haber-Bosch process that snatches nitrogen out of the air, eliminating some of the hidden environmental and geopolitical costs of today's fertilizer industry. The hitch here is that renewable energy sources currently provide roughly 2 percent of our electricity supply (excluding hydropower). There's no guarantee that a clean power system will be in place in time to protect American farmers from rising fossil fuel and fertilizer prices.

If you ask McCauley where future solutions lie, he'll tell you in better seed technology—including in more sophisticated genetic engineering. A new crop of "miracle seeds" is on the way that could be coded for radical efficiency—producing, for instance, crops that require 50 to 70 percent less nitrogen, and plants that can expand available farmland by growing in drought-ridden areas. Monsanto and the Gates Foundation are already pouring billions into this research, hoping to create a Green Revolution 2.0.

There are strong humanitarian arguments for this new generation of farming technologies, but certainly many Americans will find small comfort in a future of hyperengineered and chemically doused foods. If you count yourself among them, you may find yourself becoming ever more supportive of local organic farms, community-supported agriculture, and farmers' markets—or even thinking about expanding your vegetable garden into something that can regularly supplement your diet. First Lady Michelle Obama pushed this trend in the spring of 2009 when she planted an organic vegetable garden at the White House—the first since Eleanor Roosevelt sowed her Victory Garden during World War II—promoting homegrown food as a source of both personal health and patriotism.

If we can't get to Philpott's vision of 100 percent organic farming in the foreseeable future, Americans are now more empowered than ever to support a shift in that direction. And while I'm all for working toward our own personal and local solutions, we also have to keep our eyes on the bigger picture—accelerating the development of a clean energy network for transportation and electricity that can take the place of fossil fuels before it's too late.

We are witnessing the end of the golden age of globalization.

—Marc Levinson, economist

6

Chain of Fuels

THE STORY OF A 20,000-MILE SPINACH SALAD

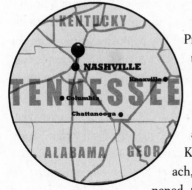

Preparing a dinner salad would seem to be a guilt-free exercise for even the most conscientious consumer, but this routine task left me nerve-racked one winter evening soon after my visit to Ken McCauley's Kansas farm. It started with the spinach, a velvety green bundle that I happened to notice came from Cal-Organic farms—located on Golden State soil some 2,000 miles from my Nashville, Tennessee, grocer. Then I spotted "Mexico" in fine print under the "Ripe Now!" sticker on my avocado. My other ingredients—carrots, red bell pepper, cucumber, mushrooms, garlic, olive oil, and lemon—hailed from as far away as China, Chile, Italy, and Honduras. It wasn't hard to deduce, with the help of a dusty high school atlas, that my salad fixings had traveled a combined distance of more than 20,000 miles to my chopping block, via fuel-hungry trucks, ships, and airplanes spewing untold pounds of CO_2. My organic salad suddenly looked like a globetrotting energy boondoggle.

I had talked with McCauley about the fossil fuels consumed by fertilizer and farming equipment, but here was another, even bigger piece of the agricultural picture. What happens to our food after it leaves the farm? Growing and harvesting crops only accounts for 20 percent of the total energy consumption of our food system, as I'd learned from Michael Pollan—the rest is used in processing, transportation, and storage. So how much fuel does it take to ship, truck, and fly the contents of a standard meal to our dining room tables?

A quick Web search brought up some alarming statistics. There's actually a term, "food miles," that refers to the distance food travels from where it's grown to where it's consumed. One recent study by Iowa State University's Leopold Center for Sustainable Agriculture found that on average, produce sold in the Midwest travels about 1,500 miles "from farm to fork." The study examined the travel routes of different fruits and vegetables from farms throughout the continental United States to the main food market of Chicago, Illinois. Only pumpkins and mushrooms traveled fewer than 500 miles to reach the market, while broccoli, cauliflower, grapes, peas, spinach, and lettuce were all trucked more than 2,000 miles to market.

In the decades before Earl Butz and the agribusiness movement, American grocers were supplied, in large part, by local family farms. In those days, consumers were accustomed to eating seasonal produce that would have to be frozen, canned, or pickled for winter months. My distant relative who lives on a farm in Parrotsville, Tennessee, inherited a knowledge of the canning process from her mother and grandmother, an annual ritual she described to me: "We collected peaches, apples, tomatoes, peas, beans—really anything that can be canned. After packing the goods in Mason jars, we put the fruits and jams in a bath of boiling water, and the vegetables in a pressure canner" that kills bacteria for safekeeping over many months. To procure enough canned goods to last from November to April, this process of collecting, preparing and canning the food requires weeks of diligence in the kitchen.

The result is far more delicious than the canned goods you get at your local grocer, but few Americans choose to preserve foods themselves in an era when "summer" berries, corn, and tomatoes can be purchased all year long. Local foods account for less than 1 percent of total agricultural sales in the United States, according to government data. In recent years, the United States has imported more than 25 billion pounds of produce annually from international farms—farms primarily located in warm climates with year-round growing seasons. In 2008, imports accounted for more than half of our fresh fruit supply and nearly a quarter of our fresh vegetables. I could not find an official calculation for the average distance traveled by produce shipped and air-freighted internationally, but a conservative estimate would put it at well over 2,000 miles.

The more I read about this sprawling food distribution network, the more I puzzled over the problems inherent in my 20,000-mile salad. In a world of rocketing fuel prices and global warming, my run-of-the-mill dinner seemed outrageously lavish—absurd, even. Given the related fuel usage and pollution, how could it possibly be sustainable to source so much of our basic sustenance from places so far away? And if such consumption *isn't* sustainable, what are the practical alternatives and solutions? It's hard enough to reengineer certain industries and institutions—cars, plastics, farms, even the U.S. military—to operate in a world without oil. But how do you begin transforming the unthinkably vast and terrifically fast global network of planes, ships, trains, and trucks that deliver products at all hours of the day and night to store shelves?

What I'd begun to realize as I counted up my food miles was something obvious to any economist but imperceptible to most American consumers: the modern global economy depends on long-distance supply chains—the intricate and complex processes by which companies move materials, parts, and finished products to consumers. This never-ending flow of global goods makes economic sense thanks in part to cheap energy.

We're all complicit. In our modern American lifestyles we've grown accustomed to a dizzyingly wide selection of product choices at rock-bottom prices, a function of the so-called Walmart effect. To support this level of variety and affordability in our consumer goods, we depend on low-wage workers in developing nations whose cheap labor keeps production costs way down. The hitch, from a fuel standpoint (considered apart from the many social justice concerns raised by fair-trade advocates), is that these workers live in far-flung locales—often literally on the other side of the world—so their products quietly, invisibly consume energy as they traverse entire nations and oceans before entering our homes.

The contents of my own home told this very story. After my salad reckoning, I wandered from my kitchen through the rest of the house and noticed, with fresh perspective, that my dining table is from Vietnam, my porch swing from Malaysia, my car from Japan, my bedsheets from India, my alarm clock from Taiwan, my tea candles from Singapore, my desk from Sweden, my couch from Hong Kong. It took a global village, I marveled, to manufacture what I'd previously seen as routine, daily amenities. Gathering up these household items from the far corners of the world was a process dependent on the very same fuel that powers our wars, moves our cars, builds our plastics, and nourishes our farms.

CONSIDER THE BANANA

The first step in understanding the bigger picture—and future—of supply chains, I reasoned, was to take a closer look at my 20,000-mile salad and the food distribution system behind it. So I called up Edgar Blanco, a research scientist at the Massachusetts Institute of Technology's Center for Transportation and Logistics who is investigating the carbon impacts of food transportation. A bright and buoyant scientist in his mid-thirties, Blanco sometimes punctuated his sentences with an engaging "Yes!" (As in: "Suppose you are an apple traveling to a distribution center. Yes!" or "This is a very complex problem to model and dissect. Yes!"). He was born and raised in Bogotá, Colombia, and got his Ph.D. from Georgia Tech's School of Industrial and Systems Engineering, where he fell in love with logistics. "I was amazed when I came to the United States at how things move here, how you can get everything all the time. I wanted to see what makes that happen."

He began to focus on carbon impacts as in recent years, a growing number of MIT's corporate partners started asking the university how they could reengineer their supply chains to reduce their carbon footprints. Blanco envisions a future in which every single product we buy will be digitally tracked—like a FedEx package along its delivery route—enabling logistics experts to follow the journey of each cantaloupe, pair of sneakers, and television in real time as it moves throughout the supply chain, calculating its carbon footprint along the way. "This is a very young field," he said. "We are just discovering it."

As for my salad conundrum, Blanco gave me reasons for both comfort and concern. On the one hand, higher food miles don't necessarily translate into higher energy use. "An apple that travels from Chile by ship to my Boston grocer," he explained, "can actually require less fuel and have a smaller carbon footprint than an apple that has been trucked in from Washington State." Two reasons: First, the U.S.-raised produce may be cultivated with higher amounts of fertilizer and more machinery-intensive farming practices than the long-distance produce, canceling out some of the fuel savings that come from shorter delivery routes. Second, and most important, fossil fuel consumption depends on, among other factors, what mode of transport your food takes. The most efficient modes are train and ship, which—remarkably—burn about nine times less energy per ton-mile traveled than trucks, and thirty-three times less than airplanes.

The worrisome news was that, despite the inefficiencies of truck travel, more than 85 percent of the produce transported in America shuttles between states via eighteen-wheelers. Only about 15 percent is delivered by train. (This disproportionate reliance on trucks is one more consequence of the widespread dismantling of railways in the mid-twentieth century that fundamentally shaped America's cities and transportation systems.) If you add up the distances traveled annually in the national, regional, and local U.S. food transport systems, it amounts to roughly 3 billion miles.

And that's just for produce that travels *inside* the United States. Blanco walked me through the logistics of transporting produce to the

United States from abroad. "Consider a simple banana," he told me. The vast majority of bananas consumed in the United States are grown on hundreds of farms throughout Central America, where they are picked while still green, placed in cardboard packaging, and hauled via trucks to central distribution hubs. The bananas are immediately refrigerated to inhibit ripening, repackaged in sturdier boxes, and trucked to coastal shipping ports. Here they are loaded into temperature-controlled containers that hold nearly a thousand banana boxes each, and then sent on their ocean voyage. Once they arrive at U.S. ports, these same containers are hauled to hundreds of distribution centers throughout the United States by trucks (one eighteen-wheeler carries the cargo of a single shipping container). At these U.S. hubs, the bananas are stripped of some of their packaging and stored for four days in specially designed airtight rooms pumped full of ethylene gas (a petrochemical), which rapidly ripens the fruit to its yellow color. Shrink-wrapped and repackaged, the bananas are then driven to tens of thousands of grocers by another round of refrigerated trucks.

"In total, the bananas go through at least five different transportation legs and three different stages of packaging from farm to grocer," Blanco told me. By his calculations, when you add up the energy required to transport the bananas and to manufacture their plastic and cardboard packaging, on average the distribution of each 40-pound box of bananas produces between 20 and 40 pounds of CO_2—the emissions equivalent of burning 1 to 2 gallons of gas. That adds up—way up—given that the United States imports roughly 6 billion pounds of bananas from Central America every year.

The environmental and cultural problems posed by long-distance food delivery have captured the attention of many consumers and retailers. In recent years, a subculture of "locavores" has sprung up—consumers who limit their diets to foods that can be found within a roughly 100-mile radius of their homes. Patron saints of this movement include bestselling authors Michael Pollan and Barbara Kingsolver, who have chronicled the Amishlike existences they led while growing much of their own sustenance, slaughtering their meats, preserving their food for the winter months, and eschewing basics that are mostly grown overseas such as coffee, cocoa (chocolate), sugar, and tropical fruits. These sacrifices, as they've written, reaped big rewards, including the enjoyment of fresher and more flavorful foods, personally reconnecting to the land, supporting small-scale organic family farms threatened by agribusiness, in addition to wiping out food miles.

The trend has caught on—recent years have seen a surge in local eating, with farmers' markets almost tripling in number from only around 1,755 in 1994 to more than 4,500 in 2008. Top chefs and eateries have begun to offer local fare. Cris Comerford, the Obama family's White House executive chef, for instance, is a local foods advocate, and a popular section of the cafeteria at Google's northern California headquarters is the "150 Club," which serves food purchased within 150 miles of the campus.

These are encouraging trends—but they don't solve the problem of my own 20,000-mile salad. The truth is, I'd love to eat a low-carbon-footprint diet, if only I could find enough time and money to shop exclusively at farmers' markets, let alone to grow and preserve my own food. For better or worse, I need a more mainstream solution. So I went searching for answers at the Bentonville, Arkansas, headquarters of Walmart—the retailer that sells more goods to more Americans than any other. Surprisingly, it was there that I found the most revealing news yet about the challenges of and potential alternatives to America's long-distance food distribution system.

THE WRITING IS ON THE WAL-MART

Walmart, the world's biggest food distributor, has also recently become America's biggest purchaser of local produce. In 2008, the company sold $400 million worth of locally grown fruits and veggies, sourcing one-fifth of its produce locally in the summer months. Walmart isn't alone in this shift toward local foods. Whole Foods sources about 17 percent of its produce annually from local farms. Consumer demand for these products is soaring, said Whole Foods' global produce coordinator, Karen Christensen, because of "a growing consciousness in our culture about both climate change and connecting to local communities in an increasingly globalized world." That growing consciousness is particularly pronounced in England, where in 2008 the supermarket chain Tesco began a "carbon labeling" program for seventy thousand products, ranging from tater tots to televisions. The labels reveal the estimated amount of greenhouse gas generated from the production, transportation, and use of each item.

I asked Walmart's director of sustainability, Matt Kissler, to explain why the retail behemoth was going local. Catalyst of suburban sprawl and employer of 1.8 million (many with gripes about schedules, pay, and benefits), Walmart personifies the corporate consolidation that has hurt local farms, businesses, and sustainable small-town living. Kissler, whose family farms potatoes and apples in Michigan, said the decision was strictly bottom-line: "It's win-win—you get better tasting products that are more affordable," he told me.

Hauling produce unnecessary distances means the grower is earning less money, Walmart is earning less, and the customer is paying more because the money is going into increasingly expensive fuel. (Walmart has its own truck fleet of more than 200,000 vehicles—it doesn't hire an independent trucking company to handle distribution—so the company is keenly attuned to fuel costs.) "You can also deliver a much fresher product to the retailer that lasts longer on shelves," Kissler added, "and reduce the labor costs of stocking those products." Local foods additionally cut down on packaging costs because produce that's trucked shorter distances requires less protective cardboard and insulation.

Before Walmart sourced all the peaches sold in its three thousand megastores from only two suppliers; now the company buys 12 million pounds of peaches annually from farms strategically located near Walmart hubs in eighteen different states. This change alone kept the company's delivery trucks from burning about 110,000 gallons of diesel in 2008, and sliced its customers' farm-to-fork distance by 670,000 miles. Another example: the cilantro for Walmart's East Coast stores used to come from California, but the company transferred the contract to farmers in Florida, where the herb can be grown just as easily (these farmers had been growing alfalfa but readily agreed to change their crop to secure the Walmart contract). That action alone slashed a quarter million food miles in a single season and saved Walmart another $1 million in fuel costs. Pleased with the fuel savings that come with shorter supply chains, Walmart is applying this strategy on a national level to corn, tomatoes, and eventually dozens of other fruits and veggies, in the hope of encouraging a greater diversity of available produce within regional food markets. It's a wonderful irony, I thought, that this consolidated retail giant would become a major catalyst working *against* consolidation— and perhaps eventually against the monocrop agribusiness model fostered by the Green Revolution.

Kissler made no bones about the fact that long-distance products will, in all probability, always comprise a large chunk of the agricultural market, since local foods are fundamentally limited by narrow variety and short growing seasons. There's no getting around the fact that the soil of northern states such as Minnesota and Maine yields crops for as little as four months, while other states with wet growing seasons have limits on variety because they can't grow crops vulnerable to mold. "The consumer is boss," said Kissler. "So as long as the boss wants bananas in January we are going to have to source them from outside of the United States." The same goes for the year-round demand for springtime asparagus, summer squash, tropical mangoes, and autumnal beets. (While it would be possible to grow these products off-season in huge local greenhouses, artificially heating such indoor farms can consume more energy than trucking the products in from tropical climes.) Another Walmart executive I spoke with made the case that there are good medical reasons for importing food: "A diversity of diet and nutrients is associated with better health and longer life spans—no one wants to compromise that."

While the message I got from Walmart was heartening—local, fuel-saving foods will make up an ever-bigger portion of the national produce market—I realized, as I browsed the grocery shelves of the Bentonville Walmart Supercenter, that I had in fact only scratched the surface of the bigger food supply chain picture. Supply chains for fresh produce are incredibly simple—primitive, even—compared to those of more complex products with many component parts. Try counting up the food miles of a product made from a long list of ingredients—it's all but impossible. We live in a world where it can take dozens of nations to produce a microwave dinner, for instance, or even a single loaf of bread.

LITTLE LOAF, WHO MADE THEE?

Examine the ingredients in Sara Lee's standard Soft & Smooth White Bread and you'll find that they come from all corners of the globe. Guar gum, which maintains moistness, comes from guar plants mostly cultivated in India; calcium propionate, which inhibits mold growth, is a powdered preservative commonly sourced from the Netherlands; honey is imported from all over the world, including India, Vietnam, and Argentina; flour enrichments commonly come from China; beta-carotene, which enhances the golden color of the crust, is typically sourced from Switzerland; wheat gluten, which keeps the bread from turning crumbly, comes from France, Australia, and many countries in between. And that's just for starters.

A June 2007 article in the *New York Times*, "Globalization in Every Loaf," examined the diverse origins of these ingredients. It explored the many complexities that arise when Sara Lee has to source the thousands of ingredients it uses for all its products, from low-carb frozen pizzas to chocolate cherry cheesecake. "Today, up to a third of the hundreds of suppliers Sara Lee uses are based overseas or have foreign operations," reported the *Times*. The company must compete in a global commodities market with other major food conglomerates such as Nabisco and General Mills.

At its Illinois headquarters, Sara Lee has a "purchasing floor"—think of it as a mini version of the New York Stock Exchange—in which sourcing experts scan computer readouts of commodities prices, scouring the globe for the cheapest and most readily available ingredients. If, for instance, suppliers in Europe have a surplus of wheat gluten and can deliver the stuff just when Sara Lee needs it, sourcing experts might pounce on a bulk order. The choice of suppliers can change according to fluctuations in price and availability. It stands to reason that the chances of sourcing all of these ingredients exclusively from local or regional markets are next to none.

This sourcing process is not unlike that of the automakers and electronics manufacturers that buy component parts from all over the world. A typical Nokia cell phone, for instance, has between 500 and 1,000 components, derived from suppliers in more than twenty-five countries. Traced back to its earliest supply chain origins, this modern-day gadget begins with the raw natural resources of fossil fuels, metals, and minerals for the product itself, and trees for the packaging. Those building blocks are extracted and then shuttled to refineries to be processed (using large amounts of energy and heat) into usable materials—plastics, chemicals, aluminum, copper, iron, cardboard. These materials are then transported to dozens of factories worldwide and assembled into component parts—including printed circuit boards, cables and wires, batteries, plastic components, thin plates, electromechanical parts, screws, keyboards, and display screens. These components are bundled, packaged, and transported to ten Nokia plants around the globe, where they are assembled into finished products and packaged again. With that, they're off to store shelves—Nokia sells its products at thousands of stores in more than 150 countries. In between these many phases of production there are multiple stops at warehouses, layer upon layer of packaging, and change after change of transportation mode as the products journey across land and sea.

The banana supply chain, by comparison, looks like a stroll in the park. A recent study of Nokia estimates that the full product life cycle of a 3G phone consumes the energy equivalent of 1.7 gallons of oil (approximately 60 percent of which is consumed in the phone's manufacture, the rest in its transportation and use). Once you take into account the billions of phones circulating in the world's purses and pockets, you find that the energy equivalent of millions of barrels of fuel are consumed annually for a product consumers generally regard as disposable. The average life span of a cell phone is a scant two years, and consumers often replace their phones sooner in favor of trendier models. The hidden energy required to fuel such trends in product manufacturing and obsolescence is, in a word, breathtaking.

Even our kids unwittingly form part of the consumer supply chain problem: more than three-quarters of the toys sold in the United States are manufactured abroad. Take, for instance, Mattel's supposedly all-American Barbie doll. The product information claims that the famously buxom figurine was made in China. And yet, in an article titled "Barbie and the World Economy," the *Los Angeles Times* reported that "several other countries contributed to the making of the . . . Barbie as much as or more than China did. From Saudi Arabia came the oil that, after refining, produces ethylene. Taiwan used the ethylene to produce vinyl plastic pellets that became Barbie's body. Japan supplied her nylon hair. The United States supplied her cardboard packaging. Hong Kong managed everything." Today, the very notion of product-origin labels—"Made in the USA" or "Made in China"—is by and large misleading, if not completely moot.

Mobile phones and Barbie dolls are small parts of the sprawling universe of gadgets continually produced and consumed on a daily basis. Just think of the number of laptops, iPods, toasters, Nintendos, lawn mowers, dishwashers, power drills, automobiles, and televisions all constantly crisscrossing the earth in their various stages of completion via vast delivery networks. It's a symphony or pandemonium, depending on how you look at it.

The fact that such a sinuous and distributed system actually *functions*—let alone in a timely and economical manner—struck me as a dazzling triumph of logistics and a testament to twenty-first-century ingenuity. But I still couldn't quite get my mind around some basic questions. How did this absurdly complex system evolve? How could it possibly be efficient to cobble together a single product over such

niles in 1999." Across the board, cheaper transportation removed barri-
rs to long-distance product delivery and helped catalyze the early stages
f global investment. It helped, too, that oil prices remained extremely
ow throughout much of the 1980s and '90s.

Reagan's antiregulatory approach persisted, to a large extent, throughout the George H. W. Bush, William J. Clinton, and George W. Bush ad-
ninistrations that followed—in fact, right through the multitrillion-dollar
collapse of the highly interwoven and largely unregulated international
banking system in 2008. As free market economics enjoyed a heyday in
he United States in the 1980s and '90s, it also began to spread throughout the world. After the Berlin Wall fell in 1989, billions of people living
n communist and developing nations joined the marketplace, further
spurring global investment.

President Clinton accelerated the trend by implementing some two
hundred landmark trade agreements—most notably the North American
Free Trade Agreement (NAFTA), which established a trading bloc that
included the United States, Canada, and Mexico. "The global economy
is giving more of our own people and millions around the world the
chance to work and live and raise their families with dignity," said Clinton. "We don't need to build walls, we need to build bridges. We don't
need protection, we need opportunity."

Clinton's pro-trade policies received criticism from many for allowing the export of hundreds of thousands of U.S. manufacturing jobs;
and praise from other quarters for contributing to the shift in the United
States from a manufacturing-based economy to a more high-tech, service-
based economy. While the number of blue-collar factory jobs (and their
salaries) shrank in the United States, the number of white-collar service
jobs swelled. In the 1990s, U.S. GDP shot up 35 percent, and the total
volume of global trade surged.

But for all that the combined forces of deregulation and free trade
agreements did to build global supply chains, politics only tells half of
the story. The other half is technology: The development of supply chains
would have been far more sluggish if two crucial technological innovations hadn't come into play—first the container, then the computer.

vast distances? How long can this system continue to work in an era of
unstable oil prices? And how will we unscramble this egg if the system
starts to break down?

TRADE WINS

The modern global supply chain came of age only very recently—in the
1980s and 1990s—thanks to an unusual convergence of public policy
and information-technology developments. But the political stage had
been set for this sprawling trade network nearly half a century earlier, in
the aftermath of World War II. As the war came to a close, the United
States spearheaded the effort to broker an era of global cooperation.
Buoyed by its industrial productivity during the war, the United States
was soaring into an unprecedented position of wealth and power. The
countries of Europe, meanwhile, lay blood-soaked and stripped of their
riches. Japan was in ruins. Not only had millions of soldiers and citizens
died in these nations, but the buildings, roads, factories, and infrastructure of most of their major cities were reduced to rubble. They needed
rebuilding from the ground up.

America took the lead with two initiatives—the Bretton Woods ac-
cords of 1944 and the Marshall Plan of 1947–51—that sought to recon-
struct war-torn nations and open the world to unfettered trade.

These initiatives promised big rewards for the United States, which
would suddenly have huge new markets for the export of its goods. As
William Clayton, Roosevelt's assistant secretary of state for economic
affairs, put it, "We need markets—big markets—around the world in
which to buy and sell." The constantly humming assembly lines of World
War II had helped make the United States into an industrial "Colossus," as British historian Robert Payne noted, producing vast quantities
of manufactured goods—cars, ships, airplanes, weapons, tools, chemicals, and plastics. In addition, a global free trade system would help the
United States by giving it access to new troves of raw materials for its
booming industries. Though the United States in 1945 was churning out

half the world's coal, nearly 70 percent of the world's oil, and more than half of its electricity, we were already getting thirsty for more.

This embrace of global trade represented an about-face in American economic policy. President Franklin Roosevelt had lifted the cloud of the Great Depression in the 1930s with his nationalistic New Deal, which focused on building regulated domestic industries, imposing strong restrictions on trade, and showing trade favoritism toward friendly nations. Now the United States was ready to open up the playing field, offering all nations access to markets and raw materials. The newly cooperative world leaders established the building blocks for a global economy: "convertible currencies" and fixed exchange rates that could be used as reliable vehicles for investment, trade, and payments. The U.S. dollar was established as a gold standard currency for central banks, and agencies were established to oversee exchange rates—namely, the International Monetary Fund (IMF) and the International Bank for Reconstruction and Development (World Bank), both headquartered in Washington and captained by U.S. economists.

Fast-forward to the late 1960s. The reconstruction effort and guidelines for the global trade system had been successful—so much so that Europe and Japan had regained their economic health, adding millions of jobs and attaining strong per-capita incomes. Meanwhile, inflation was driving down the value of the dollar, and pro-U.S sentiment throughout the world was deteriorating due in part to the Vietnam War. America's global leadership position was under threat. The 1970s only made matters worse. In the midst of the Arab oil embargo and a serious economic downturn, President Nixon borrowed from Roosevelt's playbook, regulating domestic industries, imposing a 10 percent surcharge on imports, and reengineering U.S. trade policy with more restrictions. But instead of lifting, the recession worsened; the economy plunged into a downward spiral that persisted through the Carter administration.

The message at the end of the 1970s seemed clear: government intervention in markets is not a panacea. Economist Herbert Stein, an advisor to Nixon, explained it this way:

We were at the end of two decades in which government spending, government taxes, government deficits, government regulation and government expansion of the money supply had all increased rapidly. And at the end of those two decades the inflation rate was high, real economic growth was slow and our "normal" unemployment rate . . . was higher than ever. Nothing was more natural than the conclusion that the problems were caused by all these government increases and would be cured by reversing, or at least stopping, them.

Thus the stage was set for an era of deregulation and unfetter[e] markets throughout the 1980s and 1990s, which in turn opened t[h] floodgates to global trade. President Reagan summed up his econom[y] strategy, dubbed "Reaganomics," this way: "Only by reducing the grow[th] of government can we increase the growth of the economy." In practic[al] terms, Reagan's approach was to cut taxes, cut government spendin[g] and slice through what he called the "vast web" of government regul[a-] tions that he felt had a choke hold on the economy. Economist Milt[on] Friedman, who strongly influenced Reagan's governing philosophy, d[e-] scribed Reaganomics as "a means to an end, not an end in itself. Th[e] end was freedom, human freedom, the right of every individual to purs[ue] his own objectives."

As president, Reagan encouraged the deregulation of the transpor[ta-] tion industries—greatly accelerating a trend that had begun in the '7[0s.] The trucking, airline, shipping, and rail industries had previously be[en] tightly controlled—their rates had been fixed, along with the amo[unt] of cargo operators could carry, the number of trips they could take, [the] distances they could travel, and the compensation they were guarante[ed.] When these restrictions fell away, even as blue-collar worker comper[sa-] tion declined, the trucking, airline, and rail industries expanded dram[ati-] cally, and the cost of cargo transportation plunged.

The air cargo industry, to take one example, "experienced 500 per[cent] growth," according to a Department of Transportation study, "increa[sed] from five billion revenue-ton-miles in 1975 to 25 billion revenue-

SHIPSHAPE

Sometimes the simplest innovations can have the most profound cultural impacts. That could be said for the wheel and the wedge, just as it could be said for the humble shipping container. Believe it or not, those big red, orange, yellow, and blue boxes you see dangling from cranes and stacked like Legos at shipping ports did not exist before the late 1950s— nor did they move into widespread use until decades later. They have since become the core of a highly automated system that has helped shrink a vast planet into a global village, invisibly shuttling products to and from virtually anywhere.

Perhaps nobody knows more about containers than economist Marc Levinson, author of the recent book *The Box: How the Shipping Container Made the World Smaller and the World Economy Bigger*. I called him up to get his insights on this remarkably primitive innovation and learned that the simple metal shipping container did more than launch a thousand ships—it launched millions. "Before the shipping container came along, transporting goods was so expensive," said Levinson, "that it didn't pay to ship many things halfway across the country, let alone halfway around the world." Containers helped to make the shipping of general merchandise so cheap, in fact, that for decades the distance a product had to travel "became pretty much of an afterthought."

Before the shipping container entered the picture, the cargo on ships, trains, and trucks was handled via a "break-bulk" system in which teams of longshoremen and dockworkers loaded and offloaded each piece of freight—barrels, drums, crates, cases, bales—by hand. Once standardized containers came into play, automated cranes replaced human labor for loading and offloading. Shippers began to charge per container, no matter what was inside, instead of charging a separate price for each commodity. All of this added up to simpler, cheaper, more rapid transportation. Before containers, transportation costs accounted for as much as 25 percent of the value of the commodity shipped; by the turn of the twenty-first century they accounted for only around 2 percent. "The world was full of small manufacturers selling locally," noted Levinson, but "by the end of the twentieth century purely local markets for goods of any kind were few and far between."

The container was born out of the same kind of practical ingenuity that gave rise to Tupperware, Levittown, and the Model T Ford. It was the brainchild of Malcom McLean, a self-made entrepreneur from Maxton, North Carolina, who, like so many of the great innovators before him, had a restless mind: "He wouldn't be able to sit still five minutes," according to one of his colleagues. "You'd either have to play gin rummy with him or discuss business with him. Can't go quail hunting with Malcolm without him betting on the first, the most, the biggest."

McLean began building his trucking empire just out of high school. A tall boy with a toothy grin, McLean sold eggs for his mother and pumped gas at a service station near his home until 1934, when he had saved up enough money—$30—to put a down payment on his first second-hand truck. Within weeks, he was using the vehicle to run his own small distribution service, hauling textiles and tobacco products throughout the South. Every dollar he made went to purchasing more trucks and securing more routes. By the mid-1950s, McLean Trucking was one of the largest trucking companies in the nation and was listed on the New York Stock Exchange. Its owner began puzzling over the increasing costs of highway congestion and poor road conditions. Then came his pivotal realization: "Rather than driving down crowded coastal highways," Levinson recounted, "why not just put truck trailers on ships and ferry them up and down the coast?"

McLean bought some old military tankers and intended to load them up with trucks. Quickly he realized that there was no need to put the entire vehicle on board—why waste the space? All he needed to transfer was the aluminum box that carried the truck's cargo. In 1956, he hired an engineer to design a modular container that could be crane-lifted off the back of a truck and neatly stacked onto the deck of a tanker in mere minutes. "McLean was never one to be carried away by the romance of ships," reported *Forbes*. "His basic thinking centers on transportation, period—how to get a box from A to B for the cheapest price."

On the first voyage, he cut the price of loading loose freight onto cargo ships from $5.83 a ton to 16¢ a ton—slashing costs by a factor of 36. He soon began to realize the other benefits of container shipping, most notably the promise of reliability: the automatic loading and unloading of ships worked like clockwork, and the containers reduced damage claims and insurance costs—huge advantages to businesses needing to guarantee that their products would arrive on time and intact.

Despite these benefits, the container trend took a while to catch on given industry inertia and resistance from dockworker unions. When the first of McLean's converted tankers with its load of fifty-eight containers set sail from Newark, New Jersey, on its way to Houston, one officer with the International Longshoremen's Association reportedly commented, "I'd like to sink that son of a bitch." But by 1970, when McLean had begun buying up railroad lines to connect to his shipping lines, worldwide containerization of cargo had jumped 40 percent. The trend continued to gather momentum, with container use increasing by more than 20 percent a year. While cargo ships carried a total of 1.9 million tons of containerized goods in 1970, by 1980 they were carrying 10 million tons. The average size of container ships grew: vessels could each hold up to 1,800 containers in the 1970s; by 1980, that number had jumped to 3,000 containers, and by 1990 to 4,200, making the cost of shipping ever cheaper.

Since then container ships have more than doubled in size—today's biggest vessels can carry 12,000 units, 6,000 containers, and megaships are being designed that can hold up to 8,000. As international shipping capacity grew at these dizzying rates, the resulting economies of scale helped further cut the cost of imported goods, which in turn contributed to the ever-increasing consumer demand for imports. Trucking and train volumes grew in concert, as merchandise traveling by sea must then voyage by land to distribution hubs and stores. Airfreight volumes, too, increased, even though air-shipped goods remained costlier for businesses and consumers; the speed of air transport enabled a new level of immediacy for time-sensitive products ranging from sushi to medical supplies. The total value of all merchandise imported to the United States during the 1990s more than doubled.

The 1990s were by far the most decisive decade in supply chain growth—not just in terms of cargo volume but also, and more important, in terms of precision timing and strategy. Enter the Digital Age. As networked computers linked producers on one side of the globe to suppliers on the other, they opened up a vast new realm of information-sharing, and round-the-clock design and inventory management. "Let's say you sell shirts," Levinson offered, explaining that with modern communications I could digitally transmit my shirt designs to a factory in Thailand, and then source, via online inventory networks, "Chinese fabric made from American cotton, Malaysian buttons made from Taiwanese plastics, Japanese zippers, and decorations embroidered in Indonesia." The finished order, loaded into a 40-foot container, could be delivered in a matter of weeks to a distribution center in Tennessee—all by pressing some buttons on a keyboard.

Computers, in other words, gave businesses a new level of transparency and control. Suddenly, businesses could access a universe of data to help them decide exactly the best and cheapest locations in which to fabricate each component of their products—taking into account constantly shifting factors such as wage rates, taxes, energy costs, import tariffs, and transit times. Digital networks made it easier to manage supply chains and increased their reliability, therefore enabling ever more complexity in the assembly and distribution of goods to come into play. That's why, today, the majority of the container boxes that are shipped and trucked around the world hold component parts, not finished products. Levinson calls these components "'intermediate goods,' factory inputs that have been partially processed in one place and will be processed further someplace else."

Information networks have also paved the way for what are known as just-in-time delivery strategies: businesses can obtain goods closer to the time they need them, rather than having to stockpile large and costly quantities of components weeks or months in advance. This offers a tremendous advantage in balancing supply and demand, ensuring that unwanted products won't go to waste. Today, every time a customer buys a car from a Toyota dealership, say, or even a bag of charcoal from

Walmart, a signal is instantly transmitted to the manufacturer telling it to produce and ship one more product. "Global supply chains have become so swift, so immediate, so routine," noted Levinson, "that when U.S. customs authorities stepped up border inspections following the [September 11] terrorist attacks . . . auto plants in Michigan began shutting down within three days for lack of imported parts."

Just-in-time supply chains also enable retailers to respond immediately to changing consumer trends. In his book *The World Is Flat*, Thomas Friedman offers another striking example: after the 9/11 attacks, the clothing retailer Zara noted that its New York City customers were in mourning and wanted to dress in somber colors, and within mere weeks had restocked its inventory accordingly.

But as supply chains have helped companies become ever more nimble, responsive, and efficient in inventory management, they've also made companies increasingly dependent on energy-intensive transportation. The more businesses divide up their products into component parts, the longer the supply chain becomes and the greater the distance their products must cumulatively travel. Likewise, the more heavily businesses rely on just-in-time delivery, the faster and more frequently ships and planes must travel to reliably meet those demands—consuming more fuel along the way.

Some logistics strategies have become grossly inefficient. The *New York Times* described one scenario in which "chicken and fish [were] crossing the ocean from the Western Hemisphere to be filleted and packaged in Asia not to be consumed there, but to be shipped back across the Pacific again." Such scenarios only made sense at a time of dirt-cheap energy prices.

"Energy is a tax on distance," economist Bud La Londe of Ohio State University explained to me. "Generally speaking, if the cost of energy rises, the distances products travel through the supply chain must shrink." Sure enough, if you look back through history, high transportation costs have acted like tariffs—taxes that raise the cost of imports, and limit them as a result. "If you're a manufacturer and your transport costs are high, your main concern will be to locate near your customer," said La Londe, "even if this requires smaller factories or higher labor and operating costs."

In the 1980s and '90s, when energy costs were very low relative to labor costs, it made good economic sense to ship mass-produced goods long distances and take advantage of economics of scale and cheap overseas wages. But this practice may no longer be sustainable in a world where oil prices more than quadrupled in the years from 2001 to 2008 and remain unpredictable. Unpredictability can be almost as troubling to companies as high costs, for it limits their ability to build fixed infrastructure and devise long-term strategies.

Levinson put it this way: "Even as Americans lament too much globalization, they are in fact on the verge of facing the opposite dilemma: too little." Levinson believes that we are on the cusp of a historic shift. "We are witnessing," he told me soberly, "the end of the golden age of globalization."

CLOSER TO HOME

In the summer of 2008, as oil prices were hurtling toward $135 a barrel, I noticed a rash of articles in the financial press with similar headlines, all of them marking a new chapter in the story of supply chains: "Oil Prices Spread Pain as Transporting Goods Get Costlier" (*Wall Street Journal*), "Small Business Crunched by Fuel Prices" (Reuters), "Oil Costs Force Business to Rethink Supply Network" (*Financial Times*).

One of these articles, "Shipping Costs Start to Crimp Globalization," told the story of Levinson's end-of-an-era assertion in stark numbers: "The cost of shipping a 40-foot container from Shanghai to the United States has risen to $8,000, compared with $3,000 early in the decade," reported the *New York Times*. "Big container ships, the pack mules of the 21st-century economy, have shaved their top speed by nearly 20 percent to save on fuel costs, substantially slowing shipping times."

By early 2009, oil prices had tumbled again, along with shipping costs. Yossi Sheffi, an MIT engineering professor who advises businesses on supply chain strategy, told me that the volatility spurs confusion among business leaders: "They're wondering, 'Should I invest in reengineering my supply chain? Or, if oil prices stay low, will that make me less competitive?'" Even as they scramble to react to these price shifts, businesses have other factors to consider, such as global warming: government regulations that levy a price on carbon emissions will also raise the cost of long-distance product distribution. Taking these issues into account, many businesses are betting that the era of cheap energy is over, and they've begun seeking ways to shrink their supply chains.

Take Procter and Gamble, the world's biggest consumer goods company, which makes everything from pet kibble to Pampers: "A lot of our supply chain design work was really developed and implemented in the 1980s and 1990s, when . . . oil was 10 bucks a barrel," Keith Harrison, head of the company's global supply division, told the *Financial Times*. "I could say that the supply chain design is now upside down." The company has begun making plans to shift its manufacturing sites closer to consumers in order to cut transportation costs. Kimberly-Clark, the parent company of Huggies and Depends, recently reengineered its supply chain to locate eight U.S. distribution centers near rail hubs in order to cut down on truck usage, and saved nearly half a million gallons of fuel in a single year. Ikea, the Swedish furniture manufacturer, opened its first U.S.-based factory in 2008—a move that shrank the distance of its product deliveries to American consumers.

Home Depot, too, has begun sourcing some of its flowers and plants—perishable products that require speedy and expensive delivery—from domestic nurseries rather than shipping in cheaper products from growers in South America. Emerson, a company that manufactures electronics, recently moved some of its production stateside, after years of manufacturing in Asia, to save on transportation costs.

On a smaller scale, consider an outfit like Go Airport Express, a shuttle service with a fleet of about 130 vans that ferry Chicagoans to and from transportation hubs. The company's fuel costs jumped more than

$35,000 in a single year, which has caused it to rethink its routes for greater efficiency and to encourage customers to travel in groups rather than individually, reducing its number of trips.

SUPPLY CHANGE

There is a great beauty to the "cooperative symphony," as some economists have described it, that occurs across a supply chain. To the free-market advocate Milton Friedman, robust trade represented a kind of holistic political utopia. In one famous televised commentary, he mused about the production of a basic pencil:

> Look at this lead pencil. There is not a single person in the world who could make this pencil. Remarkable statement? Not at all. The wood from which it's made, for all I know, comes from a tree that was cut down in the state of Washington. . . . This black center we call it lead, but it's really graphite . . . I think it comes from some mines in South America. This red top up here—the eraser—a bit of rubber, probably comes from Malaysia. . . . Literally thousands of people cooperated to make this pencil. People who don't speak the same language, who practice different religions, who might hate one another if they ever met. . . . What brought them together and induced them to cooperate to make this pencil? There was no commissar sending out orders from one central office. It was the magic of the price system. . . . That is why the operation of the free market is so essential—not only to promote productive efficiency, but even more to foster harmony and peace among the peoples of the world.

Absent from these words, however, is the central role that fossil fuels played in this cooperative network. Cheap energy is the unseen force that has kept the entire supply chain symphony in motion. It's the web that made a divided global civilization into a networked global village (and in so doing, many argue, it also allowed wealthier nations to benefit from workers contributing labor with less than "fair trade" compensation). When oil was cheap, we took this invisible web for granted. Now volatile oil prices are increasingly limiting our freedom of movement and the movement of goods around the globe. Anyone who has purchased a plane ticket in recent years knows just how much the cost of transportation soared. In the late 1990s, a ticket from New York to Paris cost a few hundred dollars. In 2008, that same trip cost in the thousands. Suddenly, the global village began to feel big again.

It's hard to say how significantly the global manufacturing landscape will shift in response to oil price instability. Will we see a major restructuring of national or even regional supply chains across the board; or will we see changes only at the margins, on perishable and lower-value goods such as produce? Marc Levinson foresees continued growth of international trade, but he believes it will be much slower than in decades past. As transportation consumes an ever-greater portion of the cost of a product, many goods—especially those that are heavy and cumbersome to ship—will be manufactured closer to home. But it's hard to imagine a scenario anytime soon in which complex, costly, and labor-intensive products such as cell phones, flat-screen TVs, and laptops are sourced and assembled—start to finish—in the United States.

Even if supply chains do radically shrink down to regional markets, they could boomerang back to the distributed systems of the 1980s and '90s as soon as new, reliably cheap forms of clean transportation and renewable energy hit the mainstream. Until then, chances are good that American consumers will feel a reversal of the Walmart effect that we've come to take for granted. That means accepting that, like cheap gasoline, the huge variety and rock-bottom prices of our consumer goods will likely become (if only temporarily) a thing of the past.

See what I mean? The grid is running out of room. There's just no space down here to put more copper. You can't put ten pounds of baloney in a five-pound bag.

—Dennis Romano, Con Edison employee

7

Short Circuits

WHY A HIGH-TECH SUPERPOWER
HAS A THIRD-WORLD GRID

Thursday, August 14, 2003, began as a typical late-summer day in Manhattan: buildings hummed, subways rumbled, shoppers bustled, taxis honked. But at 4:09 p.m., the great churning chaos of the city was brought to a spectacular halt when the power lines feeding the island went dead. Almost at that exact moment, hundreds of other cities and towns lost power—including Newark, Albany, Cleveland, Baltimore, Detroit, and Toronto—in what was to that point the single largest blackout in history. More than 250 power plants and thousands of miles of transmission wires had shut down across an area spanning some 9,300 square miles that was home to 50 million people.

Thousands of businesses were without power for a period of roughly twenty-four hours. Banks and ATMs were closed. Cable Internet service was shut down. Television stations were knocked off the air. Millions of mobile phones had no reception. (Even though most service providers had backup power generators, the sudden influx of emergency and other calls overwhelmed the airwaves.) Train transportation was disrupted. Car and truck traffic was snarled without stop lights. Gas pumps were inoperable without electricity. Air traffic was halted, with flights grounded or cancelled at the major international hubs of JFK, LaGuardia, Cleveland-Hopkins, Baltimore-Washington, and Toronto Pearson airports.

All at once, the entire supply chain of that region—the computers, trucks, trains, ships, and planes that had been moving countless millions of products throughout the northeastern United States and southeastern Canada—was frozen. The proverbial symphony of capitalism had been silenced. The systemic paralysis and economic loss caused by the blackout of 2003 was more immediate, unpredictable, and widespread than any loss that could have been caused by a shutdown of the region's oil pipelines.

This had happened because of an implausibly small and remote accident in that mysterious system known as "the grid"—the sprawling network of power plants, cables, poles, towers, transformers, substations, and stanchions that creates and carries electricity into our homes and offices. The accident resulted from two decidedly unassuming culprits: an old, overheated transmission wire in a distant field in Ohio and a cluster of overgrown trees.

On that hot August afternoon, transmission lines all across the Northeast had been straining at high capacity as people cranked up their energy-hungry air conditioners to beat the heat. (On consecutive days of 85-plus-degree heat, cities use roughly double the amount of power required to keep them running on cooler days.) Coursing with heavy currents, some transmission cables began to overheat and droop. Typically, this wouldn't be a problem: the lines are in fact designed to sag when carrying heavy electricity loads and then contract when conditions ease. But the cable known as the Harding-Chamberlin Line had no room for movement: the trees beneath it had grown lush without proper trimming by FirstEnergy, the Ohio utility that owned and maintained it.

When the sagging copper cable touched the high tree canopy, the leaves began to smolder and the line short-circuited, immediately shutting down—an automatic safety precaution to prevent a fire. The 345,000

volts it was carrying tried to find a new path, instantaneously jumping onto neighboring lines. Already under stress, those cables couldn't
handle the extra burden, in turn shutting down and causing a massive
cascading effect. With each successive line that tripped off, ever-bigger
currents tried to shift onto other parts of the power grid across state
lines, strangling and shutting them down. Amazingly, the grid is so old
and primitive that its alarms malfunctioned, and its operators weren't
aware of the line failures until it was too late to respond.

The resulting collapse of the Eastern Interconnection was widely reported as a harbinger of crisis in the U.S. grid—a challenge that parallels
our oil crisis but shouldn't be thought of as one and the same issue. Oil
does not, in fact, generate much electricity in the United States (it fuels
only about 2 percent of power plants nationwide). Half of American
power plants burn coal, a fossil fuel with many of the same constraints as
oil: it's polluting, it's finite, and it's increasingly costly to burn and extract.
The electrical grid does not get as much media attention as oil, in part
because we are more self-sufficient in supplying our power than our petroleum: we own, install and maintain our own power plants and wires,
and America has vast domestic reserves of coal. But the closely linked
issues of global warming and soaring digital-era electricity demands
have made the challenge of upgrading our electricity supply system no
less urgent than the task of breaking our addiction to oil—maybe even
more so.

Anybody who's been in a blackout knows that electricity is almost
as vital to modern survival as air, food, and water (in fact, it often filters the air, preserves the food, and pumps the water). Fuel shortages
may limit transportation, inflate grocery bills, and slow the economy, but
electricity shortages cause total system shutdown. From the moment
our alarm clocks ring to the time we switch on the morning news, run
the dishwasher, charge our cell phones, download our podcasts, jump on
the subway, assemble our PowerPoint presentations, FedEx our reports,
shop on Amazon, visit the ATM, hit the supermarket, and unwind in
front of late-night TV, we draw from the continuously pulsing grid of
electricity that flows across the country.

But while American consumers have worried about fluctuating gasoline prices in recent years, most of us have turned a blind eye to our electricity bills, which have been creeping slowly upward. Some industry experts predict that electricity prices will soon take a flying leap, tripling—even quadrupling—over the next decade. Just as global oil supplies can't keep pace with increasing demand, the electrical system is buckling under the combined pressures of antiquated grid equipment and ballooning usage. The blackout of 2003 foretold this grim reality, spurring heated debate among lawmakers and electricity providers about the grid's future. How can we transform this massive aging network of plants and cables into the nimble, efficient, clean energy system of tomorrow?

DIM LIGHTS, BIG CITY

Such big-picture concerns were far from the minds of most New Yorkers that sultry August evening after the blackout. For many of us, the power failure was cause for celebration. A kind of primitive euphoria overtook the city. As I wandered around darkened Manhattan, I saw people camped out in the streets on floral-patterned sheets and in folding chairs to escape the 90-degree heat of their apartments. Some had emptied the contents of their refrigerators onto platters and were offering eclectic snack combinations—deli meats dipped in Miracle Whip, Jell-O pudding pops, and imported Brie—to neighbors and strangers alike. Candles dripped onto asphalt and beer cans bobbed in tubs of melting ice. Stoops became live music stages—I stopped for a while, leaning against a handrail, to listen to an impromptu trio of guitar, trumpet, and violin.

Walking up through the revelry of Broadway into a subdued Times Square, I felt a wave of nostalgia for a time when electricity didn't separate us from the night sky and one another, pulling us into the solitary tunnels of our televisions, cell phones, and computers, into the cocoons of our homes and offices, behind the scrim of city lights.

Apart from this primal romance, however, New York City was paying a huge price. It was one of the last areas affected by the blackout to get its electricity restored, due in part to the sheer complexity of its grid. After twenty-four hours, the city was up and running again, but its businesses had collectively been hemorrhaging about $700,000 per minute. The stock exchange had been shuttered. Grocers' refrigerators had shut down and mountains of food had begun to rot. Restaurants, movie theaters, banks, courts, law and insurance offices, ad agencies, clothing retailers, factories, and convenience stores had all been closed; subways, bus lines, and roadways were out of commission.

In that single day of darkness, the city lost nearly $1 billion in revenue. (The Northeast region as a whole suffered a $6 billion setback.) There were health costs, too: hospitals lost power (one doctor described suspending a surgery in mid-procedure, suddenly in the dark), and life-support systems had to be powered by old backup diesel generators. Three reported deaths were tied to the outage.

The experience illustrated for many of us just how central—how vital—energy is to twenty-first-century life. Hardly a single American company or institution does not depend in its daily operations on the constant presence of an electronic infrastructure. Electricity is the thread from which our entire economy dangles, but it has so deeply permeated our interactive, ever-connected lifestyles that we often fail to notice it at all—until its flow is interrupted.

How, in the first place, did our grid come to be so vulnerable, so run-down and so glutted with dirty sources of power? How, for that matter, does electricity flow throughout the United States, and who controls it? What would it take to fix this broken system?

One logical place to start looking for answers was New York City. There you will find portions of the electrical grid that are among the most antiquated and overburdened in the world—and yet nowhere are the stakes higher for keeping the energy flowing. It is, after all, the capital of global trade and home to the United Nations, a place with a greater population density and higher per-capita income than almost any urban area in the nation.

I approached Con Edison, the utility that supplies all the power for

New York City. Lou Rana, the company's president and COO, agreed to discuss his efforts to bring a twentieth-century grid up to twenty-first-century standards, and to show me the nuts and bolts of the system. I joined a team of his electrical engineers who were heading underground. We suited up in worn gray coveralls, goggles, and yellow hard hats, pulled the iron lid off a manhole, and crawled beneath the streets of New York to examine the guts of the grid.

THE WIRES

Often referred to as "the world's biggest machine," the North American electricity grid as a whole is an integrated network of generators and millions of miles of wires that crisscross the United States and Canada. It snakes across fields, over mountains, through tunnels, along highways, beneath sidewalks, under rivers and seas. If you live anywhere in Canada or the continental United States, this mega-machine "reaches into your home, your bedroom," as one writer put it, "and climbs right up into the lamp next to your pillow."

Within the North American grid are three divisions—the Eastern Interconnection, covering the eastern two-thirds of the United States and Canada; the Western Interconnection, encompassing most of the rest of the two countries; and the Texas Interconnection, covering most of the Lone Star State. Within each network, demand and supply must be perfectly synchronized—meaning, at any given instant, the amount of electricity consumed must equal the amount of electricity produced.

Most power plants create electricity by converting mechanical motion (typically a large fan spun by pressurized steam) into a charge of highly mobile electrons. Think of electricity more as an *action* than an *object*—it's not the electrons themselves, but the movement of electrons over conductive wires. "Electricity is not power," as Edison put it; "it is a method of transporting power." The current in these wires travels at virtually the speed of light, zapping from the power plant to your lightbulb in nanoseconds. Since the grid can't store electricity, it functions as the ultimate just-in-time delivery system: the power is generated exactly when it's needed. So every time you flip a switch or plug in your iPhone, a generator miles away has to churn out a tiny bit more juice. And every time you douse the lights, a generator has to slightly lessen its power production. An imbalance in the grid—too little power to meet demand, or too much pulsing into the system to be used—can trip off wires and shut plants down.

The grid is designed as a hub-and-spoke system, in which large centralized generators supply electricity to thousands of end users. Often sited far from population centers, the generators pipe power into towns and cities via long-distance high-voltage transmission lines (the cables you see strung up on those large metal scaffolding structures that look like robots). Throbbing with nearly 800,000 volts of electricity, these cables feed into facilities called substations, at which the power is "stepped down" (reduced) and then piped via smaller distribution lines that flow like rivers into thousands of tributaries running into factories, offices, and homes. All told, the U.S. grid has about 300,000 miles of high-voltage transmission lines and 5.2 million miles of local distribution lines. When one cable in a network short-circuits, others nearby will automatically pick up the burden. But if the surrounding cables are also overstressed, they too can fail, causing a cascading effect that can knock out major portions of a network.

In recent years, the U.S. power grid has become increasingly prone to such interruptions. Average temperatures have risen, homes have gotten bigger, and so have air-conditioning demands. Our use of electronic gadgetry has also jumped—think of the power-hungry laptops, flat-screen TVs, cell phones, PDAs, iPods, and digicams that have become so commonplace in the last decade. Thanks to our technology-rich lifestyles and the inefficiency of our buildings and power plants, Americans consume, per capita, at least 50 percent more electricity annually than the citizens of Europe and Japan.

But we don't have the infrastructure to support our lavish habits. We've seen almost no expansion or evolution of the grid that struggles to sustain our skyrocketing digital demands. Former energy secretary Bill

Richardson has explained the problem this way: "We're a major super-power with a third-world electricity grid."

Just as our cars generally run on the same combustion engine technology first mass-produced a century ago by Henry Ford, just as our highways and public transportation systems have scarcely been upgraded since the days of President Eisenhower, and just as a high percentage of our military machines are based on the same energy-guzzling models used in World War II, our power plants and all the other components of our electricity grid are shockingly outdated. It's a humbling paradox that the Internet we now use so regularly—a dynamic, super-resilient, ultrasmart information network—is itself kept alive by a primitive network of cables "not much smarter than sewage pipes," as one engineer described it to me. The average age of the equipment that makes up our grid infrastructure is more than forty years, and many components were designed and installed before World War II. The tired bones of this inefficient system lead to costly glitches: the Department of Energy estimates that routine blackouts and line overloads cost Americans roughly $150 billion annually—about $500 apiece.

I don't mean to downplay its magnificence: the North American grid has enabled colossal growth and innovation over the past century and has rightly been hailed as one of the greatest engineering achievements in history. But compared with the rest of today's technology, it's an old, lumbering dinosaur built to sustain a machine-age economy, not equipped to bear the heavier loads or adapt to the complexities of the digital age.

The grid was engineered with mechanical switches, which means it can't quickly respond to fluctuations in demand or block off sections to protect them from cascading effects. Its high-voltage lines were designed before planners had any idea that power would flow so far across state lines, so the copper cables are inefficient and vulnerable to overload. They lose about 5 to 10 percent of the current they carry over long distances, wasting an estimated $12 billion on electricity annually.

And since the grid primarily pumps electricity like water in one direction—from the producer-hubs to the consumer-spokes—it doesn't respond well to currents that go against the flow. That makes it difficult for large numbers of homes and offices to install solar panels or other small, distributed systems that could flow electricity "backward" into the grid, contributing clean power.

All of this adds up to a system that's tragically outdated in an era of climate change. More than a third of all greenhouse gas emissions produced in America come from electricity generation, and 80 percent of these pollutants come from power plants that run on coal—the cheapest and most carbon-intensive form of energy. Today, many coal plants in the United States are more than fifty years old and have reached the end of their effective life cycles. They need to be replaced—but with what? There is a huge opportunity here to introduce clean, renewable energy sources such as wind, solar, and geothermal—technologies that could jump-start the American economy and become the next great global industry. But we won't see a major shift toward greener, more reliable power sources without a simultaneous upgrade in grid transmission technology.

I got a firsthand look at the challenges our power industry is facing when I climbed inside the New York City grid with Rana's engineers. Like everything in that city, the grid is a crazy marriage of new stuff and old stuff—here a solar-powered green skyscraper, there a state-of-the-art superconductor, and over there a massive snarl of ancient tubes and wires held together with duct tape.

GOING UNDER

Con Edison's chief of underground grid maintenance, Dennis Romano, had agreed to accompany me down below with his crew of electrical engineers to explain what I was seeing. A jovial man with a permanent five o'clock shadow, Romano seemed amused if a bit baffled at my excitement over this brief trip.

In spite of what I'd learned about the grid's fragility, I had a fanciful notion of what I'd encounter below the streets of New York: a vast, orderly chamber 50 feet underground containing thousands of gleaming wires all labeled and mapped according to the neighborhoods and buildings they fed, gauges glowing to indicate the volumes of current coursing on each line—as clean and intricate as the innards of the world's biggest iMac.

Instead, my descent into a manhole on lower Broadway lasted all of 17 feet—and the shallow tunnel I crouched through opened onto a chamber roughly the size of an average walk-in closet. The floor was covered with a murky pond of street runoff, crumbled asphalt, and garbage fragments, and the air was clammy and foul. The walls revealed a gory cross section of the grid: emerging from dozens of cement ducts was a spaghetti-like tangle of grimy wires pulsing with so much electric current I could see them vibrate, like hoses with liquid gushing through them.

The New York City grid encompasses more than 80,000 miles of cable—enough to circle the globe four times. Peel back the sidewalks of Manhattan and you'll find a larger concentration of copper than anywhere else on the planet—more, in fact, than in the world's largest copper mine. All that metal can be found within 15 feet below street level, sandwiched in with water mains, sewage pipes, and telephone lines. (These pipes and tubes are constantly in need of repair, so they have to be placed close to street level for speedy access.) There is no large central chamber where all the wires are organized, labeled, and monitored; instead, there are some 260,000 manholes throughout the city, each one providing access to the wires feeding just a handful of buildings.

Many of these cables are over fifty years old. As the wires age, they degrade under a battery of stresses. The combination of sweltering heat in the summer and freezing cold in the winter causes them to expand, contract, and weaken. The constant vibrations of the city and its underworld—rumbling subways, feet pounding on pavement, incessant traffic—can wreak havoc over time. When water mains break and sewage lines overflow, they can soak and erode grid equipment. When salt is scattered on snowy streets, it often eventually drips into street cracks and manholes, eating away at the cables' insulation. Equally common is a nick in a cable from a construction worker's jackhammer or backhoe. Any one of these burdens can overstress and shut down a wire.

An even greater stressor on the system is the relentless demand for power. A decade ago, electricity use typically subsided at around 10:00 p.m., but now everything is running 24/7—banks, shops, clubs, cafés, office buildings, libraries. And people are charging many more gadgets overnight, so some of the wires and equipment that need to cool down never get a rest. When a wire shorts out, maintenance teams have to perform a kind of crude surgery: they tie a piece of rope to the end of a cable, which is as thick as a beer can, yank it out, then stuff a replacement cable back inside the duct using a metal rod. It's a laborious and dangerous process, as occasional leaks in underground gas mains can migrate into the manholes with the potential to ignite from sparking wires, resulting in "hot flashes" that yield temperatures so high that synthetic fabrics have actually melded to human skin and contact lenses have affixed to eyeballs.

But the biggest challenge facing New York City is its outsized electricity demand, which is growing at a rate of nearly 2 percent a year. That doesn't sound like much, but it translates to an additional annual load of 200 megawatts—enough to power nearly a quarter million homes or a midsized city. "It's like moving Albany onto the New York City grid every year," Lou Rana later told me. That's a big challenge when you have a system as congested as Con Ed's.

"See what I mean? The grid is running out of room," Dennis Romano said as we huddled in the dank manhole, gesturing at a mass of wires so dense it was like a Friday afternoon traffic jam at the mouth of the Holland Tunnel. "There's just no space down here to put more copper." The lines, he added, can only carry a finite amount of electricity: "You can't put ten pounds of baloney in a five-pound bag." Romano was describing gridlock in the most literal sense—the grid in its current form is reaching a physical threshold, meaning it can't be built out any further.

The problem, said Romano, isn't just that the grid's wires are reaching critical mass; it's that there's no available space for building new power plants. This becomes a question of real estate, since it's next to impossible to find new locations for generators in a city as densely developed as New York. Even when potential sites are charted, many community groups in New York City are justifiably opposed to adding more polluting facilities to neighborhoods already afflicted with high asthma rates. And the suburbs around New York City tend to be too densely populated and concerned about property values to allow Con Ed to string new transmission lines across their backyards. That's why few new transmission cables—and for that matter, few new power plants—have been installed in or around the city in recent decades.

In fact, most large metropolitan areas in the United States—San Francisco, Los Angeles, Miami, Chicago, Houston, and Boston, among other cities—are grappling with similar challenges of overburdened grids, limited space for new plants and transmission wires, and community resistance to the construction of polluting facilities. In the foreseeable future, even smaller cities such as Nashville will face these problems as they become more densely populated.

"At the rate our demands are growing," Romano said, "we could outgrow the grid in under ten years." When we ventured back up to street level, I could see why: New York was voraciously guzzling power. Bank machines were whirring, flat-screen monitors were flickering, and an Old Navy store had flung its doors wide open, sending a misty plume of air-conditioning out into the stifling 90-degree heat. Across the way, Banana Republic and Bloomingdale's were doing the same. "That right there," said Romano, nodding toward the open doors, "is why the grid gets hammered in summer months. People assume we can air-condition the streets. They just don't think about it." For someone like Romano, who feels the pulse of electrical currents in his hands every day, it's hard to believe that New Yorkers are so blindly marching toward the brink of crisis.

WATT IN THE WORLD

New York City's love affair with electricity began not in crisis but as an opportunity unprecedented in history. It was less than 2 miles from the

dank, grimy manhole I'd visited that the world's first power plant was unveiled with great ceremony on September 4, 1882. Thomas Alva Edison,
having debuted his lightbulb four years earlier, was about to test his riski-
est and boldest invention yet: at a facility he'd built at 255 Pearl Street,
he planned to fire up a hulking coal-powered generator. Electricity from
that generator would—he hoped—instantly illuminate the buildings of
Wall Street a few city blocks away. It was "the biggest and most responsible thing I had ever undertaken. It was a gigantic problem, with many
ramifications," Edison later told reporters. "All our apparatus, devices
and parts were home-devised and home-made. . . . What might happen
on turning a big current into the conductors under the streets of New
York no one could say."

Edison was already a superstar, albeit an unlikely one. Born in Milan,
Ohio, in 1847, Edison was viewed as "addled" by an early teacher and
was removed from school to be homeschooled by his mother. This education was by all accounts a success; he had read works by Shakespeare
along with Gibbon's *Decline and Fall of the Roman Empire* by age twelve.
Edison worked selling newspapers on passenger trains and as a telegraph
operator. He lost an early job working the night shift on the Associated
Press wire for Western Union when one of his many experiments with
a battery led him to spill sulfuric acid, which dripped through the floor
and onto his boss's desk below. But his unconventional education and
persistence—"Genius is 1 percent inspiration and 99 percent perspira-
tion," he later said—paid off with his invention of an improved "quadru-
plex telegraph" in 1874. Proceeds from the sale of this device allowed
him to set up his famed research laboratory in Menlo Park, New Jersey,
where he turned out a succession of pivotal inventions, including the
microphone and the phonograph—recording and playing back voices for
the first time in history.

The world was enthralled, above all, with Edison's lightbulb: "There
was the light, clear, cold and beautiful," wrote one reporter. "The intense
brightness [of gas lanterns] was gone, and there was nothing irritating
to the eye. . . . It glowed with the phosphorescent effulgence of the star
Altaire. . . . It seemed perfect."

Perfect—except that the bulbs couldn't light up without electricity, and no mechanism yet existed to pump power into offices and homes, where the illumination could actually be used. Edison had tried the approach of installing on-site coal-powered generators in several buildings, including the home of J. P. Morgan, the tycoon who bankrolled many of his Menlo Park inventions. But not only were the small-scale generators expensive, they were loud and unsightly, and they belched out foul fumes. They were also prone to breakdowns and needed constant monitoring.

People wouldn't buy bulbs, Edison reasoned, without a mechanism that could conveniently light them up—quietly transmitting electricity from afar. So the "Wizard of Menlo Park" had made a daring pledge to the public and investors: he would "produce a thousand—aye, ten thousand—lights from one machine."

The day had come to make good on his promise. Edison had spent years testing and installing the components of six steam generators at 255 Pearl Street. To power these electric "dynamos," men shoveled coal into furnaces that boiled water to form steam, in turn creating pressure that rotated their turbines to produce electrical energy. Each machine was the size of a reclining elephant, earning them the name "Jumbos" for the lovable star of P. T. Barnum's circus. Edison's team of workers had also dug 18 miles of underground brick tunnels laid with copper wire to channel this electrical current, connecting Pearl Street to the financial district. They had then run smaller wires from these main channels into twelve hundred light sockets they'd installed throughout many Wall Street buildings—including fifty-five sockets inside the offices of the *New York Times*. The inventor had lobbied tirelessly to win approval from politicians for his risky experiment, obtaining fire insurance and municipal permits in spite of looming questions no one could definitively answer in advance: Would the wires catch fire? Would people get electrocuted? Would the sockets fail?

Edison spent the morning down on the streets with the Irish laborers he employed, double-checking that everything was in place. Known in the press as a folksy man of the people, Edison was often seen in his rumpled work clothes, with uncombed hair and unwashed hands. But by early afternoon, he had cleaned up, dressed in his best suit and derby hat, and joined a coterie of moneymen who had gathered in J. P. Morgan's Wall Street offices to witness the event. The air was tense with anxiety. Investors had sunk the equivalent of $10.6 million in today's dollars into Edison's experiment, and not a dime had yet come back to them. The media had pounced on Edison for the project's repeated delays.

"One hundred dollars they don't go on," quipped Edison's chief engineer, standing next to him.

"Taken," muttered Edison, staring at his watch, which he had synchronized with those of his staffers. With that, the clock struck three and Edison's team on Pearl Street threw open the main circuit breaker. Inside Morgan's mahogany walls, Edison closed the switch. Hundreds of incandescent bulbs lit up simultaneously in a five-block radius—and Morgan's office erupted with cries of "They're on!"

"Edison was vindicated and his light triumphed," reported the *New York Herald*. Elsewhere the bulbs were barely noticed in the glare of the afternoon sun, but as night fell, the spectacle of lights was dazzling. The *New York Times* reported the next day that by night the electric "light was soft, mellow, and grateful to the eye. It seemed almost like writing by daylight to have a light without a particle of flicker and with scarcely any heat to make the head ache." Some 200 customers had purchased 3,500 bulbs from Edison within two months after the Pearl Street demonstration; by October 1883, more than 500 customers were burning 10,000 bulbs.

The early foundations of the nation's power grid were built, simply, as a vehicle for light: customers would buy bulbs, and with these bulbs came a surcharge for the installation of electric lines and sockets. Edison scrambled to find enough workmen to lay new wires. By the end of the decade, more than a million bulbs glowed in homes and offices throughout the country.

Entrepreneurs were soon competing with Edison to manufacture bulbs and construct the infrastructure to feed them. Lacking standards, however, this mad rush to build an industry created chaos. The voltage on electricity lines, for instance, varied depending on who installed them and provided the service. You could plug a lamp successfully into a light socket in your home, but walk across the street to your neighbor's house and the plug might not fit or the voltage might not be compatible with your bulb.

The power industry at the turn of the century was still small and struggling. The Pearl Street experiment had cost triple what Edison predicted, and investors worried that electricity would serve only a luxury market. Despite the public's enthusiasm for incandescent bulbs, they illuminated less than 5 percent of lamps nationwide, competing with Standard Oil's far cheaper petroleum-based lamp oil.

But within just a few decades, the fledgling electricity business would become a massive industry. To understand how this shift came about, I spoke with Richard Munson, author of two books on the subject—*The Power Makers* and *From Edison to Enron*. The industry's doubt and disorder, he explained, gave rise to an era of consolidation. Just as Rockefeller had bought up refineries nationwide to standardize the cost and distribution of fuels, in 1892 J. P. Morgan arranged for the consolidation of the major illumination companies under the name General Electric. This included, most notably, Thomas Edison's company, which had itself incorporated a number of smaller entities in 1890. One company, Chicago Edison, resisted the grip of GE under the leadership of Sam Insull, Edison's former right-hand man at Menlo Park.

A dapper native of London with a tidy silver mustache and round wire spectacles, Insull had a keen mind for economies of scale. Instead of manufacturing and selling lamps and wires, Insull saw that selling the *service* of electricity—not by the bulb, but by the kilowatt-hour—would define the future of the industry. The power distributed this way could provide cheap, reliable light—posing a tremendous threat to Rockefeller's lamp-oil industry. It could also be used by any of the countless other energy-hungry innovations that would soon enter the market, from electric coffee percolators and vacuum cleaners to irrigation pumps and the tiny motors that turn factory conveyor belts.

At a time when most businesses owned their own generators and de-

livery systems, Insull saw the opportunity to build centralized power stations that could serve thousands of customers. Such an approach would
ramp up production and ratchet down cost. The bigger the power plants,
the more efficient they would be, the more customers they would serve,
and the greater their profits. In the early 1900s, said Munson, electricity
was 50 percent more expensive than Rockefeller's lamp oil. Insull knew
that the only way to compete was to slash prices. In 1897, he cut his
customers' rates in half, from 20¢ per kilowatt-hour to 10¢. As he pulled
in ever more customers and built ever-bigger generators, he continually
dropped prices until, in 1909, he was selling his electricity at 2.5¢ per
kilowatt-hour.

Now the industry was ready for real growth. "Insull's customer base
grew from about 10,000 at the turn of the century to 200,000 in 1913,"
noted Munson. "Following his lead, the industry as a whole grew at the
spectacular rate of 12 percent per year for the first two decades of the
century." These were the bold beginnings of the hub-and-spoke power
distribution system we have in place today.

PLANTING BULBS

By the mid-1930s, electricity was pumping through the vast majority of
the nation's urban homes. Centralized power plants and their inexpen-
sive product had given rise to electrified trolleys and subways; elevators
and skyscrapers; radio and television; household amenities such as refrigerators, dishwashers, and washing machines; and a surge in industrial
growth thanks to automated manufacturing, mass production, and the
extension of the American work day by electric light.

But these amenities weren't available to all. Nine out of ten rural
homes and farms still had no access to the grid. The country was starkly
divided between the plugged-in urban elite and the powerless rural
masses.

The crusade to bring power to rural America in the 1930s played out
like a political boxing match. On one side of the ring was private power,

the now massive force built by Sam Insull, J. P. Morgan, and advocates of big business such as Republican president Herbert Hoover. On the other side was public power, a new, untested concept backed by a ragtag coalition of rural Americans and progressive populists.

At the time, more than half of the U.S. population lived on farms, but the political influence of this group was extremely limited. For nearly two decades, farmers had been petitioning big utility companies to run power lines into their communities, traveling individually and in delegations to corporate headquarters to plead their case. Their food and dairy products were rotting before they could be hauled to market, they said; they needed refrigeration and electrical appliances to help streamline their grueling daily work and household chores. Time and again, the companies refused, claiming that it was too expensive to run cables costing up to $5,000 (about $75,000 today) per mile to rural areas. Some farmers offered to raise funds to help with installation costs, and even to relocate their homes closer to convenient transmission locations. Power company officials reasoned that they could have hundreds of customers per mile of cable in cities rather than a mere handful per mile in the country. Farmers, they argued, couldn't afford standard electrical rates and the latest energy-hungry appliances. Why serve this market?

For one thing, the power industry had for years been advertising its product as the very source of human dignity, targeting this message at housewives in particular. Between 1922 and 1932, GE spent nearly $50 million (well over half a billion in today's dollars) on advertising for electricity. Best known among the ads was their "Any Woman" campaign, which equated electricity with freedom for the average housewife. Slogans included "Any woman who turns the wringer—any woman who irons by hand—any woman who beats a rug—any woman who cooks in a hot, stuffy kitchen is doing work which electricity will do for a few cents per day," "Any woman who does anything which a little electric motor can do is working for three cents an hour," and "The wise woman delegates to electricity all that electricity can do." The campaign's not-so-subtle implication was that women without electricity were foolish and destitute. Other GE ads implied that women without electricity were bad mothers: "This is the test of a successful mother—she puts first things first," read one. "She does not give to sweeping the time that belongs to her children."

One female activist in the 1920s proclaimed, "Housewives must come up from slavery" and embrace the use of electric power. Historian William Leuchtenburg observed that by the 1930s electricity had divided America into "two nations" living in two different centuries: Farmers "toiled in a nineteenth-century world; farm wives, who enviously eyed pictures in the *Saturday Evening Post* of city women with washing machines, refrigerators, and vacuum cleaners, performed their backbreaking chores like peasant women in a preindustrial age."

While it was the private utilities that propagated this image, it was progressive politicians who built it into a crusade, arguing that electricity by the mid-1930s had become not just a privilege but a basic civil right. Soon after he was elected to succeed Herbert Hoover in 1932, Franklin Roosevelt toured rural Tennessee and was astonished by the region's dire poverty, attributing much of it to the lack of electricity. He established the Rural Electrification Administration (REA) in 1935 to provide funding and technical assistance to farms and rural communities to help them build and run their own electrical power systems.

"Cold figures do not measure the human importance of electric power in our present social order," said Roosevelt. "Electricity is no longer a luxury, it is a definite necessity." Texas representative Sam Rayburn, a staunch supporter of the REA who worked to push the plan through Congress, gave voice to a neglected constituency: "We want to make the farmer and his family believe and know that they are no longer forgotten people . . . yea, they are the bulwark of the government."

Although private power advocates railed against the plan, progressives had a compelling economic argument. Roosevelt saw expanding the electricity grid as an opportunity, in the midst of the Great Depression, to create new jobs and stimulate manufacturing. The agricultural sector would prosper, too. The increased profits that farmers would earn from products such as milk and eggs once refrigeration reduced spoilage would pay many times over for the cost of their power. Irrigation with electric pumps would open more land for cultivation. And as farmers became wealthier, they would consume ever-greater amounts of electricity, eventually benefiting the power sector.

The progressive argument won out, and within two years the REA had helped electrify some 1.5 million farms across the country. The cost of installing rural electric cable plunged. Virtually all American farms were wired by the mid-1950s, priming them for the Green Revolution that was soon to follow.

The long-awaited moment of connection to the grid for rural communities was breathtaking. "Farmers, their wives and children, would gather at night on a hillside in the Great Smokies," wrote Leuchtenburg, "in a field in the Upper Michigan peninsula, on a slope of the Continental Divide, and, when the switch was pulled on a giant generator, see their homes, their barns, their schools, their churches, burst forth in dazzling light."

CURRENT AFFAIRS

As with everything else about the 1950s—the era of chrome-finned cars and suburban utopias—America's embrace of electricity use in that decade was wholehearted and lavish. Electrical appliances began to pervade American homes. Window air conditioner units appeared, with sales escalating from 74,000 in 1948 to 1,045,000 in 1953. Vacuum cleaner models were upgraded from costly, clunky units rarely found in homes to powerful, svelte, standup devices. New and improved refrigerators were billed as "quiet marvels of convenience." Television came of age: in 1946, fewer than 1 percent of American homes owned TV sets, while by 1962, that number had reached 90 percent. Specialized widgets were invented for virtually every domestic chore imaginable, from electric can openers to plug-in baby-bottle sterilizers to chafing dishes advertised as "perfect for Welsh rarebit." "In 1960, the average American home sported a dozen electric appliances," noted historian Richard Munson, "double the number of a decade before."

Despite its promise to liberate women from housework, the Golden Age of electricity created so many more household tasks that at first it actually tethered them closer to home. Betty Friedan wrote of this constraint in her 1963 bestseller *The Feminine Mystique*, in which she chronicled the "nameless, aching dissatisfaction" of the American housewife in the 1950s: "Millions of women lived their lives in the image of those pretty pictures of the American suburban housewife, kissing their husbands goodbye in front of the picture window, depositing their station wagons full of children at school, and smiling as they ran the new electric waxer over the spotless kitchen floor. They baked their own bread, sewed their own and their children's clothes, kept their new washing machines and dryers running all day."

Rather than creating a female workforce, electricity initially served to create a female *consumer* force—vastly multiplying the number of products available, while television ads lured housewives to shopping malls. It wasn't until the 1960s, when the novelty of electrified goods had worn off, that the social benefits of electricity began to emerge. Between 1955 and 1975, women's participation in the workforce jumped from 40 percent to 55 percent. This was a revolution tied in part to social trends, including the feminist movement Friedan helped to foster, but technology also played a role: the average amount of time women spent on domestic chores in the 1960s and '70s dropped "from 52 hours a week in 1965 to 45 hours per week in 1975, and stayed relatively constant after that," according to a study by researchers at the University of California, San Diego.

Electricity also led to a shift in the types of positions held by working women. The number of people employed in housekeeping jobs dropped precipitously throughout the 1950s and '60s, a result both of the increased use of home appliances and the emergence of better-paying skilled jobs in the booming manufacturing sector. The possibilities created by the power industry in this era of change seemed limitless. The sixties brought with them the Beatles, Jimi Hendrix, Chuck Berry, and other legendary wizards of electric guitar—the instrument that launched an ecstatic age of rock-and-roll and recorded electronic music.

Munson walked me through the social and political impacts of electricity usage in the second half of the twentieth century. During these midcentury decades, he explained, electricity enabled the growth of multiple trends—not just in appliances and entertainment but also in subways and skyscrapers, mechanized agriculture and manufacturing. As a result, America's electricity consumption nearly quadrupled between 1950 and 1970—"blindingly fast growth rates," said Munson. In the decade of the 1960s alone, the size of the average power plant more than doubled. This increased size made these new, mostly coal-powered generators ever cheaper to operate and able to pump out more electricity for every unit of fuel burned.

As electricity rates hit their lowest point of the century, engineers dreamed of electrifying nearly every function of life, from compacting the garbage to brushing teeth. Utilities began expanding their market into household appliances that had long been powered by oil and gas, promoting the use of electric cooking ranges and water heaters. "Power companies spent millions convincing consumers that they would, as the motto went, 'live better electrically,'" Munson recounted. "They started promoting an all-electric Tomorrowland."

An energy expert from Cal Tech told *Popular Mechanics* that "a power plant the size of a typewriter" would become available and that the future held an era of "universal comfort, practically free transportation, and unlimited supplies of materials." The scientist Buckminster Fuller drafted plans for a giant glass bubble that could enclose all of Manhattan, with electric filters and air conditioners to maintain fresh air at constant temperatures. The power industry, which had come so far from its humble Pearl Street roots illuminating lightbulbs, seemed indomitable. No one could have predicted the unusual series of events on the near horizon that would threaten to bring the industry to its knees.

SMOKE ON THE WATER

On June 22, 1969, in Cleveland, Ohio, the Cuyahoga River caught fire. The river burned for thirty minutes, with flames rising as high as five stories, before the blaze was extinguished by local firefighters; in that brief time it caused $50,000 in damage. The most likely cause of this extraordinary event was that sparks from a passing train had ignited an oil slick on the river's surface, one of many pollutants from local industries. Perhaps the most extraordinary aspect of the fire was that it was not, in fact, viewed as extraordinary at all by those who lived and worked in the area. William E. Barry, the chief of the Cleveland Fire Department, described it as "strictly a run-of-the-mill fire." Industrial pollutants in the river had caught fire before, in 1868, 1883, 1887, 1912, 1922, 1936, 1948, and 1952. But this most recent fire got the attention of *Time* magazine, which stated in an article on August 1, "Some river! Chocolate-brown, oily, bubbling with gases, it oozes rather than flows. 'Anyone who falls into the Cuyahoga does not drown,' Cleveland's citizens joke grimly. 'He decays.'"

By the late 1960s, the nation's romance with the energy industry was beginning to sour. Rachel Carson's *Silent Spring*, first published in 1962, became a national bestseller and aroused widespread public concern about the environment, specifically addressing the impacts of chemical pesticides on plant and animal life. In November 1966, an air quality emergency alert in New York City raised fears about a potential "killer smog" phenomenon like the one that had killed an estimated 12,000 people in London in 1952. In January 1969, an explosion at an offshore drilling site 5 miles off the coast of Santa Barbara, California, led to the worst oil spill in the nation's history to that point. Around 200,000 gallons of crude oil poured from the site operated by the Union Oil Company (now Unocal), generating an oil slick that ran along 35 miles of the coast and proved devastating to marine life. The Cuyahoga River fire, coming as it did less than six months later, further galvanized the public's awareness of environmental issues.

Gaylord Nelson, a World War II veteran and U.S. senator from Wisconsin, cited the Cuyahoga fire as a contributing factor to his decision to organize the first national Earth Day. "Pollution was rampant in all kinds of rivers all across the nation," he later said. "There was bound at some time to be a dramatic fire and this turned out to be the most dramatic one of all." Nelson had been working on environmental legislation behind the scenes in the Senate since he first joined in 1963, sponsoring the Wilderness Act and the Wild and Scenic Rivers Act, but he was troubled by the slow pace of progress. "All across the country," he recalled, "evidence of environmental degradation was appearing everywhere, and everyone noticed except the political establishment." He announced the date of a huge grassroots protest in September 1969, planning it for April 22, 1970.

In the meantime, another national politician had become aware of the rapidly shifting tide of public opinion on environmental matters. President Richard Nixon was an unlikely hero for the environmental movement, and his motives continue to be debated to the present day (with J. Brooks Flippen arguing, in his book *Nixon and the Environment*, that Nixon aimed to "take the initiative away from the Democrats").

Whether he was a true believer or a politician to the core, there's no denying that Nixon perceived this issue to be of substantial and growing concern to the public. He chose to make it a centerpiece of his first State of the Union address in January 1970, saying, "The great question of the '70s is: shall we surrender to our surroundings or shall we make our peace with nature and begin to make reparations for the damage we have done to our air, to our land, and to our water?"

On April 22 of that year—the first Earth Day—no fewer than 20 million people marched through towns and cities across America demanding legislative change in Washington. "The reason Earth Day worked," Senator Nelson recalled, "is that it organized itself. The idea was out there and everybody grabbed it. I wanted a demonstration by so many people that politicians would say, 'Holy cow, people care about this.'" Soaring air pollution rates caused largely by unregulated power plants were at the top of the activists' agenda. Protestors voiced outrage over the escalating levels of air pollutants including sulfur dioxide (which causes acid rain) and smog-forming particulates (linked to asthma and emphysema).

The Earth Day event mobilized activists nationwide—in high schools, universities, synagogues, churches, and community groups. Environmentalists organized debates, rallies, and sit-ins. They distributed instruction manuals such as *How to Challenge Your Local Electric Utility: A Citizen's Guide to the Power Industry.* "The electric power industry is out of control," read the manual, asserting that eventually "the entire United States [could be] covered with nothing but power plants." It outlined instructions for blocking sites for new generators and for opposing rate increases that would enable the industry's continued expansion. "While this manual does not offer a panacea for the problems of control and use of energy in our society," its authors concluded, "we hope that it has at least given you an idea of where to start." Utility CEOs were getting nervous, knowing that lawmakers would have to respond to the public sentiment.

In July 1970, Nixon authorized the creation of the Environmental Protection Agency. In the months and years that followed, he signed off on sweeping environmental legislation that sped through Congress, most notably the Clean Air Act and Clean Water Act. These laws had serious implications for the once untouchable power industry. Any new power plants built in the United States now had to comply with air quality standards, adopting innovative new filtration technology to "scrub" or remove a certain percentage of pollutants before their exhaust was released. Industry leaders argued that such regulations would cripple their profits, but their protests barely registered next to the tectonic shift of social change.

POWER STRUGGLE

In 1973, three years after the first Earth Day, another historic transition rocked the power sector in the form of the Arab oil embargo. At the time, nearly 15 percent of the nation's power plants were fueled by petroleum, a resource that had been so cheap and abundant in the postwar decades of the 1950s and '60s that it could compete with coal as a power plant feedstock. When the embargo caused the price of oil and natural gas to quadruple overnight, electricity prices followed. Congress soon enacted a law that banned the use of oil in new power plants.

Utility executives remained calm in the panic that ensued over America's dependence on Middle Eastern oil, for they believed they possessed a way to address both environmental concerns and the embargo: nuclear power, dubbed "the peaceful atom" by industry publicists. Nuclear energy, its advocates proposed, could replace both polluting coal and prohibitively expensive oil.

Power companies went so far as to resurrect the utopian vision of earlier decades, arguing, as Munson told me, that "the time had come to fully electrify America in the name of energy independence." Utilities wanted to displace the oil used in transportation with electric power and to vastly expand America's generating capacity. President Nixon and his vice president and successor, Gerald Ford, promoted "Project Independence," which included a plan to build 200 nuclear reactors throughout the country and 150 new coal plants. (Air pollution concerns had diminished against the backdrop of an oil supply shock.)

But here again, utility executives encountered an unwelcome shift in public opinion. On March 28, 1979, at the Three Mile Island nuclear generating station near Harrisburg, Pennsylvania, an accident led to the partial core meltdown of a nuclear reactor. Radioactive contamination leaked into surrounding communities. Though no deaths or injuries could be attributed to the event, it caused a sensation, in part because of the uncanny timing, three weeks earlier, of the release of the movie *The China Syndrome*, an unsettling drama starring Jane Fonda and Jack Lemmon, about safety issues at a nuclear power plant. Public outcry after the Three Mile Island accident led to a tightening of federal restrictions that made the building of new reactors prohibitively expensive.

Meanwhile, as nuclear energy was losing support, American consumers had begun to do something novel: they were conserving electricity. America had not seen a decline in consumer power demands since World

War II, but soaring costs and a rising environmental consciousness were now motivating consumers to turn out their lights and turn down their thermostats. The rapid 8 to 10 percent annual growth in electricity use in previous decades tapered off, sinking below 1 percent.

President Jimmy Carter, who took office in 1977, spurred along the trend. "Every act of energy conservation is more than just common sense," he stated. "I tell you it is an act of patriotism." Carter signed legislation regulating energy efficiency at federal buildings, and established temperature restrictions for commercial structures. And though he was not opposed to coal—in fact, he promoted it as an important alternative to oil and "our most abundant energy source"—Carter was also a strong advocate of renewables, funneling millions into clean energy research and development. He proposed that solar panels, which had been used to power U.S. satellites since the late 1950s, could satisfy 20 percent of America's energy demands by 2000. Many in the energy industry dismissed these goals. The editor of the trade publication *World Oil*, for instance, famously predicted that solar power would have as much impact as "a mosquito bite on an elephant's fanny."

Ronald Reagan helped turn the political tide back in favor of the utilities. He pulled the solar panels off the White House, lowered appliance efficiency standards, and ended the renewable energy fund. And while he couldn't eliminate the air pollution laws enacted under Nixon, his administration came to be known for its limited enforcement of them. American consumers began to resume their lavish habits, with power demand quickly climbing throughout the 1980s.

But while the outlook was brightening in some ways for the power industry, the massive coal plants built in the 1950s and '60s were beginning to break down, requiring costly maintenance. Power companies didn't dare pull these behemoths offline and build new facilities, because the required scrubber technology to remove pollutants "had nearly doubled the building cost of a new plant," said Munson, "from roughly $800 million to $1.5 billion." The old plants had a notable advantage: they were exempt from more recent pollution-control laws. The environmental movement was now a well-funded litigation machine, and quick to bring lawsuits against new power plant construction that violated existing law. This meant that even if the government was lax on enforcement, utilities could still face expensive lawsuits waged by green groups against potentially noncompliant plants.

As a result, the 1980s and 1990s became decades of stasis and stalemate in the electricity industry. Not only did the industry do everything possible to keep its highly polluting old plants online, merely replacing their aging parts bit by bit, but it substantially ramped up their usage in order to meet rising electricity demands. Though the plants had been built to run for twelve-hour cycles with time to then cool down and rest, the industry began to run them round the clock. "Virtually the only new plants built between 1980 and 2000," Munson told me, "were small-scale natural gas plants" that could be fired up at a moment's notice during demand surges but were too costly for long-term use. The grid's distribution system, meanwhile, remained largely unregulated. There was little federal oversight to ensure that the old copper cables would be updated or even reliably maintained.

GRIDLOCK

After my trip underground with Con Edison engineers to witness New York City's overburdened power grid, I contacted Kurt Yeager, the former president and CEO of the Electric Power Research Institute and former director of energy research at the EPA. I wanted to hear his opinion on the severity of America's grid problem. It was hard to believe that our growing digital-age power demands are supported by such archaic technology. But Yeager confirmed what I'd learned. "It's dumbfounding," he said, "that the electricity industry—which has enabled the digital economy to arise—is one of the last industries to go digital."

A grid with nanosensors and electronic controls—rather than the mechanical levers and switches that operate the current system—would enable utilities to predict and prevent problems. Such systems have already been implemented overseas, explained Yeager: "We have one of the least reliable grids in the developed world. The average U.S. consumer experiences power outages several hours a year. If I'm in Japan, it's several minutes a year. In Singapore, it's several seconds. These are smart systems and they never overload the lines." In the coming years, the U.S. grid in its current form will only become more unreliable as demand grows and equipment continues to age.

The problem is not just the wires, Yeager added, it's the plants themselves. Most power plants today function just like Edison's 1882 dynamo on Pearl Street: they boil water using fuel (be it coal, natural gas, or the energy from nuclear fission), and the pressurized steam that's created rotates a turbine, which in turn creates an electrical charge. "Nearly all our coal plants are more than thirty years old, and many of them are twice that or more." The worn power plants that utilities have been nursing along for decades "are like Grandpa's ax, the one with the handle that's been replaced seven times. We just keep retrofitting these old dinosaurs."

After decades of exposure to unrelenting heat and pressure, the iron casings and steel components of these generators invariably erode and crack. There's a limit to the number of times cracks can be patched, gaskets tightened, or turbine blades replaced. Many experts say that between half and two-thirds of the coal plants in America need to be replaced within the next decade.

As the plants' components gradually erode over time, they waste ever more heat and become ever less efficient. "The coal plants we have today operate at 30 percent efficiency," said Yeager, "which means 65 percent of the energy that goes in just goes up in the stack or is wasted heat." In other words, for every three lumps of coal that goes into a power plant, only one lump's worth of electricity comes out.

As fossil fuel costs rise, this becomes a losing proposition for utilities—and for the consumers who unwittingly pay for the system's inefficiencies. Moreover, these outmoded plants make the American economy in general less efficient than the economies of its global competitors. Amory Lovins, who runs the energy-efficiency think tank Rocky Mountain Institute, has written that "The waste heat discarded at U.S.

power stations alone amounts to 20 percent more energy than Japan uses for everything." (Much of that waste heat can be captured and recycled into clean power, and new technologies are in development to do just that.)

If half or more of all U.S. power plants need to be replaced in the next decade—a massive and costly undertaking—what will they be replaced with? If left to its own devices, it appears that the industry would opt for more of the same: after California's rolling blackouts in 2001 and the Northeast outage in 2003, the drumbeat for new coal plants, which had been subdued since the 1970s, began to sound. Industry leaders began to warn of the looming crisis in the electricity supply system, and argued for fast-tracking the development of new coal-powered plants. The terrorist attacks of September 11 also aroused widespread political support for coal amid the outcry for energy independence. "America is the Saudi Arabia of coal," went the refrain on Capitol Hill, and research and development programs were introduced to commercialize methods for liquefying coal into a gasoline alternative. The George W. Bush administration, too, threw its support behind coal, expediting the licensing process for new coal plants.

But coal projects are increasingly facing an uphill battle in the United States. Much like there was during the early 1970s, there is today a rising tide of consumer opposition to this energy source, which is the most polluting of all fossil fuels. Environmental groups estimate that the impacts of coal through global warming, mountaintop mining, lung disease, and the disposal of fly ash (the by-product of burning coal) cost society more than $400 million annually. Congress and the Obama administration have also been working aggressively to pass a federal cap on greenhouse emissions that will significantly increase the cost of producing power from coal, and accelerate the shift toward cleaner-burning plants and renewable energy.

More often than not, when we consider our energy challenges in the United States—whether it's rebuilding the grid or expanding offshore drilling and wind energy development—the emphasis tends to be on building out more supply, not cutting back demand. The often neglected reality is that saving fossil fuel is a lot cheaper—and better for an ailing economy—than is purchasing it from suppliers either foreign or domestic.

Fortunately, digital microchips are making appliances ever "smarter" and increasingly energy efficient. Americans are, however, using so many more devices for so many more hours per capita that we're canceling out some of the benefits of this growing efficiency. The average Internet user is online at least twenty-seven hours a month. And with the proliferation of cell phones, laptops, PDAs, digital cameras, and music players, each American owns at least five external power adapters, according to the Environmental Protection Agency. Overall, consumer electronics account for roughly 10 percent of household electricity use, and the figure is rising. Plasma TVs in particular have become the home's equivalent of a Hummer; some models can consume as much electricity per year as an in-room air conditioner.

Moreover, according to government data, 10 percent of household electricity in the United States is wasted in powering devices that are not in use. Your televisions, computers, stereos, microwaves, DVD players—many gizmos with a microchip—continue to sip power when they're switched off so they can keep running their internal clocks and retain their settings. Amory Lovins calculated that the total amount of energy lost annually in the United States when appliances are dormant is "equivalent to the output of more than a dozen 1,000-megawatt power stations running full-tilt." Such preventable energy waste costs Americans hundreds of billions of dollars a year. But like the customers I watched walking in and out of the Old Navy store as it air-conditioned the hot streets of Manhattan, most of us just don't think about it.

GET SMARTER

Lou Rana, Con Ed's president, did offer some encouraging news about the direction of the energy industry when we discussed his plans to renovate New York City's complex, aging grid. Sporting a crisp black suit with a bronze silk tie in his windowless conference room, Rana was a far cry from the folksy, rumpled, man-of-the-streets persona of Thomas Edison. But he had the inventor's passion for his subject. For nearly two hours, Rana excitedly discussed the "smart grid," a concept he described as a "high-tech, superefficient, ultrareliable, self-healing, . . . clean, green electricity machine." It represents a radical overhaul of the way in which energy is generated, distributed, and consumed, he told me—an overhaul with a potential impact on the energy industry that could rival the Internet's impact on communications.

Con Ed has already been experimenting piecemeal with some components of a smart grid, which Rana mapped out for me, drawing squiggly lines on a whiteboard. First, he's been testing superconductor wires that carry far bigger loads than do the current copper cables and reduce the energy lost in transmission from 10 percent to less than 2 percent. Rana's engineers are installing nanosensors like those Yeager described that can monitor electrical current flows remotely, allowing grid operators to track and contain power surges before they begin to cascade. Con Ed is also installing "smart sensors" that detect human activity in buildings, enabling the grid to automatically power down lights, elevators, and appliances when they're not in use. "Price monitors" are being tested that can alert residential customers when demand is low and electricity is therefore cheapest for energy-intensive household chores such as running the dishwasher or dryer. Rana is also working to bring solar-powered office buildings online that can draw power from the grid on cloudy days and feed energy back into it on sunny days when the solar panels produce a surplus. He is developing a plan to obtain 20 percent of New York's City power supply from small-scale distributed power sources— solar panels and clean-burning microplants fueled by natural gas, for instance—installed on apartment and office buildings. This would help address the problem of building big new power plants and transmission lines on extremely limited real estate.

Spurred by a state government order, Con Ed has also decoupled its revenues from its power volume—meaning it no longer gets paid for the volume of electricity that pulses through its wires. As a result, the company is encouraging customers to cut their demand with measures such as switching to efficient lightbulbs and appliances.

None of these ideas can be implemented on a large-scale basis without a major investment. A full smart-grid conversion would cost tens of billions of dollars for New York City alone. It remains to be seen who, if anyone, will be willing to pay for such a change. New York consumers famously resist rate hikes, and the state's coffers are running low. Even with sufficient funds, it's not clear whether the system could be installed in time before the grid's demands finally outgrow supply, as ever more of its aging components collapse under pressure. The easier path would be to continue replacing the grid piecemeal, copper wire by copper wire. But this won't do in the long run. Without the smart grid, more and bigger blackouts could lie ahead as demand grows in a system with limited capacity for expanded supply.

CHAINED LIGHTNING

Edison's Pearl Street experiment gave rise to an electricity generation industry that today encompasses nearly 5,000 energy providers serving 124 million residential customers, 17 million commercial customers, and 770,000 industrial customers. With a staggering $300 billion in annual revenue, the power industry is one of the biggest in the nation. And the United States is expected to see a 30 percent growth in electricity demands between now and 2030. In that same time frame, global demands are expected to almost double as developing nations come online. Electricity usage in these nations is growing even faster than is petroleum consumption.

These numbers don't take into account a vast new market that could open up: the age of electric vehicles. As hybrid cars are growing in popularity and new plug-in models are soon to be introduced, the futurists of today are envisioning—just as their counterparts did decades ago—a century in which all transportation is powered by electricity. The whole energy system, they believe, will be unified under the flow of electrons.

This seems almost laughable given the current fragility of the U.S. electricity supply system. How, I wondered, can we confidently move toward an all-electric future if we're operating on a Third World electricity grid? One way or another, by necessity if not by choice, the archaic system of plants and cables has to be rebuilt. Will it be replaced with the same old twentieth-century fossil fuels, mechanical switches, and copper wires? Or will we opt for a smart grid and usher in a generation of clean, sustainable technologies?

"The mind can not conceive," said Thomas Edison in 1916, "what man will do in the twentieth century with his chained lightning." Now, as we face an era of even greater reliance on electricity, it's time to explore this question again—what will the United States do in the twenty-first century with its power?

Two

GREENER PASTURES

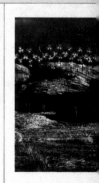

The Dawn of the New Energy Era

Say you have a wind turbine on your land that does $20,000 per year. It's just blades on a post—simple machinery, easy maintenance—but it'll be blowing out there for, oh, a century, maybe more. Over time it generates $2 million for your family, that single machine. It's a whole different scenario than an oil well that produces for a few years, and then it's done for.

—T. Boone Pickens, energy entrepreneur

8

Earth, Wind, and Fire

HOW RENEWABLE ENERGY WILL
DETHRONE THE POWERS THAT BE

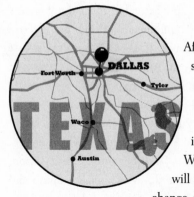

After months of exploring the great successes and sobering realities of our energy past and present, I was ready—desperate even—to map what lies ahead. What will America's postpetroleum future look like? Who is working to build it? And how will we get from here to there? Such a change will require rethinking the largest and the smallest details of a landscape wherein fossil fuels flow constantly through our cars, homes, businesses, factories, and food supply systems. It will require rethinking nearly every aspect of who we are, as individuals and as a nation.

In his book *The Beer Can by the Highway*, John Kouwenhoven contemplates "what's American about America" and concludes that a "distinctive blend of technology and . . . democracy" is what makes this country unique. He writes that our "inventiveness, adaptability, and other qualities which tend to foster industrial productivity (and hence abundance) are directly traceable to drives inherent in the democratic ideal."

In my journey through America's highways, farmlands, military frontiers, and drilling territories, I had witnessed this phenomenon again and again. I had seen how a combination of technological ingenuity and the American dream in people ranging from Croatian salt miners to Norwegian American farmers propelled our culture forward into the modern age, from Anthony Lucas's Spindletop gusher to the breakthrough inventions of John Rockefeller, Thomas Edison, Henry Ford, William Levitt, Earl Tupper, Malcom McLean, and Norman Borlaug.

I'd come to realize, moreover, that this was not solely a good thing. American ingenuity had consequences both greatly beneficial and perilous in the twentieth century. It built our mechanized military, engineered our oversized cars, sprawled out our cities, invented our plastics, nourished our bumper crops, extended our supply chains across the globe, and connected the millions of miles of our electric grid. And along the way, it got us roundly hooked on fossil fuels. Just the same, American ingenuity may still have the power to reverse the negative consequences of our fossil fuel past and kick our costly habit.

I had a bullish sense of hope on this last point, and set out to explore the conviction more deeply. I wanted a firsthand glimpse of our emerging clean-energy landscape, where every kind of natural element—wind, sun, ocean tides, the warmth of the earth, even human exercise and sewage—can be transformed into electricity. No area of our economy more vibrantly reflects American ingenuity today than the development of "clean technologies," a catchall phrase that refers to nonpolluting energy sources and the new generation of buildings, transportation systems, and factories that will use them. (A portion of these energy sources is also commonly called "renewable"—those that are cyclically replenished by nature, not permanently lost when used as are fossil fuels.) It's also true that a daunting range of obstacles could stall the development of these innovations and threaten their success—barriers I wanted to examine as well.

WAR WITHOUT GUNS

My venture into America's energy future began with a trip to a flat, dry, and mostly barren town in the gusty prairies of west Texas that had become one of the leading frontiers of America's clean energy development. The very same region of the country that offered up a seemingly limitless supply of oil in the days of Spindletop had been selected as the site for the world's largest wind farm. In June 2007, billionaire investor T. Boone Pickens announced a $10 billion bet to build a 4,000-megawatt wind facility centered in the town of Pampa, Texas. I made arrangements in the summer of 2008 to travel west with Pickens from his Dallas office to Pampa, a few hundred miles away, where 200,000 acres of this massive wind experiment had been marked for construction.

I was, if anything, more curious about Pickens himself than about the project he was planning. Pickens's recent decision to advocate renewable energy struck me as a promising development for our energy future. As Pickens goes, I thought, so goes America: he is the oil titan who comes the closest (in both his achievements and his inherent contradictions) to being a modern-day John D. Rockefeller, and his choice represented a 180-degree shift in energy policy.

Pickens made his billions as an oil entrepreneur overseeing the production of more than 200 million barrels of petroleum from wells throughout the world. As a corporate profiteer, he famously tried to snatch Gulf Oil from the grip of Chevron. As a political booster, he has spent fifty years prodigiously supporting Republican campaigns. He spent millions to fund the Swift-boat ad campaign that damaged John Kerry's bid for the presidency in 2004—a stance for which he voices no regret to this day. Pickens is, in short, an unlikely hero for the modern-day environmental movement.

But in the summer of 2008, Pickens announced what he called the "Pickens Plan," a massive push to free America from its dependence on foreign oil and to develop clean alternative sources of energy, including natural gas, wind, and solar power. "I've been an oilman my entire life," he said, "but this is one emergency we can't drill our way out of." To some traditionalists in the oil industry, Pickens is a defector, a turncoat of sorts. Others—environmentalists and progressives included—consider him an enlightened soul, a converted sinner, a man in his twilight years (Pickens is in his eighties) who saw his industry hurtling toward doom and sought salvation in the clean-energy gospel.

Along with the Pickens Plan came a massive PR blitz to promote it, a $58 million media campaign including print ads and TV and radio commercials as well as appearances on every program from *Charlie Rose* to ABC's *The View*. In this, Pickens has aimed to create an unusual new embodiment of the 1970s grassroots environmental campaign, calling his crusade "a war without guns" and recruiting citizen-soldiers in a video posted on his Web site: "We have to have an army and you're part of the army. Bring your friends, family, your church group, everybody together that you can think of." More than a million and a half Americans have signed on to fight for his agenda.

I met Pickens on what turned out to be a fitting day for a discussion of wind energy—a blustery morning that, by the time I arrived at his office in downtown Dallas, was quickly evolving into a violent electrical storm. Wind, Pickens assured me, is a long-term investment that promises huge financial rewards: "I'm out to hunt elephants—I don't get off the trail for rabbits." His office attests to a man who hunts big game: it's filled with luxurious leather couches and mahogany desks; cast-iron sculptures of bison; giant gilt-framed oil paintings of sunsets, running horses, and the Wild West; and photos of Pickens smiling alongside Republican presidents from Richard M. Nixon to George W. Bush, with jaunty personal notes from each commander in chief displayed.

I was surprised by Pickens's appearance and manner. He wasn't the ruthless, swashbuckling figure of legend who'd inscrutably stared from the cover of *Time* in 1985 above the headline "The Takeover Game— Corporate Raider T. Boone Pickens." Instead, he was approachable and good-natured, even paternal in his manner. A fit, sturdy man with a tidy

crop of silver hair, Pickens works out daily with a team of personal train-
ers and appears younger than his years, his age belied only up close by
a discreet hearing aid, a slight hesitation in his walk, and an occasional
tired droop to his steel blue eyes.

Pickens has brought an unexpected cowboy swagger to the clean-
energy agenda, describing the renewable industries, for instance, as
"hotter'n a three-nutted tomcat." He took polite but firm exception to my
mention of Al Gore's climate change initiatives. Though Pickens does
see global warming as "a serious concern," he makes no pretense to being
an environmental do-gooder: "Don't think I'm Al Gore. I'm not going
to do a major investment in wind farms for the environment first and
money second. I'm a fella who thinks with my wallet."

Pickens believes that the world has already reached peak oil. "The
industry can at a maximum produce 85 billion barrels a day, while the
global demand is 86.4 billion barrels," he told me. "You bet your ass
that demand is gonna keep rising, along with the price." While he has
cheered proposals to open protected lands and offshore areas to drilling, he said that's only one small element of a sound long-term national
energy strategy.

Pickens believes that America can cut its oil dependence in half
by shifting our transportation sector to natural gas (for heavy vehicles)
and electricity (for passenger cars) while substantially increasing wind
development to supply 20 percent of our electricity demands in under
a decade—up from roughly 1 percent today. "We also need efficiency,
solar, nukes, any alternatives we can get," he said, but stressed his belief
that wind has the biggest growth potential.

For all his emphasis on financial pragmatism, Pickens confessed to
having a sense of "mission" in his current work. "I think I was put here
for two reasons," he told me. "To make money and be generous with it,
and to find good ideas and get them into play. I think I've got a solution
to peak oil. The only other plans I've seen are to roll over and die."

BLOWIN' IN THE WIND

As we boarded Pickens's Gulfstream jet to Pampa, the storm had improved only slightly; Dallas's Love Field airport had grounded many flights to wait it out. But Pickens, a man who makes his own rules, instructed his pilot to take off as scheduled. Married four times, Pickens craves change and seems to like controversy even more, readily embracing a paradigm shift in his industry while other tycoons resist it. He welcomes rather than avoids risk.

He credits his father, Tom, for this life lesson, dating it back to the night of his birth in Holdenville, Oklahoma. It was a difficult delivery, and at one point the doctor informed Tom, "You can save your wife or your baby, but not both." Pickens's father refused to choose, instead urging the doctor to attempt—successfully, as it turned out—the hospital's first-ever Cesarean section. "I've always said I'm the luckiest guy in the world," Pickens often muses. "I certainly think I was on that particular night."

Tom Pickens leased mineral rights for oil companies, earning a comfortable but modest living through the Great Depression. Pickens loves to recount stories of his own early days as a newspaper boy in Holdenville, when he grew his clientele from 28 customers to 156 by muscling his way into the paper routes of other local boys—testing a strategy of "rapid expansion by acquisition" that he later mastered as an entrepreneur. After graduating from Oklahoma A&M (now Oklahoma State), Pickens accepted a job as a geologist at Phillips Petroleum, but he eventually grew impatient with the company's bureaucracy. At twenty-six, he was married with two children and a third on the way, and if he left Phillips he'd be turning his back on a guaranteed $500 a month. "No one gave me a prayer of succeeding, except my parents," he recalled.

But he decided to set to work as an independent geologist, applying $1,300 from his Phillips retirement account toward a Ford station wagon in which he traveled the Texas Panhandle, consulting and arranging oil-drilling deals. His early successes led him to the founding of Petroleum Exploration Inc., later Mesa Petroleum. As he continued his bootstrap "expansion by acquisition," he gradually amassed a vast independent oil empire and later founded BP Capital Management, a private equity firm that invests mainly in energy ventures.

Now, even as he is betting huge on wind and promoting solar and biofuels, Pickens says he'll continue to make money off the energy past that generated his fortune for as long as he can—after all, the rising demand for oil is a good formula for near-term profit.

Unlike Chevron's Paul Siegele, Pickens doesn't believe that technology breakthroughs will open up new frontiers for oil. "The oil just isn't there—no technology can change that. And with China and India pushing up the global demand, new discoveries just can't keep up with it." Pickens is personally offended by the fact that his native country has become so thoroughly dependent on Middle Eastern oil. "It's sheer lunacy that we're sending our money to the Arabs," he told me, adding that "in the era of peak oil, the USA is sucking hind tit." Translation: we're the runt that will suffer most when the oil runs low. "The U.S. spends $700 billion annually on foreign oil—money that's going largely to our enemies. Our attitude has been 'Send us oil, never mind the price or the conflict.' That's gonna cripple the hell out of this country. I can't stand around and watch that happen."

As the storm cleared and the turbulence died down, Pickens settled comfortably into his monogrammed leather seat, satisfied that the skies had finally cooperated with his plan. He looked out the plane's window at the vast ranchlands of western Texas where he envisioned building his fleet of 2,700 wind turbines, each standing 300 feet tall with blades half the length of football fields. I followed his gaze out to the flat, distant prairie only occasionally punctuated by farms the size of postage stamps, with two-lane highways strewn like loose threads across the landscape.

The economics of energy is radically changing, Pickens told me, moving his bony hand through the air like a wizard waving a wand. His formula for the long-term shift to renewables is simple: The more global oil demands increase, the more oil prices are bound to increase over time, the more pollution will increase, and the more money we'll send to our enemies overseas. Conversely, the more the demand for renewables increases, the more their costs will decline, the more pollution will decline, and the more jobs we will create at home. "The shift to homegrown energy sources is patriotic, it's economic, it's diplomatic, it's environmental—it's win-win-win-win."

That's the argument Pickens presented to a room full of hundreds of ranchers and farmers in the Pampa Civic Center—our destination once we landed. He had come to host a town hall meeting to pitch his plan to make this rural community in the Texas Panhandle—population 17,800—the "wind capital of the world." And he had come to answer questions about how such a fate would change the lives of the individuals whose land would be underneath his turbines. Even though Pickens had already assembled leases for 200,000 acres of turbines—enough to start construction on his project—he wanted to double that area and still had some local politicking to do.

The Pampa Civic Center is a spare, one-room building surrounded by strip malls, a narrow highway, and flat prairie as far as the eye can see. Inside I found cinder-block walls, concrete floors, and metal fold-out chairs with seats bowed from years of use. Dressed in khakis, tasseled loafers, and a hunting vest, Pickens stood out against the sea of overalls and cowboy hats. He stepped onto the plywood stage to a hesitant smattering of applause. Pickens is a controversial figure in these parts—while some see him as a folk hero, many resent him for buying up water rights in western Texas and controlling precious public resources through his Mesa Water, Inc. But the room quickly warmed to him as he outlined the financial incentives of his wind farm. "Wind will revitalize rural America, starting with Pampa," Pickens vowed.

His turbines would create fifteen hundred new jobs, and landowners could continue farming and running cattle with an added revenue boost from harvesting wind. Five turbines would be located per square mile, and those who housed them on their property would get royalties of up to 7 percent on the energy produced, said Pickens. That works out to an average per turbine of $10,000 to $20,000 per year—in an area where the average per-capita income is below $18,000.

"Why aren't you putting them on your land?" one audience member asked Pickens as he milled among the crowd after the event, referring to Pickens's nearby 68,000-acre private ranch. The magnate answered directly and honestly that he'd rather not look at them. Another audience member commented that this shouldn't be a concern—most Texans have spent their lives looking at vistas of oil derricks, "which aren't any prettier'n windmills."

Would the turbines make noise? Yes, answered Pickens, but "it's the sound of money in your pocket." The best part, he noted, is that wind is a perpetually renewed resource with no declining production curve, meaning it doesn't have the price volatility of oil. "Say you have a wind turbine on your land that does $20,000 per year. It's just blades on a post—simple machinery, easy maintenance—but it'll be blowing out there for, oh, a century, maybe more. Over time it generates $2 million for your family, that single machine. It's a whole different scenario than an oil well that produces for a few years, and then it's done for."

To Pickens, who spent his career weathering the cyclical booms and busts of the oil markets—that problem dating back to Spindletop of instability caused by supply gluts and deficits—the promise of an energy industry fueled by sources that can be infinitely replenished is a dazzling novelty.

Pickens's grand vision has since hit some sizable roadblocks. Amid the credit crisis and the sharp decline in oil and natural gas prices at the end of 2008, the tycoon's $1.3 billion private equity fund reportedly plunged in value by a staggering 97 percent. In the economic downturn, Pickens also lost the backing on his $10 billion wind project, preventing him from building the high-voltage transmission line that would have delivered his rurally produced power to faraway population centers. Pickens set about modifying his proposal, putting the Pampa project on hold and exploring the possibility of dividing it into several smaller wind farms in states including Oklahoma, Kansas, Wisconsin, and Texas. But he maintains his strong support for renewables, noting that the 2008 plunge in energy prices was only temporary—"the price is going back up"—and that the fundamentals of the Pickens Plan are still sound: "I haven't changed anything; you don't shoot for the moon from the start."

TELLTALE SIGNS

Is it actually possible, as Pickens claims, that renewables such as wind could become the new American fuels, eventually replacing coal and oil entirely in powering our cars, homes, and industries?

Texas, which has among the strongest and steadiest wind patterns in the United States, lies at the base of a "wind corridor" that produces reliable wind speeds of between 14 and 18 miles per hour. The corridor stretches up the center of the country through the heartland and into the Dakotas, fanning out to Illinois and Wisconsin in the east and Montana in the west. The raw capacity is there: "The total potential wind development throughout this corridor is 400,000 megawatts," as Pickens explained. (One megawatt powers roughly a thousand average-sized homes.) "That spells opportunity, serious, big-deal opportunity."

The "big-deal" raw capacity is also there in solar. Enough solar energy beats down on the surface of the earth every hour to supply 100 percent of the world's energy needs for a year. In theory, the state of Nevada alone could produce enough electricity through solar technology to power all American businesses and homes.

America's most successful energy innovator, Thomas Edison, was hopeful about the future of these power sources long before terms such as *renewable energy* existed. "This scheme of combustion to get power," he said in 1931, "makes me sick to think of it—it is so wasteful. . . . You see, we should utilize natural forces and thus get all of our power. Sunshine is a form of energy, and the winds and the tides are manifestations of energy. Do we use them? Oh, no! We burn up wood and coal, as renters burn up the front fence for fuel. We live like squatters, not as if we owned the property."

The challenge is figuring out how best to harness these natural forces. Once they are harnessed, the challenge becomes one of transporting huge new volumes of clean electricity great distances across our already overburdened and inefficient grid—the 900 miles from Pampa, say, to power the lights in my home in Nashville.

Pickens is a key leader in the American effort to develop renewable energy sources and an improved electric grid, but he's far from alone. Hundreds of big investors and businesses across the globe are currently making plans for—and in some cases betting their very economic survival on—the future of renewables.

In 2008, $47.5 billion in wind turbine investments were made around the world—producing 27,000 megawatts of clean electricity or enough to power approximately 20 million American homes. Roughly a third of those new megawatts were produced through projects based in the United States. Joseph Romm, a former high-level official at the Department of Energy, predicts that, given the increasing global demand for clean power, "wind investment will quadruple within ten years, even in a down economy." That's the good news. The bad news is that this money may go primarily to overseas manufacturers.

American companies helped invent and produce the first modern commercial wind turbines more than sixty years ago—but most manufacturers are now European, thanks in part to government mandates and industry incentives that have been widely implemented in Europe. Today, six of the world's ten leading wind turbine manufacturers are located in Denmark, Germany, or Spain. We've lost the competitive edge in solar, too: more than 90 percent of solar-electric panel manufacturing happens outside of the United States. "Up until recently," explained Bracken Hendricks, an energy policy expert with the Center for American Progress, "Congress allocated a smaller annual budget for clean energy R & D than the pet food industry did for inventing new kinds of kibbles."

This long-starved budget received a sudden, historic windfall in February 2009 when Congress committed $70 billion over the next ten years to the development of clean technologies, smart grid upgrades, and fuel efficient vehicles. President Obama's mission for such an investment—part of his American Recovery and Reinvestment Act—was to take "a big step down the road to energy independence and lay the groundwork for a new green energy economy that can create countless well-paying jobs."

Some critics fear that such massive public investments will stifle ingenuity, but similar efforts in the past have done just the opposite—

pushing the mass production of new technologies into high gear. Federal investments during World Wars I and II encouraged innovations in cars and trucks, airplanes, oil pipelines, and plastics; the rural electrification program under Roosevelt created hundreds of thousands of miles of the electric grid. Government agencies such as NASA and Pentagon programs such as the Defense Advanced Research Projects Agency (DARPA) created many of today's technologies—microwaves, GPS navigation systems, microchips and computers, the Internet, and cell phones, among countless others. Private companies went on to make billions of dollars through refining, adapting, and improving these technologies for mainstream use, in turn benefiting the daily lives of millions of Americans.

THE COLOR OF MONEY

Calling it a step in the right direction, Pickens widely promoted the federal recovery plan, as did dozens of other big business leaders who stood to benefit from the energy-industry incentives. Pickens, in fact, is one of many industry leaders who have begun shifting their allegiance from conventional fossil-fuel-based technologies to renewables in recent years, pushing to expand the frontiers of clean energy even as the overall economy has slowed. I spoke with executives at companies ranging from Walmart to General Electric and Duke Energy, all of whom have made significant investments in renewable power sources and energy efficiency.

"Increasingly for business, 'green' is green," said GE's CEO, Jeffrey Immelt, explaining that sustainable strategies are becoming ever more financially rewarding. Immelt reminded me that GE, founded by Thomas Edison, gave America the first lightbulb, the first power plant, and the first jet engine, so "now it's a very natural progression" for the company to pioneer next-generation technologies—wind turbines, solar panels, cleaner-burning coal plants, and energy-saving lightbulbs and appliances. "The role of business is to solve the world's toughest problems," Immelt said, "and make money while we're doing it."

GE is also the only major U.S. player in the wind industry. "For a hundred years," one of the company's top engineers told me, "GE has made stuff that spins, from aircraft engines to gas turbines, so we've learned a heck of a lot about how to shape blades aerodynamically." The company bought its wind division from a bankrupt Enron in 2002 for $350 million and by 2008 had grown it into a $6 billion-a-year business—one of the best-performing divisions in the conglomerate.

Among the many forces pushing Immelt toward green strategies are the clean energy demands of his clients in Europe and Japan: "We're not going to design a gas turbine for three different markets. We're only going to design one gas turbine," he told me, "so we have to shoot for the highest standards. If our customers in Europe and Japan want clean and efficient [products], that's what we're going to build." At a recent *Wall Street Journal* conference titled "ECO:nomics," Immelt responded to criticism by some conservative journalists of his support for clean technology and a tax on carbon emissions, insisting that he was "not an environmentalist," but rather that his green initiatives represented the best way for his company to remain competitive in a fast-paced global economy. "I don't need to be lectured by anybody in this room," he declared after one particularly heated exchange, "about how to compete."

One of Immelt's best customers is Jim Rogers, an Alabama-born executive who has been dubbed "the green coal baron." Rogers runs Duke Energy, a massive utility in the Southeast and one of the three largest corporate emitters of greenhouse gases in America. Rogers has devised a plan to "decarbonize"—eliminate his company's carbon emissions—by 2050 that attracted much skepticism and controversy in the industry. Sooner than that, by 2030, Rogers plans to cut his coal use by more than half—from nearly 70 percent of his company's portfolio to under 30 percent—and he's increasing solar and wind to 11 percent of his supply. He's also devised a "save-a-watt" program that enables Duke to benefit—rather than financially suffer—from improving the energy efficiency of the homes, offices, and factories of its customers.

Rogers believes these actions will save him money in the long term, as global competition will invariably drive up the price of coal and as federal restrictions on carbon emissions come into play. "Mandatory caps [on carbon] are essential," he told me, "because a lot of companies like ours are making big decisions right now: Do we build coal plants? Do we build nuclear or natural gas or renewables? How much do we depend on energy efficiency? These are investments that last for half a century or more."

One of Rogers's best customers is Walmart—another, even more improbable new green leader. The food supply chain issues I had examined at Walmart were only one part of a much larger overall strategy. Walmart has an annual greenhouse-gas footprint comparable to CO_2 emissions of New Hampshire, and is the biggest private consumer of electricity in the United States. So in 2004, after policy discussions with founder Sam Walton's son and company chairman Rob Walton, then-CEO H. Lee Scott decided to trim his energy bills.

He started with an experiment in Aurora, Colorado—a superstore that from a distance looks like any of the thousands of other unsightly big-box stores operated by the retail giant. But on closer look, the store's sprawling parking lot is made of recycled asphalt from a local airport runway. Its heating oil comes from the grease pans of its own deli. Its electricity is powered by rooftop solar panels and an on-site windmill. Nearly every component of this structure, in fact—from its waterless toilets to its superefficient lighting, refrigerators, and air-conditioning system—represents the cutting edge of clean energy design.

Scott built the Aurora Walmart as the retail outlet of the future—a pilot project that he says will serve as a model for new store development. And green stores are just the beginning of Scott's grand plan, which has since been reiterated by his successor, Mike Duke. It includes goals so audacious they border on the fantastical: Walmart's multinational empire, Scott claims, will eventually run on 100 percent renewable fuels, emit zero waste, sell only sustainable, energy-wise products, and require its domestic and international suppliers to uphold strict green principles. Along the way, Walmart will save billions on energy costs.

Scott, like Pickens, Immelt, and Rogers, was up-front about the big dollar incentives for his company: "As I got exposed to the opportunities we had to reduce our impact, it became even more exciting than I had originally thought. It is clearly good for our business," he told me. "We are taking costs out and finding we are doing things we just do not need to do, whether it be in packaging or energy usage, or the kind of equipment we buy for refrigeration in our stores, that there are a number of decisions we can make that are great for sustainability and great for bottom-line profit."

GE, Duke, and Walmart are just a few among dozens of old-school industrial behemoths—DuPont, Dow Chemical, Caterpillar, Alcoa Steel, General Motors, and Archer Daniels Midland included—now acknowledging to varying degrees that the challenges of global warming and energy independence can become uniquely American business opportunities.

SOMETHING VENTURED

Not just corporations but also leading American banks and investors are staking a claim on renewable energy markets. Warren Buffett, CEO of Berkshire Hathaway, and Bill Gates of Microsoft are among a coterie of billionaires placing big bets on wind. Goldman Sachs has pumped over $2.5 billion into renewable energy sectors and is working to significantly reduce its corporate greenhouse gas emissions. Citigroup has pledged $7.5 billion to fund renewable energy projects, with a goal of $31 billion in the next decade. Bank of America has committed $20 billion to financing green innovation. The list goes on, with numbers so huge that they begin to lose their shock value, but it adds up to a spectacular cascade of investment.

Smaller-scale venture capital investments to develop start-up companies in the clean energy sector also surged by the billions in recent years. In 2001, companies working on clean technologies attracted over $500 million of venture capital investments in major world markets: North America, Europe, India, and China. In 2008 that figure was more than $8 billion. According to leading venture capitalist John Doerr, an early investor in Google and Amazon who has since shifted his dollars to the energy sector, "The field of green-tech is the largest economic opportunity of the twenty-first century."

It's no coincidence that Silicon Valley is a hotbed of energy start-ups. Doerr is one of many big guns who helped launch the dot-com boom and are now bringing their high-tech investment savvy to the world of energy. Steve Case of AOL, Bill Gates of Microsoft, and Vinod Khosla, cofounder of Sun Microsystems, have all put millions into clean energy investments. Intel, Hewlett-Packard, and IBM have all established handsomely budgeted renewable energy divisions. "We're seeing in green technology what we once saw in the Internet," Khosla told me, describing his investments in areas like biofuels, batteries, and bioplastics. "But in this case the end result isn't buying a book or getting a date online, it's saving the future of civilization."

No dot-com giant is doing more for the clean energy battle than Google. The company's philanthropic arm Google.org recently announced a plan to make renewable energy cheaper than coal—and to do it in "years, not decades." Dubbed "RE<C," the program aims to produce one gigawatt of electric generating capacity—enough to power the city of San Francisco—from green sources. Google's founders, Sergey Brin and Larry Page, have earmarked hundreds of millions of dollars for this project.

Brin and Page are thirtysomethings who started their business in a Menlo Park garage in 1998 and now rank among the thirty richest people in the world. They have worked to create a distinctive corporate culture with a dash of idealism that they believe fosters creativity. "We have a mantra," Page has said. "'Don't be evil,' which is to do the best things we know how for our users, for our customers, for everyone." Google openly acknowledges that its idealism and self-interest are mutually compatible: "Being 'green' is essential to keeping our business competitive," reads the company's Web site. "It is this economic advantage that makes our efforts truly sustainable."

POWER TO THE PEOPLE

The enormous investments of these companies won't pay off, however, financially or for the environment, until they reach the daily lives of most Americans—until the wind turbines, solar panels, and clean cars they are funding actually begin to shape how we move, work, eat, and play.

A similar burst of investment, after all, first drew the energy from fossil fuels to create our current American landscape of cars, plastics, fertilizers, and supply chains. But just as it took years and sometimes decades before Americans began using coal and oil in such diverse everyday applications, it will be years before we see clean energy innovations become commonplace in our lives. Already, however, promising signs of a shift in this direction are emerging. I found one of today's most creative experiments in green living less than two hours' drive from my home, just off Interstate 75 in central Tennessee. This new tract of houses, located near a midsized Ford dealership, a Krispy Kreme doughnut shop, and a Piggly-Wiggly supermarket, quietly testifies to one hopeful possibility for the future generation and use of energy in America.

Sited on tenth-of-an-acre lots, these neat, one-story colonials looked so ordinary on the surface that even with a clearly marked road map I initially drove straight past them. Built by volunteers of Habitat for Humanity, the houses had narrow front porches, painted shutters, and carefully trimmed lawns strewn with baseball bats, tricycles, and kickballs. But this middle-American subdivision named Harmony Heights in Lenoir City is much more than another quiet avenue in a pleasant town. It is a revolutionary testing ground for the zero-energy home of the future—an experiment that could have implications for American daily life almost equal to those of Edison's Pearl Street dynamo.

Zero-energy home (ZEH) is the term coined by researchers at the Department of Energy for a house, like those in Harmony Heights, that annually produces at least as much energy as it consumes. But a ZEH is not strictly self-sufficient—it is linked to and plays a constant game of give-and-take with the grid, pumping energy into the grid when its rooftop solar panels convert more light into electricity than is needed by the home, and drawing energy from the grid when the panels are inactive at night and on cloudy days.

It's not much of a challenge to build a ZEH if you have an unlimited budget. Even a sprawling mansion could technically be zero-energy if an expensive solar array was purchased that was large enough to cancel out its high energy needs. But consider the challenge—one that's crucial to today's economic climate—of building a zero-energy home for under $100,000. That's the goal of Jeff Christian, a building researcher at the Department of Energy's Oak Ridge National Laboratory. A lanky man with straw-colored hair and angular features, Christian has been working on ZEH development for nearly a decade. He hopes to have a mass-market prototype in place by 2011—a model home that is 70 percent more energy efficient than a standard new home its size, supplying what little electricity it needs from a 2.5-kilowatt solar roof and costing no more than its conventional counterpart.

Given that solar is still more expensive than grid energy, Christian has to make up for the house's added energy and technology costs with extreme efficiency measures. He aims to get the cost of the home down to the point where the additional monthly mortgage payments for its green features will be no higher than the total monthly savings on its heat and utility bills. The Department of Energy teamed up with the Tennessee Valley Authority and Habitat for Humanity to make these Lenoir City houses into a living laboratory in which it could test new ways to curb domestic energy use.

The first house I visited in the Harmony Heights development belonged to Adam Indrajaya and Lina Kinandjar, a landscape worker and pastry decorator, respectively, who moved to Tennessee from Malaysia nine years ago with their two young children. Lina is proud of the immaculately kept home. Wearing a bright pink blouse and a ponytail, she buzzed around the house opening up cupboards and closets to show me the hidden wires and sensors within the house's "surveillance system."

The families who applied for these Habitat homes agreed to let Oak Ridge Laboratory engineers install dozens of sensors monitoring all aspects of their electricity consumption, presumably seeing this as a small price to pay for rock-bottom energy bills. Though the home was modest in size, it was not lacking in creature comforts, with its large TV, surround-sound stereo, DVD player, Dell computer, and two aquariums, in addition to a standard-sized refrigerator, washer, dryer, dishwasher, and microwave. The couple had never heard of energy efficiency or solar panels before Habitat selected their house "to get a special treatment," as Adam put it.

Quickly, the family began to see benefits. "We got paid!" Lina marveled as she showed me her latest energy bill, bearing a credit of $35 in one month. "We are like a little power plant." The local utility buys electricity from the family during the day when the children are at school, the parents are working, the lights and appliances are shut off, and the solar panels are producing a surplus. "That's just the time when the grid is cranking at peak load," explained Christian. "It's a perfect symbiosis— homes use the least power when businesses use the most."

Since the previous month had been a particularly sunny one, the utility had purchased more electricity from the house over the course of the billing period than the family had purchased from the grid. Not all months yielded this kind of payback, but on average, the family's energy "deposits" to the grid were balancing out its energy "withdrawals." This was a far cry from the nearly $200 per month the couple had been paying for utilities at their old apartment.

The more efficient the home is, the bigger the payback. Standard energy-saving features in Lina and Adam's home include heavily insulated walls to prevent the loss of heat or air-conditioner-chilled air, and ultraefficient lighting and appliances. But the house is also fitted with more specialized features, including the added wires that allow for the family's back-and-forth exchange with the grid; exterior paint made of light-colored materials that deflect heat rays in order to bring down air conditioner use; a geothermal heating and cooling system buried in the front yard; a mechanism that captures the heat from shower water after it goes down the drain; and even a device that captures the warmth that comes off the coils behind the refrigerator.

Compared to the potential smart grid that Con Edison's Lou Rana had described to me, this was a decidedly low-tech electrical laboratory—bearing no smart features that could automatically dim lights and power down appliances. "We'd love to use fancy controls and a lot more aggressive technologies," Christian told me, "but we have to work within the realm of financial reason that makes sense for our occupants." (When I asked him if he was developing the Prius of solar homes, he replied drolly, "No, it's more the Yugo.")

Eventually, he said, smart house features will become standard in all homes. Christian, now in his fifties, believes he will live to see a future in which homes will function as the gateways between an improved, intelligent electric grid and battery-powered electric cars. Smart homes will monitor the real-time price of energy, selling energy back to the grid when the demand (and payback) is highest and drawing energy from the grid for household chores and charging car batteries when energy is cheapest.

It will all be automatic—the home's networked computer system will determine the best time to run the dryer, knowing it would cost, say, $5 per load in the middle of the day, but only 45¢ per load in the middle of the night. Likewise, the house will automatically cycle off air-conditioning and "hibernate" appliances according to the same pricing trends. It will even monitor the whereabouts of its occupants, switching off the lights and the TV when someone leaves a room and turning them back on when she returns.

The most promising immediate impact of Christian's experiment in Harmony Heights was one that both he and Lina mentioned separately: how rapidly the family's conservation habits were improving, spurred by the promise of financial rewards on their electric bill. Lina's neighbor Kim Charles told me that she now tries to wait for days of bright sunlight to do her household chores: "It's just like gardening—you notice what the sun is giving today."

On cloudy days, Charles runs the dryer and dishwasher late at night when electricity is cheap. Even the children have begun taking the initiative in turning out the lights, knowing a penny saved is a penny earned—perhaps for a trip to the toy store. "It has become like a game to them," she said. "We used to think that electricity is like air—a given," she added. "We never really spent much time thinking about it."

GOOD DAY, SUNSHINE

This small but profound revolution in the lives of a few families has the potential to spread well beyond Lenoir City. Eventually, millions of people in America could live in zero-energy homes. But one crucial factor must change first: clean power has to get cheaper. A race is under way in corporate, academic, and government laboratories nationwide to develop and improve clean-energy technologies while ratcheting down their costs.

Today's Thomas Edisons, John D. Rockefellers, and Henry Fords are chasing success with innovations ranging from towering windmills to batteries made from tiny living organisms. There won't be one big discovery, I learned—no "silver bullet" solution. As energy insiders often say, it will be a "silver buckshot," a vast array of complementary innovations that will together dramatically improve energy efficiency in our individual and collective lives and diversify our power sources.

Solar and wind applications are by far the best known and most advanced of renewables currently in development. Enough solar energy—this fact bears repeating—beats down on the earth's surface every hour to supply all the world's energy needs for a year. But what's the best, most effective way to channel it?

Traditional solar panels convert light energy to an electrical charge through the medium of silicon, a semiconducting material also used in the microchips embedded in all digital devices. Silicon channels electricity with precision and high efficiency. In microchips, the silicon enables electrons to flow through and excite memory blocks of stored information; in solar panels, the silicon enables photons (the light particles in sun rays) to flow through and excite chemicals in the panel's membrane that then release electrons. (Note that silicon should not be confused with silicone; they are derived from the same basic raw material, silica, but are combined with different additives.)

The invention of the first photovoltaic cells (the industry term for devices that turn sunlight into electricity, often called "PV" for short) was in fact a happy accident of scientific research. In 1953, when Gerald Pearson, an American physicist at Bell Laboratories, was experimenting with silicon for electronics applications, he accidentally doped a batch of silicon with chemicals that gave it the ability to convert sunlight into notable amounts of electricity. Within a year, Pearson and two Bell colleagues had developed a model that could power electrical equipment. The space industry seized on the discovery, and four years later, a solar array was hitched up to the U.S. satellite *Vanguard 1,* providing electricity to the vessel's radios. It wasn't until the 1970s that solar power was introduced into commercial applications, ranging from the tiny power sources on calculators to President Carter's rooftop panels on the White House.

From the beginning, the solar industry has grappled with two central challenges: cost and conversion efficiency. The latter refers to the percentage of sunlight that can be absorbed into the silicon and converted into electricity. The solar panels of the 1970s were able to convert only about 6 percent of the light absorbed into electricity. Today's commercial crystalline silicon panels generally get about 15 to 20 percent efficiency—a substantial leap, but still leaving much room for improvement. Likewise, the cost to install solar capacity decreased more than 30 percent between 1998 and 2007—a big drop, but solar power is still much more expensive than conventional grid energy for consumers.

The persistent challenges of cost and efficiency in traditional solar devices have spurred creative experimentation among young innovators seeking to design cheaper, more powerful, easier-to-produce solar applications. A company called Energy Innovations, for instance, is developing ways to vastly increase the efficiency of solar panels with optics technology that concentrates the rays of the sun to levels that are hundreds of times more powerful than diffuse sunlight. It's the same general principle applied by children using magnifying glasses to channel the sun into laser beams hot enough to singe weeds and ants. To test his theory, Energy Innovations' founder Bill Gross and his team developed a product called the Sunflower—curved glass dishes, resembling stadium lights, installed in rows atop solar panels. Embedded with microprocessors, they digitally evaluate and follow the path of the sun. Like the flowers after which they are named, the dishes raise their faces to the east as dawn breaks, tracking and concentrating the sun's light as it moves across the sky, and then bowing down again as the sun sets.

Gross's team used ultrapowerful solar cells that have almost double the efficiency of standard silicon cells but are also many times more expensive; the concentrator enabled them to offset much of the cost increase from this material by shrinking the size of the panels. It's a promising concept, but still there are wrinkles to iron out—above all, continuing to lower the cost.

Other innovators, such as the company Miasolé, are working to mass-produce thin films, a new generation of cheap, flexible photovoltaic materials that can be mounted almost anywhere. While conventional silicon, which is brittle, must be framed in heavy, solid panels on a roof, the thin-film solar material is slender and bendable and can actually *become* the roof—either directly embedded into shingles or produced in adhesive sheets that can be peeled off a backing like a sticker and smoothed on top of metal roofs. It's consumer-friendly innovation, and cheaper to manufacture than many solar products, but still its conversion efficiency is significantly lower than that of silicon, meaning it requires a great deal more surface area to produce the same amount of electricity.

The northern California firm Nanosolar has developed a new kind of thin film that converts sunlight into electricity using nanotechnology— the science of engineering on a molecular level. The company contends that these minute nanoparticles can offer a spectacular increase in efficiency so that the panels produce five to ten times more electrical current than conventional thin films. Better yet, this so-called "nanoink" can be "printed" onto almost any surface as a coating no thicker than paint. It could be integrated into the rooftops of cars and trucks, suspended in liquid and painted onto house exteriors, applied as a transparent coating to windows, or even integrated into fabrics. (In theory, your blue jeans, for instance, could double as a cell phone charger.) It could also eventually be mass-produced—for "about a tenth of what current panels cost," according to *Popular Science*, "and at a rate of several hundred feet per minute."

Again, there are still hurdles to surmount. Will the nanoparticle films be durable and reliable? Can the costs come down as much as promised? The company is pressing forward and has begun shipping to select customers a product called SolarPly solar cells that are as thin as aluminum foil. Martin Roscheisen, CEO of Nanosolar, says his facility will create 430 megawatts of solar cells a year, enough to power more than 300,000 homes.

Even stranger and more far-flung concepts in solar technology are currently in development, including "hairy" solar panels made from a material that resembles tiny blades of grass and is "grown" from organic light-absorbing carbon nanotubes, and massive solar "space fields" (still early in the design process) that could be launched into the stratosphere to beam down a constant stream of energy—uninterrupted by clouds or night—via concentrated microwaves. The scope of innovation is astonishing, but there's no guarantee of successfully moving many of these technologies out of the lab and into production.

HOT HOT HOT

It's exciting to think about the vast creativity behind today's clean energy innovation, but in fact the most promising technology in the solar arena is also the most basic. Joseph Romm calls it "the technology that could save humanity." It's a method that is so uncomplicated, so primitive and even obvious that it smacks of the proverbial elephant in the living room. "I like low-tech," said Romm. "Simple is good."

Today, the majority of our power plants still function according to the same general system used by Edison's Pearl Street generators—burning fossil fuels to heat water, which creates steam that spins turbines. Why not find a way of using our most formidable heat source, the sun, to boil the water that powers the turbines? A flock of up-and-coming companies—Solel, BrightSource, and Solargenix among them—have not only put this concept into action, they've attracted billions of dollars of investment and won contracts to build more than 500 megawatts of solar-thermal installations.

"Concentrating solar power" (CSP) is similar to the Sunflower's solar-concentrator concept—except the challenge here is concentrating the sun's heat energy, not its light rays. One form of CSP is known as the parabolic trough system: giant curved mirrors magnify the effect of the sun, heating a thick fluid—either oil or molten salt—to a temperature of 700 degrees. The fluid then circulates in pipes that boil vats of water, creating steam that drives turbines. Another system—the "power tower"—uses massive flat mirrors angled to reflect the sun's rays onto a receiver at the top of a tower that, in turn, heats the fluid and drives a similar process.

A key advantage of CSP is that it offers solar the critical opportunity to become an on-demand base load power source that can be relied upon night and day. The solar heat is retained by the viscous liquid for hours after the sun has gone down and can drive the turbines late into the night—the time when electricity demands tend to subside. At this point, a natural gas generator can kick in to keep the fluid hot, turning off automatically when dawn breaks and the sun begins to flood the mirrors and do its job again.

Numerous CSP plants have been built. One in Nevada operates round the clock using only 2 percent natural gas—and 98 percent solar power. According to Romm, solar thermal plants spanning an area of 92 square miles in the Southwest—where clouds are scarce and solar resources remarkably consistent—could, in theory, generate electricity for the entire United States.

The drawback of CSP is the same as that of wind or any other centralized power generation: rather than being a distributed form of electricity produced directly on someone's rooftop, this form of solar needs to be piped over very long distances via transmission lines. Advocates argue that this is a reasonable trade-off for the reliability it affords.

Solar technology in all its various forms and stages of development has the potential to completely redraw America's energy map. It could erase the old, troubled lines of connection from Saudi Arabia to our gas pumps and from coal mines to our households, and replace them with lines connecting rooftop panels to our household appliances, and a new network of transmission cables originating in the Southwest and spanning the country.

ON WITH THE WIND

Wind energy could have an even bigger impact on the shape of our energy map. By far the most mature and cost-competitive of all the renewable markets, wind is even cheaper than coal as a power source in some states that offer credits and subsidies for clean electricity. Iowa, Minnesota, New Mexico, and Oregon already get 3 to 6 percent of their power from wind; Texas has a total installed wind power capacity of 7,900 megawatts—the production equivalent of about a dozen coal plants. In 2008, the DoE released a report showing that the United States as a whole could get 20 percent of its power from wind by 2030. That would slash carbon dioxide emissions from electricity generation by a tenth and create nearly 10 million American jobs.

Two major challenges stand in the way of ramping up wind. Number one is distribution. It stands to reason that wind blows most powerfully and consistently in places that are far removed from population centers. (Most people prefer not to live in constantly gusty locales, with the exception of the Windy City, Chicago.) Which means that the best wind resources in the country's central corridor will need to be piped over long distances to population centers via a network of superconducting transmission lines advocated by alternative energy proponents such as Pickens. Already, Texas state officials have approved a $4.9 billion plan to build transmission lines for wind energy throughout the state. It's a regional example of what needs to happen on a national scale—the growth of a coast-to-coast superconducting grid system as comprehensive as

Eisenhower's interstate highways. The DoE's 2008 report showed that a nationwide transmission system supporting a 20 percent growth in the wind industry could be installed at a cost of 50¢ per month per American household—under 2¢ a day.

The other challenge is energy storage. Even in the reliably gusty central corridor of the United States, wind simply can't be trusted to blow all the time—or when it's convenient. It may blow at night (when electricity demands are negligible), or dwindle during peak demand, or not blow at all for days. Engineers are scrambling to develop better, cheaper batteries and storage devices to reserve wind energy when there's a surplus, and dispense it on demand. But no good large-scale solutions exist just yet.

Companies are experimenting with various options, including plans to convert wind energy into underground vaults of compressed air that can later be released to provide electricity on demand. Others are finding ways to use wind energy to pump water into chambers that can then be opened as needed to run hydroelectric generators. Researchers at the Massachusetts Institute of Technology have even found ways to bioengineer viruses that can hold an electric charge, potentially creating a new category of high-powered batteries that can be as thin and flexible as Scotch Tape.

Even without fully developed solutions to storage and transmission problems, countries such as Spain and Denmark have already successfully integrated between 12 and 20 percent wind power into their grids. Pickens stressed that the United States can't wait around for storage solutions to mature before laying the groundwork for our renewable energy future. "We can't have a 'ready, aim, aim, aim' approach to this crisis," he told me. "We've got to fire."

WAVES OF THE FUTURE

Wind and solar are currently the best-known clean energy sources, but innovators and investors are also working to capture power from many other natural elements—the ocean tides and waves, the warmth of the earth, even human exercise and waste. Consider the unsung area of innovation known as geothermal—every bit as logical and promising as the solar thermal field that's drawing so much investment. Just as heat beams down onto the earth from the sun, it also rises through hot granite toward the surface of the earth from molten chambers deep within its crust. This is the same subterranean heat that causes havoc in ultradeep oil-drilling operations such as the one I visited on the Cajun Express.

Unlike the warmth of the sun, this underground heat is constant—it doesn't get interrupted by weather conditions or night. Tapping geothermal heat is a fairly straightforward operation: drillers burrow wells deep into the earth—anywhere from 1 to 2 miles down—and tap the heat in one of two ways. Either they capture the steam that's created in natural hot pockets or they inject fluids into the wells that absorb the heat and flow back up to the surface to drive steam-generated power plants.

The best geothermal resources in the United States are concentrated in the West, particularly in California and Nevada, as well as areas of Alaska, Hawaii, Idaho, and Utah. The technology is proven, low-risk, and fairly affordable to install, and has great capacity to scale up. The United States currently has about 2,500 megawatts of geothermal power (the equivalent of the energy produced by three coal-powered plants), but a recent U.S. Geological Survey study found that we could generate at least fifteen times that amount. In response, federal agencies have recently opened vast new areas of public land to geothermal development.

There's another notable form of geothermal energy that's somewhat humbler at first glance but an extremely worthwhile investment for anyone with a backyard. This technology doesn't generate electricity per se, but it does wonders to slash electricity demands by using the temperature of the earth to warm a building's interior in the winter and cool it in the summer. Dig a trough about 10 feet below your home, and you'll find that no matter what the season, the earth at this shallow depth stays at a year-round temperature of between 45 and 75 degrees. Geothermal household systems like the one buried in Lina and Adam's yard in Lenoir City simply harness these temperatures. Pipes are embedded a few dozen feet below ground, carrying fluid that absorbs the ambient temperature and is pumped back into the home, keeping it the same temperature as the earth beneath it.

This takes a great deal of pressure off heating and air-conditioning units—even in the dead of winter or the summer's scorching heat, you only have to heat or cool your home from that steady baseline temperature averaging about 60 degrees. Geothermal, in other words, offers the dual benefit of reducing the need for oil, gas, or electric heating in the winter while also slashing the high electricity demands of air conditioners on hot summer days. Geothermal heat pumps can cut home greenhouse gas emissions by more than half and save enough on heating and cooling bills to pay back their initial cost of several thousand dollars in less than five years.

Another vast untapped power reserve is the ocean. According to green energy advocate Bracken Hendricks, "Since water is a thousand times more dense than air, its movement packs a much greater wallop and allows us to create vast amounts of energy with a smaller mechanical footprint." Capturing ocean energy is in fact a form of both "lunar power" and wind power: the gravitational pull of the moon is conveyed through tides, explained Hendricks, and the force of the wind through surface waves. These oceanic forces are relatively constant. The California Energy Commission has stated that the wave power off California's coastline could power up to a quarter of the state's electricity demands. The technology for capturing wave power is relatively simple: big buoys compress water or air as they bob up and down, and then use that fluid or air pressure to drive an electrical generator. The ocean itself, however, is unruly at best. "The ocean's tides and swells can destroy almost anything man-made," explained Hendricks. "It's extremely difficult to design materials or machinery that can withstand their awesome force."

Among the challenges is finding a way to keep the cords connecting the buoys to the compressors from snapping. Innovators are scrambling to meet the task, and Hendricks walked me through some examples. The New Jersey–based company Ocean Power Technologies is developing digital software that can gauge the force of waves and automatically position the buoys to buffer that force. The Canadian start-up Finavera Renewables is experimenting with a technology called AquaBuoy that connects the compressors to the buoys through cables designed like huge muscles that flex and relax according to wave pressure. These and other companies are looking to develop wave-power projects off U.S. coasts, but they are still in trial phases. "Ocean power is where wind power was a decade ago," said Hendricks, predicting that this resource could eventually generate anywhere from 6 to 20 percent of America's electricity.

POWER OF ONE

Compared to the movements of the ocean, the force generated by a lone man on a treadmill is negligible, but it is power nonetheless—and yet another form of renewable energy that innovators are looking for ways to capture. Every time you step on a Stairmaster or turn the wheels of a stationary bicycle you are creating kinetic energy that can be harnessed and converted into electrons. Imagine this: you head for your one-hour workout at the gym, and before you mount the elliptical machine, join a spin class, or lift weights, you swipe a "kilowatt card" that tracks the amount of energy your muscles can muster.

Doug Woodring, an extreme-sports fanatic and renewable energy entrepreneur, has debuted the concept at a chain of exercise clubs throughout Asia. It's an intriguing marketing gimmick, to get people counting their kilowatts along with their calories. According to the *Wall Street Journal*, "The gym chain has rigged up 13 machines at one of its clubs, and when all of them are in use, the power generated amounts to about 300 watts—roughly enough to run three 27-inch television sets, five 60-watt light bulbs or several hundred video iPods."

This is peanuts in the grand scheme of electricity production, but it's part of a growing trend in harvesting human energy. The Florida-based company ReRev.com is now making elliptical machines for U.S. gyms that can power iPods and cell phones during workouts, and are particularly popular on college campuses. The U.S. military is working to develop "heel strike generators" mounted in soldiers' boots that can help charge up their walkie-talkies and reduce the battery load troops must carry in the field. A London design firm called Facility: Innovate has been developing flooring materials that collect kinetic energy from throngs of people walking through busy subway concourses.

A more substantial human-derived energy source is something we generate in great abundance: garbage. In 2006, the Atlanta-based company Geoplasma debuted a new method of vaporizing trash—blasting it with heat at temperatures higher than those of certain parts of the sun. Two by-products come of this process: a rocklike slag that can be used in road construction, and a synthetic, combustible gas that can be burned to run turbines and create electricity. A prototype plant in Florida has experimented with vaporizing 3,000 tons of garbage a day, and in turn producing 120 megawatts. About a third of that powers the vaporizing process itself, but the rest is sold to a neighboring Tropicana facility to run its production equipment.

Geoplasma says that its process is emissions-free and the slag is nontoxic—although the technology is so new that these claims haven't fully been tested. A central concern with waste gasification is that garbage itself is not exactly a renewable resource. Introducing to an already wasteful society the notion that trash can simply "disappear" could encourage still more waste and discourage essential recycling.

There's one form of human waste, however, that *is* inherently renewable—the kind we flush. Better known in more lighthearted energy circles as "poop power," sewage electrification is still in its early stages of development, but the concept is proven. There's no burning involved— instead, an industrious breed of bacteria is unleashed upon the feces in an anaerobic digester, and as they eat through the waste the bacteria release methane, a natural gas that can be used as fuel for generators. There is still a long way to go in refining the process, but one pilot project at a sewage treatment plant in Washington State has managed to harness bacteria-produced electricity to power a portion of the plant itself.

In theory, this waste-generated power could be transferred to the grid, where it would become part of an increasingly eclectic mix of renewably sourced electrons.

THE MEANING OF CLEAN

In looking at clean technologies, one highly controversial subject should be mentioned: "clean coal." It's almost impossible today to watch an hour of television without hearing that term come up in advertisements, often paired with a backdrop of verdant meadows. Oil, coal, and natural gas are nonrenewable energy sources, and all of them produce polluting greenhouse gases when burned. Natural gas is certainly the cleanest of fossil fuels, producing much less carbon dioxide than oil or coal—that's why Pickens has proposed it as part of his clean energy strategy. It is, nevertheless, a mined fossil fuel that comes with serious environmental consequences, and one that we heavily import from overseas.

The coupling of the term "clean" with "coal" is even more controversial. Burning coal intrinsically produces pollutants that cause smog and acid rain in addition to global warming. Scrubber technology can be applied that eliminates some of these toxins before they are released into the atmosphere. Engineers have also devised a way to gasify coal by subjecting it to tremendous heat and pressure; they then remove the carbon dioxide from the gas before it is burned to produce energy. But these cleansing technologies are expensive, and they have not yet been proven to work at a commercial level. Questions also loom about how to safely store large amounts of carbon dioxide once it has been captured.

In other words, despite the coal industry's widespread promotion of "clean coal," at present there is no fully clean way of generating energy from this most carbon-packed of all fossil fuels. "Clean coal is like healthy cigarettes," Al Gore has said. "It does not exist." Gore and renowned climate scientist James Hansen of the NASA Goddard Institute for Space Studies have called for a moratorium on all new coal plant development, and have been leaders in a growing wave of activism against coal.

In March 2009, Hansen organized a peaceful protest in Washington, D.C., to stop all coal-fired power plants, stating, "Let's use this as a rallying cry for a clean energy economy that will protect the health of our families, our climate, and our future." Thousands of protesters marched from the Spirit of Justice Park to the U.S. Capitol's own coal-fired power plant. Carrying signs that said "Closed for Climate Justice," "Coal-Blooded Killer," and "No Coal Is Clean Coal," they blocked the entrance to the power plant, preventing workers from getting in. Marchers protested coal's unparalleled climate impacts, and voiced concerns about the disposal of waste ash from burned coal, the smog and mercury emissions from coal plants that pose public health risks, and the strip-mining method (known as "mountaintop removal") that is often used to harvest the fuel. The D.C. march was not the first protest of its kind: residents of Austin, Waco, and Dallas, Texas, held a candlelight prayer vigil in 2006 that helped block nineteen new coal-powered plants proposed in their region.

The opposition seems to be working. As the public outcry intensifies and federal restrictions on greenhouse gas emissions loom on the horizon, investors are growing leery of coal. The Energy Information Administration predicted in 2008 that more than 100 gigawatts of new coal-fired power plants would come online by 2030. But in 2009, as an increasing number of investors began pulling their financing from coal projects, the agency's prediction had plunged by more than half to 46 gigawatts. That's still a substantial amount of new coal-generated power—a 10 percent increase over the quantity we have online today—but it shows a precipitous shift in the landscape of the power industry.

"Last year there was nearly four times more wind power added to the grid than coal power," said David Hawkins, director of the Natural Resources Defense Council's Climate Center. "That says a lot about the direction we're headed." But he stressed that the fight to curtail the use of coal in the United States will be a politically bloody one due to the coal industry's considerable lobbying heft.

There is a strong argument to be made for pushing ahead with efforts to create, if not *clean* coal technologies, then *cleaner* coal technologies that capture CO_2 and sequester it underground. Coal currently accounts for half of electricity production in the United States. Until we develop a stronger, more reliable grid and long-term storage solutions for intermittent wind and solar resources, these renewables will not be able to fully replace coal's supply of large-scale, on-demand power.

Even if the United States were to eliminate all coal from its energy mix, China and India are building the generating capacity of two conventional coal plants every week. "In the last five years alone, China has built the equivalent of the *entire* U.S. coal fleet," Hawkins told me. "If we don't figure out a way to burn coal responsibly, all the plants scheduled to come online by 2030 will send 30 percent more CO_2 into the atmosphere than all the coal that has ever been burned in human history. It would doom any efforts to stop global warming."

Hawkins is one of many environmentalists who have supported the development of carbon sequestration technology by companies such as GE. Certainly the old coal plants must be taken offline, and the newer ones need to be retrofitted with pollution-scrubbing filters. Hawkins believes that coal can be significantly reduced from supplying half of the U.S. energy mix to a quarter by removing the oldest, most inefficient plants.

Nuclear power, which supplies 20 percent of America's electricity needs, is another controversial energy source that has been promoted as "clean." It is clean, in the sense that generating nuclear power produces no greenhouse gases. But it carries with it heavy financial, political, and environmental burdens: The capital costs of building nuclear reactors are significantly higher than those of coal plants, and the reactors take many years to construct and bring online. The uranium that powers nuclear plants is a nonrenewable resource that we largely import from abroad, undermining our goal of energy independence. Nuclear plants also produce radioactive waste that can be dangerous to plant and animal life for periods of tens of thousands of years. Right now there are no large-scale, politically viable solutions in the United States for the safe disposal of used uranium. Given these constraints, it seems that the challenges of dramatically scaling up renewable energy are less formidable than those of expanding our nuclear plants.

In the long term, do we have enough guaranteed renewable energy sources to eliminate the need for fossil fuels and uranium altogether? Eventually, yes. Add up the massive potential of the innovations I've outlined—wind, solar, tidal, geothermal, and power from human motion and waste—and it seems abundantly possible that we can find a way to power our electricity needs with renewable energy alone.

In reality, that fairy-tale image of Oz—glittering solar panels, whirling wind turbines, crystal clear skies—is most likely a long way off. But if we can't move to 100 percent renewable energy in a single decade, perhaps we can do it in three or four.

THE ROAD AHEAD

The big bursts of innovation we are seeing in wind, solar, and other clean sources of power today are in essence a modern form of wildcatting—a thrilling, heroic, and high-risk game of ingenuity that will, like the early days of Spindletop, lead to more losers than winners, and more investment dollars lost than gained. "There will be dozens of failures for every one success," said the energy expert Bracken Hendricks, referring to the flurry of new activity in clean energy. "But in that one success lies a key to the future of humanity."

A sea change transforming our politics, our economy, and our cultural identity is in the works. Even though America has been lagging behind, we can still leap to the head of the race. As we jockey for position, it's useful to look back at the pivotal moments of change that defined our successes and missteps over the last century: the birth of wildcatting and the global oil industry; the stratospheric growth in industry that surrounded World War II and pushed aviation and automobiles into the mainstream; the meeting between FDR and King Ibn Saud on USS *Quincy* that shaped decades of foreign policy, and Roosevelt's electrification program, which transformed rural America; Eisenhower's highway system, which gave birth to suburbia and fueled the American dream; the Green Revolution, which tripled our farming productivity;

the Arab oil embargo, which carried lessons about oil dependence we failed to heed; and the supply chains and dot-com boom, which have woven us into the vast web of the global economy.

We are, as author John Kouwenhoven writes, "a society which puts a premium upon adaptability, mobility, and a willingness to forgo the security of fixed status and the certainty of dogmatic absolutes." We are capable of tremendous innovation and sudden technological shifts. We live in a time when grand-scale change is possible—and becoming ever more so.

It's a crucial moment for federal leadership to lay the foundations for the explosive growth ahead. One of the many sobering consequences of the recent economic recession has been that billions of dollars in private investment have been pulled from the clean technology sector. Anthony Lucas's discovery at Spindletop would not have been possible without his Pennsylvania-based investors any more than Edison's power plant experiment would have been possible without J. P. Morgan's backing. Quite simply, there can be no invention or innovation without the money to pay for it. The 2009 Recovery and Reinvestment Act has helped blunt the impact of the recession on the renewable energy sector—keeping new solar and wind companies in business that would otherwise have folded. But still more action is needed.

I spoke with business leaders, scientists, and innovators about the policy options at our disposal. Their top priorities include a grand national plan, Eisenhower-style, to install a new smart nationwide energy grid—fast. We need both new high-power transmission cables crisscrossing the country and much more reliable distribution cables at the local level. (The Recovery Act did include some funding toward that end, but a comprehensive plan and significantly more resources will be needed to properly execute the project.)

A renewable energy standard requiring utilities to get at least 40 percent of their power from renewable sources by 2020 would directly connect millions of American households to clean power sources. A massive energy-efficiency program for buildings and appliances could help offset 85 percent of the projected increased demand for electricity by 2020.

And while the Obama administration took a significant step in establishing the position of "energy czar" in the White House (an advisory role to coordinate energy- and climate-related activities among the agencies), a new, permanent cabinet-level position—a Secretary of Smart Growth—could infuse principles of sustainability and innovation into all limbs of the executive branch with deeper, longer-term impacts.

The most significant policy change will be the federal limits on greenhouse emissions that loom so near. These "cap and trade" regulations will put a cap—or ceiling—on greenhouse emissions across many industries and enable companies to buy and sell the right to pollute. A dirty company that produces more emissions than the law allows, for instance, can buy "carbon credits" from a clean company that performs better than the law requires.

Such a system is designed to impose a price on carbon dioxide that reflects the environmental cost of burning fossil fuels—a price that will greatly accelerate the shift from America's energy past to our energy future.

As Joseph Romm told me, even the most expensive renewable energy, solar panels, "would quickly become cost-competitive with coal by 2015 if coal producers had to pay a price for the pollution they emit." Carbon limits will dramatically hasten efforts to develop affordable carbon-scrubbing technology for coal plants, and step up the competition among entrepreneurs and utilities to develop the best, greenest, cheapest, and most efficient replacement technologies, eventually phasing out fossil fuels altogether.

SIMPLE GIFTS

What sets the clean, independent energy network of tomorrow apart from the fossil fuel system of today is, above all, its utter simplicity. In a sense, we're not so much barreling into the future as we are reconnecting to the past. Romm walked me through some ancient applications of the renewable technologies Americans are now exploring. Two thousand years ago, Chinese inventors used simple windmills to pump water. The innovation spread to Persia and the Middle East, where it was used to grind grain. More than ten thousand windmills operated in the Netherlands in the 1800s, powering innovations in farming and engineering. Even in the early part of the twentieth century in America, windmills dotted the farmlands and oil fields of the West, providing electricity for isolated farmers and wildcatters, pumping water for irrigation, and operating derricks.

Solar power, too, dates back to 212 B.C., when Archimedes advised Greek soldiers to use their curved bronze shields to concentrate beams of sunlight on Roman ships. The solar concentrator concept was also sketched in Leonardo da Vinci's notebooks as a means of generating heat; he saw it as an alernative to burning wood that could help conserve forests.

Perhaps the most ancient form of renewable energy is geothermal. As early as 10,000 years ago, North American Paleo-Indians settled around hot springs and used their warmth for heating and cooking. The first residential geothermal system was installed in the fourteenth century in Chaudes-Aigues, France, where engineers tapped the heat of boiling underground wells and piped it into homes. The system has worked so well that it's still in use today.

Renewable energy channels natural elements so simple, so essential, that they almost seem too obvious or commonplace for a society like ours, which craves complexity and challenge. But today, at a time of so much environmental, economic, and geopolitical turmoil, simple solutions are profoundly opportune.

The car will no longer be an autonomous machine that gets filled up with gasoline once a week. It'll be a component of a larger intelligent network—a traveling battery, essentially, that uploads and downloads electricity to and from a grid that's constantly growing smarter, greener, more decentralized. . . . As the grid gets greener, the cars get cleaner.

—Dan Reicher, Google employee

9

Autopia

DETROIT DOES THE ELECTRIC SLIDE

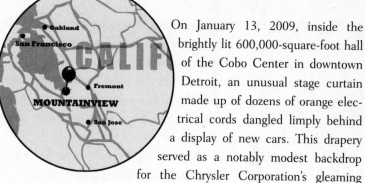

On January 13, 2009, inside the brightly lit 600,000-square-foot hall of the Cobo Center in downtown Detroit, an unusual stage curtain made up of dozens of orange electrical cords dangled limply behind a display of new cars. This drapery served as a notably modest backdrop for the Chrysler Corporation's gleaming prototype vehicles at the 2009 North American International Auto Show. Spaced roughly 3 inches apart, the orange power cords with their black plugs varied in length from 3 to 6 feet and gave the overall effect of threadbare party streamers left up a day too long. But these peculiar decorations weren't an oversight—they were a deliberate choice, mirrored by a power-cord symbol encircling the Chrysler logo. The vehicles on display ranged from sleek red sports cars to boxy black sedans and white SUVs, and all shared one feature: in place of gas tanks, electrical sockets.

I had heard predictions of a future of all-electric cars in my discussions with grid experts and renewable energy advocates. But it was surprising—shocking, actually—to see that future acknowledged and even embraced by a humble Detroit. Ever since the roaring 1950s, Detroit's automakers had been loudly proclaiming the virtues of large gas-guzzling cars, SUVs, and pickup trucks.

In fact, just twelve months earlier, at the 2008 Cobo Center show, the Chrysler Corporation had unveiled a muscular new off-road vehicle in front of a high-tech waterfall that spelled out "Jeep" in falling sheets of sparkling liquid. It had hired *Sex and the City* star Kim Cattrall to stroke the curves of a glamorous Mercedes SUV prototype. The company had even wrangled a herd of 130 Texas longhorn cattle to accompany its latest Dodge Ram pickup truck (rated at a dismal 15 mpg) as it lumbered toward the showroom. The building had buzzed with more than 700,000 car lovers enjoying light shows, hip-hop beats, misty swirls of dry-ice-fog, and models dressed in sequined gowns.

The crowds at the 2009 auto show were smaller. Models in somber black dresses stood alongside electric cars, unveiling them with the lifting of plain white sheets as though revealing cadavers. What was most remarkable about the 2009 show, in fact, was the quiet, unremarkable way in which it heralded the end of the gasoline-powered combustion engine as we know it—the engine that has propelled vehicles across America's highways for over a hundred years.

It wasn't just Chrysler. For the first time in the show's history, almost every major automaker debuted cars with electrical sockets and made serious, long-term commitments to producing plug-in vehicles. The cars on display represented a wide range of engineering approaches, each illustrating a different stage in the industry's transition from petroleum-powered to all-electric engines.

General Motors displayed the Chevrolet Volt, a sleek battery-powered sedan, and the Cadillac Converj, an electric variation on the all-American land yacht. Chrysler showed the Dodge Circuit all-electric sports car, along with its Jeep plug-in Patriot EV (for "electric vehicle"), the Jeep Wrangler Unlimited EV, and the Chrysler Town & Country EV. Mercedes-Benz, meanwhile, introduced BlueZero, a battery-only prototype that looks like a cross between a car and a scooter. Ford and Toyota also announced plans to begin offering plug-in cars in their U.S. dealerships between 2010 and 2012.

Somberly and resolutely, GM's then-CEO Rick Wagoner stood before the TV cameras to declare his company's "commitment to the electrification of the automobile." Ford chairman Bill Ford likewise announced his company's intention to "speed the acceptance of electric vehicles." The message was clear: whatever may happen in the near term to Detroit's top automakers, the "heartbeat of America" in the future will almost certainly be pumped by electrons.

And yet few in the media or the public greeted the 2009 trade show as the dawn of a new automotive era. Doubt still surrounded Detroit's every move in the wake of its financial collapse the previous fall. Rising gasoline costs had led to a panicked flight of consumers from Detroit's standard gas-guzzling models in favor of the more fuel-efficient foreign competitors. Congress had agreed to grant the Big Three automakers a provisional $25 billion lifeline, but only after scolding their executives for clinging to that 1950s business model in which size and design frills rather than fuel efficiency were equated with profits. The head of the Senate Banking Committee described the car companies as "seeking treatment for wounds that, I believe, were largely self-inflicted." The global recession only compounded this underlying problem.

Chrysler's humble display of orange power cords was designed to send a message to Congress and the public that Detroit's leaders were ready to forego frivolity and get down to business: "Any superfluous show business," GM vice chairman Bob Lutz told the media, "I think that's all a thing of the past." The companies were—albeit belatedly—beginning to embrace a lean, responsible clean-car future.

There was, however, a hitch. The electric vehicles on display were almost all "concept cars," still in the early design phase. In addition to the looming question of whether Detroit's automakers would (or even should) survive long enough to bring these cars into dealer showrooms, many critics pointed to the sizable barriers that still stand in the way of electric vehicles.

OIL CHANGE

It is an odd feeling to walk around a plug-in car from front to bank, running a hand along its smooth, glistening exterior in search of its charging port. I'd had the opportunity to do so a few months before the Detroit auto show, when I was introduced to the Chevy Volt, a midsized sedan that can go 0 to 60 in under 8.5 seconds. At first glance, it looked like any other new car on the road, with its bubbly futuristic styling. But the primary fueling mechanism on this vehicle was not the familiar screw-top gas-tank receptacle on the rear panel. It was three clean prongs—exactly like those you'd find at the back of a kitchen appliance—located just in front of the driver's door under a discreet sliding cover.

This Volt was on display at a conference held in the summer of 2008 in Washington, D.C. Hosted by Google and the Brookings Institution, the conference was examining the future of plug-in vehicles in America— specifically, the role that Washington could play in fast-tracking their progress into the driveways of middle-class consumers.

There weren't any flashing lights, waterfalls, or even modest curtains of electric cords at the event, which was held in the dimly lit ballroom of Washington's Regency Hyatt hotel. Instead, half a dozen concept cars were simply, squatly parked beneath gaudy chandeliers on a floral polyester carpet. They looked absurdly out of place, like giant futuristic bugs hunched on an old frilly tablecloth.

I'd gone there to find out how plug-in cars work and what obstacles still stand in their way. Are electric cars really green if they're powered by coal-fired plants? Is the battery technology powerful enough for long-distance driving, not to mention safe and affordable? What if you're a harried mom or dad who forgets to plug in your car at night and then gets stranded during carpool when the juice runs low, with no way of charging up at a local gas station?

The cars in the Regency ballroom represented a range of approaches to tackling these questions. The Volt and a plug-in SUV from Ford were the only vehicles on display made by Detroit. The rest had been designed by a young crop of automotive visionaries. The California-based company Tesla Motors, founded by dot-com billionaire Elon Musk, was showcasing a glamorous, all-electric roadster powered by a hive of thousands of connected cell-phone-style batteries. University of Delaware students were showing a reengineered Toyota Scion truck: they had removed its mechanical guts and replaced them with a homemade state-of-the-art electric engine that could travel 150 miles on a single charge. Google itself was displaying a standard Toyota Prius that had been "hacked into" by engineers who boosted its battery capacity and implanted special software into its controls system, enabling the car to "upload" and "download" power to and from the grid.

At a time when Detroit was under fire for its failure to innovate, here were enthusiastic engineers who represented a young, scrappy, band of rebels. The small group of speakers and guests at the conference was as eclectic as the cars, wearing everything from starched suits to sneakers. In addition to two prominent representatives of Detroit—GM North America's president Troy Clarke and Ford America's president Mark Fields—there were also utility bigwigs, including Peter Darbee, chairman and CEO of the California utility giant Pacific Gas and Electric. And there were plenty of little guys with big ideas—renegade car hackers, upstart battery makers, eco-activists, and grassroots organizations such as Plug-In America and CalCars that promote the development of green cars.

The most surprising speakers at the convention were the politicians. They represented a strikingly broad spectrum of bipartisan support for electric cars, ranging from Republican Orrin Hatch of Utah to Democrat Jay Inslee of California. James Woolsey, the Republican former CIA chief who had spoken to me several months earlier about the dangers inherent in our dependence on foreign oil, described electric cars as essential to our national security. Cars currently account for roughly 40 percent of America's total oil consumption. Plug-in vehicles, said Woolsey, have the power to end the dominance of oil in the world economy: "We can, we should, and we must, as a major national priority . . . absolutely, totally, completely destroy oil's monopoly," he thundered from the podium. Even Orrin Hatch, a rock-ribbed conservative hardly known for going against the grain, touted the promise of plug-in cars. "This technology is nothing short of revolutionary!" he enthused. "The thing about the electricity grid I really like is that it's domestic—you won't see the president of the United States flying to Saudi Arabia and begging them for electrons."

Milling among the guests, I managed to clear up a few of my basic questions. Is it really accurate to classify these vehicles as "clean"? Coal produces more greenhouse gases per unit of energy than oil, so in theory an electric car battery charged by a coal plant could be even worse for the planet than an oil-powered engine.

In actuality, given the amount of waste heat lost in combustion engines, among other factors, it turns out when you crunch the numbers that "driving a car charged from a standard wall socket is much better from a global warming standpoint than driving an average vehicle filled with oil," said David Sandalow, an environmental scholar with Brookings and assistant secretary for Policy and International Affairs at the DoE. "In fact, if we moved America's entire vehicle fleet from gasoline engines to plug-ins today, we'd cut greenhouse gas emissions by up to 35 percent."

And that's given the current energy mix. Dan Reicher, director of climate change and energy initiatives at Google.org, noted that the beauty of moving cars toward electricity is that "as the grid gets greener, the cars get cleaner." In other words, swapping out coal plants with renewable energy technologies will slash emissions from both the automobile and the power industries in one fell swoop.

In reality, I learned, the promise of an affordable all-electric car for the mainstream market—and the infrastructure needed to support those vehicles—still looms years out of reach. What has great potential in the very near term are intermediate technologies—plug-in cars that combine gasoline and electric engines and can fuel up either via the grid or at standard fuel pumps. These plug-in hybrid electric vehicles, or PHEVs, provide the link between the gasoline infrastructure of yesterday and the electric system of tomorrow.

BRIDGING THE GAP

The Chevrolet Volt that I had the thrilling and mundane pleasure of plugging into the hotel wall socket actually has two fueling mechanisms—the electrical charging port and a standard gas tank. This is the most notable of the plug-in hybrids now on the drawing board, as it's the first one slated to hit the American market. General Motors plans to mass-produce the Volt by 2010 with a sticker price of roughly $40,000. Its hybrid electric-gas engine is a total redesign of the Prius engine, with a battery quadruple the weight (about 400 pounds), enabling it to function on electric mode at all times and travel up to 40 miles on a single charge. Since the average daily round-trip commute in America is 30 miles, the car could conceivably never visit a gas station again with regular charge-ups. On trips longer than 40 miles, the backup gas-powered generator will kick in to keep the battery juiced. Fuel is never actually used to propel the car (as it is in the standard Prius); it's just there to recharge the battery when it's running low. The average fuel economy? More than 100 miles per gallon.

When first developing their fuel-efficient Prius for U.S. markets in 2000, Toyota executives made the judgment call that freedom-loving Americans would never embrace a car they would have to plug into a wall socket to charge up like an appliance. The standard Prius maximizes fuel efficiency by using both an electric motor and a smaller-than-usual gas-powered engine to turn the transmission that rotates the wheels. The electrical power in the Prius is never provided by connection to the grid—the battery is charged up by the gasoline combustion engine. Which means that while it's more fuel-efficient than conventional cars, the standard Prius still relies on gasoline to generate the power that propels it forward.

By contrast, GM's plug-in Volt (this bears repeating) can run *entirely* gas-free depending on the distance you have to travel. "You're not just using less gasoline, you're displacing gasoline with electricity," electric-car advocate Felix Kramer told me. "The plug-in hybrid represents disruptive change."

The Volt and other plug-in models in development still have several technological hurdles to overcome—most notably refinements in the battery technology. The implications of this breakthrough are nevertheless enormous: no American innovation today would be more patriotic, penny-wise, or planet-friendly than an affordable, high-performance car that gets 100-plus mpg—connecting the petroleum era to the electric age.

CAR HACKERS

Designs for plug-in vehicles have in fact been floating around commercial and academic laboratories for decades. They were initially sketched out after the Arab oil embargo, as engineers began experimenting with multiple concepts for fuel-efficient cars. But it was a small group of car hobbyists, software designers, environmentalists, and techno-geeks working in a garage in Palo Alto, California, that first put plug-in hybrids on the road. Starting in 2003, this group had begun forming in Internet chat rooms that discussed the possibility of reengineering a standard Prius so it could be charged by a basic wall outlet. Felix Kramer, who was a particularly active member of the chat room, had been inspired in part by a blueprint for the first hybrid that could be charged via the grid drawn up by Dr. Andy Frank at the University of California, Davis. Kramer—who had recently sold an Internet startup company and had time to kill—proposed that a group get together to build a real-world version of this blueprint. Ron Gremban, a retired electrical and software engineer, offered his Prius as the test bed.

Every few weekends over the course of a year, half a dozen self-appointed "car hackers" would meet in Gremban's sunny, cluttered garage, throw open the hood of the Prius, and get to work tinkering with the car's mechanical and computerized innards. By early 2005, they had patched in a 60-pound heap of used lead-acid batteries collected from local gas stations and built a car that could drive on all-electric mode for 25 miles before its gas engine kicked in. While other plug-in concepts were in development in automotive laboratories, this jury-rigged Toyota was the first real-world plug-in hybrid to hit the American road.

"Ron Gremban and Felix Kramer have modified a Toyota Prius so that it can be plugged into a wall outlet," the *New York Times* reported on April 2, 2005. "This does not make Toyota happy. The company has spent millions of dollars persuading people that hybrid electric cars like the Prius never need to be plugged in and work just like normal cars."

The car hackers published an open-source description of their experiment online so that the formula could evolve through contributions made by others and spread widely. That's just what it did. Soon several companies were created in Silicon Valley offering custom conversions of Priuses into plug-ins for around $10,000 a pop, providing a fuel economy of between 80 and 100 mpg (roughly twice that of a standard Prius). These conversions quickly drew an underground following. Dozens of orders came in from utilities, universities, and government agencies, as well as from cities including Austin and Chicago, companies including Google, and public figures such as James Woosley and San Francisco mayor Gavin Newsome.

Watching from the sidelines, General Motors had begun experimenting with the concept in its own laboratories. When I asked Kramer about Detroit's decision to develop concepts that had first been proven by California's plug-in subculture, he cheerfully quoted anthropologist Margaret Mead: "Never doubt that a small group of thoughtful, committed citizens can change the world. Indeed, it's the only thing that ever has."

WHEELS OF FORTUNE

Over the last decade, General Motors has had a love-hate relationship with the electric car. GM introduced a limited number of small

all-electric vehicles called EV1s in 1996. Consumers didn't actually own these cars—they were leased by GM as part of a market study on plug-in technology. But five years later the company yanked the entire fleet off the road (as recounted in the independent film *Who Killed the Electric Car?*) and shredded the vehicles one by one, arguing that horsepower-hungry consumers just didn't like the notion of plugging their cars into wall sockets.

A number of forces converged to put the electric car back on Detroit's drawing boards. Oil prices began to escalate beginning in 2001 and the cost of gasoline more than doubled, war raged in the Middle East, and weather patterns in the United States became more violent, making the threat of climate change feel immediate and real to consumers and politicians. Between 2001 and 2008, the focus in Washington and Detroit pinballed from one gasoline alternative to another—first to pure hydrogen, then to ethanol made from crops such as corn and switchgrass, and only then, as a last resort, to electric. GM itself examined the full gamut of options in 2005 and 2006—including ethanol, clean diesels, hybrids, and hydrogen fuel cells—before finally circling back to the electric concept.

The resounding call for energy independence led to a brief national obsession with the promise of hydrogen. Detroit and Washington began trumpeting a future in which cars would run on hydrogen-powered fuel-cell batteries, producing an "exhaust" of nothing but pure water. But automakers encountered big economic and logistical barriers. Hydrogen fuel-cell concept cars were costing from $500,000 to $1 million apiece to manufacture, and the challenge of building a hydogen fueling infrastructure to replace gasoline pumps began to look impractical in the near term. (It's an expensive and energy-intensive process to liberate hydrogen atoms from water molecules, where they are found in abundance.) A hydrogen-car future, critics argued, wouldn't be practical for the masses until at least midcentury. This didn't kill the technology so much as shift it to the back burner. Every major automaker is still working to develop hydrogen platforms, just with a longer-range view and in a way that will work in tandem with electric car platforms.

The hydrogen hype was promptly overshadowed by the soaring popularity of ethanol, the biofuel derived from corn and other fibrous crops that is commonly used as an additive in gasoline. It can also replace gasoline altogether (as demonstrated by Ken McCauley's "corn-fed" truck). The greatest advantage of biofuels is that they integrate easily with our existing transportation infrastructure. It doesn't cost much to retrofit conventional gas stations with biofuel pumps, and little more is needed to make conventional cars ethanol-compatible than new plastic linings in their gas tanks.

Roughly half of all cars in Brazil, for instance, can be powered by pure ethanol made from sugar cane. Why couldn't the United States do the same with fuel made from domestic corn crops? In 2006, GM and Ford launched multimillion-dollar ad campaigns touting the promise of ethanol with slogans such as "Live green, go yellow," and images of golden corn cobs exploding from gas nozzles. Both companies promised to convert more than half of their fleets to "flex-fuel" gas tanks capable of burning either gasoline or E85, a blend of 85 percent ethanol and 15 percent gasoline. While Detroit was moving aggesssively in the direction of biofuels, Toyota was developing more sophisticated hybrid-electric cars.

Congress, meanwhile, increased farm subsidies to spur the development of ethanol, zealously supported by the corn lobby and politicians from America's twenty-nine farm states. But this, too, met barriers. As the price of oil skyrocketed, so did the price of corn—in part owing to the increased cost of petroleum inputs on farms and the rising crop demand for manufacturing ethanol. Critics speculated that this biofuel couldn't be practically and affordably scaled up as a significant substitute for oil. Moreover, as food shortages spread through the world's impoverished nations, many antifamine activists pointed their fingers at ethanol. They argued that an increasing amount of finite cropland production was being diverted to growing fuel rather than food, driving up food prices.

Environmentalists also noted that not all biofuels have green benefits: fossil fuels are used in the production of ethanol not just through farming practices but throughout the manufacturing process that converts crops into fuel. Corn-ethanol has by many estimates a global warming impact only slightly lower than that of gasoline, whereas fuels derived from cellulosic crops such as switchgrass (which are easier to process) can cut carbon emissions from tailpipes by up to 60 percent.

Today nearly all ethanol in the United States is corn-derived, but California has recently introduced a clean-fuel standard requiring a 10 percent reduction in the carbon intensity of transportation fuels sold in the state by 2020. The initiative was designed to propel the shift away from corn-derived ethanol to cleaner biofuels, and it's a policy that many lawmakers are aiming to apply nationally.

The Obama administration has allocated billions to the development of ethanol and other biofuels, which the president has described as "important transitional fuels to help us end our dependence on foreign oil." This description alone represents a crucial shift: biofuels have been demoted from their status as the great promise of tomorrow to "transitional" sources of energy that will enable plug-in cars to use less gasoline as America moves toward a post-petroleum future. Rather than competing with the shift toward electric vehicles, in other words, biofuels have come to be seen as compatible with it.

Part of what pushed biofuels into to a supportive role—rather than the starring role—in America's automotive future was the phenomenal success of the Toyota Prius. This egg-shaped car sold more than 1 million units in its first decade on the market, and more than half of those sales were in the United States. Before the Prius debuted, American consumers had shunned the idea of electric power in cars; suddenly they were clamoring for it.

The extraordinary success of the Prius and the rapid ascent of Toyota made Detroit executives very nervous, particularly at General Motors, which saw Toyota nipping at its heels in total sales volume. Electric cars had an unlikely champion inside GM—Bob Lutz, the company's vice chairman of product development. A cantankerous man in his seventies who owns a fleet of luxury cars and four private jets, Lutz has been the industry's loudest skeptic about global warming. In 2005, he became concerned about Toyota's surging sales as oil prices shot up. Japan had been imposing tough fuel standards and hefty gas taxes for decades, given its lack of domestic oil reserves. As a result, Toyota's fleet had grown nearly 20 percent more fuel efficient than GM's. This greater efficiency contributed to Toyota's toppling of GM as the world's biggest automaker in 2007, selling 2.35 million cars to GM's 2.26 million.

Bob Lutz fretted over what to do to counteract Toyota's surging popularity, and in January 2006 he made the case to his board of directors that "the electrification of the automobile is inevitable" and, moreover, the only way to preserve Detroit's industry. That month, Lutz got the board's green light and large amounts of funding to start designing the Volt—GM's response to the Prius. "After years of avoiding the future, [GM] finally understood oil prices were not going to return to earth, global warming was a de facto political reality, and Washington was serious about imposing tougher fuel economy rules on his industry," explained David Welch in *BusinessWeek*. "GM would have to live green or die."

BATTERIES NOT INCLUDED

If plug-in cars are all they're cracked up to be, why has it taken them so long to hit the road? One word: batteries. Hardly a sexy technology when compared to devices like the iPods and cell phones they power, batteries are perhaps the single most important technology to watch as automakers shift toward an electric future.

Without certain limits to battery technology, America's automobiles might have been battery-powered from the industry's earliest days. In 1889, Thomas Edison engineered a car with a rechargeable battery, and by 1896 America's first car dealers were selling mostly electric cars. At the turn of the century, the auto industry was split among technologies—40 percent of cars were powered by electricity, another 40 percent by steam, and the rest by gasoline. In 1908 even Henry Ford bought his wife an electric car. Billed as quiet, clean alternatives to their combustion-engine siblings, electric vehicles were targeted at female consumers. One 1912 advertisement for a car dubbed the "Silent Waverly Limousine-Five," came with the tagline: "Delicate Gowns Not Marred in This Roomy Electric," and promised "no smoke—no odor—no spattered oil . . . at half the expense of most gas cars."

For all the advantages of electric power, eventually gasoline won out because it offered the inherent benefit of longer journeys: an electric battery would lose its charge far more quickly than a petroleum-fired engine would run out of gas. And it took longer to charge up a battery than it did to fill up a gas tank, making fueling stations more convenient than charging stations.

Today's battery engineers are still grappling with many of the same concerns. The invention of the first rechargeable battery dates back to French physicist Gaston Planté in 1859. Since then, the evolution of battery performance has been sluggish compared to that of other technologies (computer chips, for instance, have doubled in performance roughly every two years). Part of the problem is that the very task batteries are trying to perform—capturing and storing energy in an ordered, ready-to-use state—runs contrary to the laws of physics. It contradicts the second law of thermodynamics (known to anyone who tries to keep a reasonably uncluttered house), which holds that nature favors higher disorder. Electricity itself is by definition a restless force that's difficult to contain. The good news is that in the last fifteen years we've seen great strides in battery development, thanks largely to the proliferation of wireless electronic devices.

Batteries, in essence, create reactions between metals and chemicals that store and channel electricity. Between 1800 and 1990, the two most common rechargeable battery technologies on the market were lead-acid batteries, which powered the first electric cars, and nickel-cadmium batteries, which came into use in the 1960s, powering gadgetry like electric toothbrushes, hair clippers, and power tools. The limitations of these batteries were their weight, their bulk, and the amount of charge they could carry.

As companies such as Sony, Toshiba, and Black & Decker began to sell more cordless devices, battery engineers started testing new combinations of metals and chemicals that could produce lighter batteries with bigger, longer-lasting charges. Nickel-metal hydride batteries appeared on the scene in the early 1990s. They quickly became the battery of choice for powering portable electronic devices. Toyota engineers seized on this technology as the battery of choice for the Prius, and have used it for years in these cars with great success. The problem is that these batteries are too costly, heavy, and bulky to provide the main source of power for a car. They perform well in a backup role, assisting a gasoline engine, but their energy capacity deteriorates with each charge, and they don't have the endurance and efficiency to power a high-performance engine on their own.

Enter the latest star in the rechargeable battery market: lithium-ion. Lithium is the lightest metal in the world, and these batteries carry roughly twice the power of a nickel-metal hydride battery of the same size. You're already familiar with rechargeable lithium-ion batteries—they're those slender battery packs on your cell phone. Having enabled the wireless-communications revolution, lithium-ion batteries now show promise of paving the way for electric cars. Not only are these batteries lighter and more powerful; they can endure thousands of charge cycles without wearing down, and they recharge quickly.

But the technology still needs work. "They still need to get cheaper, easier to recycle, more energy dense, and longer-lasting," says Bill Reinert, the national manager for advanced technology vehicles at Toyota. "And above all, they need to be proven safe." What concerns automakers most is that lithium-ion batteries are prone to malfunctions if they overheat, overcharge, or get knocked around. In 2006, a manufacturing glitch caused some batteries in Sony laptops to catch fire, forcing the company to recall millions of products. Imagine the explosion that could occur in a battery that carries a thousand times more energy. Tim Spitler, a battery engineer, told the *Wall Street Journal* that automakers must be wary of producing vehicles that could "ignite and burn up grandma and two kids sitting on half a ton of batteries in the car."

In other words, a lot is at stake if car companies try to fast-track their plug-in vehicle technology. Detroit must consider the liability issues—not to mention consumer confidence—if it prematurely introduces into the market battery packs that could have dangerous defects.

That's reportedly part of the reason why Toyota changed its plan to use lithium-ion batteries in the 2009 Prius, which would have given the car a considerable boost in fuel efficiency; the automaker said it wasn't comfortable with the risk.

The other challenge is cost. These batteries are expensive, in part because they're tricky to produce. They must be assembled in immaculate temperature-controlled facilities—even particles of dust could contaminate the batteries and cause a malfunction. They also require expensive and sophisticated cooling systems. And since their internal structure has many different cells linked together, engineers have to devise intelligent software to monitor, calibrate, and manage the cells. "The software will say, 'Okay, if I see a cell that is getting too high a voltage or with a temperature peaking from the rest, I will take action,'" said Craig Rigby, an engineer with Johnson Controls–Saft.

Just as innovators are jockeying to develop the best solar panels, the race is on to see who will be the first to optimize lithium-ion battery chemistry and design—a competition that pits big Asian battery makers such as Mitsubishi, Toshiba, Panasonic-Sanyo, Hitachi, AESC, Lithium Energy Japan, and Blue Energy, against smaller U.S. developers, including EnerDel, Kokam, KD Advanced Energy, Johnson Controls–Saft, Altair Nanotechnologies, and A123 Systems. The two biggest buyers— Toyota and GM—are being extremely secretive about their selection processes. General Motors staged a competition between battery makers from around the world to see which could best meet its requirements; it is testing the products of the top contenders on machines that run the batteries to the hilt day and night, charging, discharging, and exposing them to a constant stream of stresses.

Investors are dazzled. By 2015, "the worldwide market for hybrid-vehicle batteries will more than triple to $2.3 billion," the *Economist* reported recently, summarizing the prediction of an industry expert. "Lithium-ion batteries . . . could make up as much as half of that."

Still other innovators in the field of nanotechnology are vying to develop products that could leapfrog current approaches to lithium-ion. They believe that nanotechnology can be used to dramatically improve energy conductivity inside a battery, perhaps giving batteries many times more powerful and enduring charges, and better durability. Altair Nanotechnologies claims that its batteries have three times the power of conventional batteries, last significantly longer, and charge in mere minutes.

This flurry of research and development proves not just Edison's theory that necessity is the mother of invention but also that invention in turn generates a positive feedback loop: one successful technology breakthrough creates the need for another, and another. It's reminiscent of the early days of the electricity industry, when Edison's invention of the lightbulb gave birth to a nationwide network of cables and power plants and a vast array of electrified devices.

INFORMATION SUPERHIGHWAY

Once the problem of electricity storage is solved, another hurdle looms large, colossally so: the grid. The challenge involved in moving an entirely fuel-dependent industry to electricity is vastly more complicated than purchasing a slew of extension cords. "The smart grid is to the success of hybrids what the elevator was to the success of the skyscraper," said Pacific Gas and Electric's Peter Darbee. "In its current form our electric grid can't support a massive infusion of plug-ins."

Think back to that ancient snarl of cables and wires I encountered beneath the streets of New York and it's not hard to understand why the grid would buckle if a large number of commuters arrived home from work and plugged in their cars at roughly the same time for a recharge. Yet if an intelligent computer system embedded within the grid automatically staggered the charging patterns of those vehicles throughout the night, when demand is low, "it would benefit the consumer, who would get cheap power, and the producer, who would be able to sell otherwise wasted electricity," Google's Dan Reicher told me.

Google is a leader in the effort to create a more reliable and dynamic power grid. Remember that its philanthropic arm, Google.org, is putting millions into RE<C, a program to create energy from renewable sources more cheaply than from coal inside a decade. Its smart grid investments represent yet another front in the clean technology battle. Hoping for a better understanding of Google's role in America's automotive future, I visited Reicher at the company's headquarters in Mountain View, California.

The Googleplex campus is a twenty-first-century Oz: sprawling lawns and organic vegetable gardens, a sizable fleet of bicycles and electric scooters available to employees for short trips, superefficient office buildings flooded with natural light and built from recycled and local materials, rooftops covered with more than 9,200 solar panels—one of the largest corporate solar installations in the United States. Every imaginable amenity is provided on campus for Google employees, including foosball and pool tables, laundry, dry-cleaning and car-washing services, on-site doctors, and subsidized massages. And Google's dozens of cafés serve up free, homemade food—much of it organic and locally grown. Widely reputed to be the best corporate fare in America, it is rumored to cost the company no less than $75 million a year.

It's all part of a strategy to keep Googlers working longer hours and focused on the job at hand rather than shuttling off campus for lunch dates and errands. "We've designed our campus around as little use of cars as possible," explained my Google tour guide, a blonde, chipper woman in her thirties who emanated a glow of good health. Creating an entirely self-sufficient campus saves fuel. The more amenities that are immediately available, the less driving employees have to do, which cuts down on the company's overall carbon footprint. Google pays to offset not just the massive amounts of electricity consumed by the data centers that sustain its search engines, but also the daily commutes of its staffers. Many Googlers opt not to drive to work, instead using a fleet of Google-owned commuter shuttles that run throughout the San Francisco Bay area and are powered by homegrown and locally processed biofuels.

Over a breakfast of portobello frittatas and salsa made with ingredients fresh from the Google gardens, Reicher described a future in which cars will increasingly function like computers—intelligent, software-rich machines made of lightweight materials and powered by silent, pollution-free electric motors. In this future there won't be gas pumps, but instead "smart garages" at your home and office outfitted with special power cords that plug into your car and quickly charge it. You will also be able to "upload" to the grid the electricity that you don't plan to use on a given day, and get paid for it.

This elaborate smart grid concept is, as Reicher conceded, a huge, costly, and complicated leap from today's unintelligent grid. He used a communications metaphor to spell out the challenges: "The grid was designed for one-way conversation between power plants and consumers. Now it needs to be redesigned for a conference call going on twenty-four hours a day, with ultimately millions of people at the conference."

Google hopes to play the role of an intermediary at that conference. Reicher noted that Google's founders, Larry Page and Sergey Brin, believe we're entering "a golden era of innovation in American cars" after nearly a century of technological paralysis, decades of lagging behind Japan and Europe, and years of sluggish sales. Reicher—who's been working for environmental causes since the age of seven, when he convinced a clothing company to discontinue the use of pelts from endangered wolverines to line its jackets—agrees. He said, "We've asked ourselves: what would it really take to dramatically reduce the contribution of the automobile sector to climate change and foreign oil dependence? And for us it became pretty clear that moving vehicles to electricity is really the smartest thing to do."

To illustrate why, Reicher walked me across campus to one of Google's "solar carports," an open-air garage with a sweeping canopy of photovoltaic panels. In the center of the carport were a dozen converted plug-in Priuses, with decals of electrical plugs and the phrase "Recharge IT!" in the rainbow colors of Google's logo. With battery packs nearly double the size of normal Prius batteries, these cars got an average fuel economy of almost 90 miles per gallon. Reicher yanked on a 110-volt extension cord dangling on a pulley from the rafters above and connected it to one of the cars. "Have you ever ridden in a solar-powered car?" he asked me, gesturing skyward at the solar panels. "These cars get charged by free, distributed, zero-carbon energy." And since the cars are mostly used for local trips (they're available to employees who commute to work by shuttle and need a vehicle to get to meetings), they often travel with a carbon-free footprint.

The evolution of the car and the grid are on parallel paths, said Reicher. Neither the shift toward electric cars nor the one toward clean power will be immediate. Auto technology will progress through a series of steps, moving from hybrid electric-gas engines toward plug-in versions of this technology, with the electric battery component growing ever larger as the gasoline component shrinks. Eventually, as we get the electric grid infrastructure in place and consumers adjust to the idea of driving fully electric cars, the gasoline engine will be eliminated. Meanwhile, the grid will travel a complementary path from dirty fossil fuels to clean, renewable energy sources.

V2G

In the long run, it's neither the cars nor the power sources that interest Google from a product development standpoint; Reicher does not foresee any "Gcar" or "Gsolar" products on the company's horizon. Instead Google wants to play a behind-the-scenes role in these technologies, pursuing them from an investment and research standpoint and working with corporate partners to fund their innovations.

Google hopes to reap big returns on its investments when these renewable energy and automotive technologies go mainstream, but better yet, it sees these products as the gateway to an even bigger and more exciting technological frontier—the intelligent grid system that will link millions of cars, smart garages, and homes with the new clean power sources that will sustain them.

Con Edison's Lou Rana, the Department of Energy's Jeff Christian, and T. Boone Pickens had all emphasized the need for an improved electric grid. Google is working to develop the software components of this grid. "There's a big opportunity to increasingly inject digital technology

into vehicles, but I think the bigger opportunity is going to be inject-
ing this digital technology into the grid," Reicher told me. "We want to
be developing the information products that will successfully integrate
millions of smart vehicles and homes into a vast, smart electricity web."
Google is using the Recharge IT! fleet to test just such concepts.

Reicher went on to describe a "rolling power plant" system I had
previously heard outlined by Peter Darbee of Pacific Gas and Electric—a
so-called V2G or vehicle-to-grid system in which cars not only withdraw
electricity from the grid but also deposit it back into the grid during peak
hours when demand is highest. In a V2G system, drivers could recharge
their batteries in their garages at night, then sell the surplus electricity
unused by their commute back to the grid during the day (in the hours
of highest demand) when plugged into sockets in their office garages.
A car's internal software would be in direct communication with the
power company, determining when to sell and buy power, and ensuring that the batteries maintain enough charge for the evening commute
home.

The car owner would make money from this arrangement. The utilities would also benefit, especially during peak load times, when businesses are humming and air conditioner demands are highest. Firing
up fossil-fuel-powered "peaker plants" is expensive and cumbersome for
utilities, so it would serve them well to be able to draw from the collective power of electric cars functioning like millions of tiny power plants.
"Smart" car batteries could help to resolve some of the energy storage problems of intermittent renewable sources like wind and solar—
downloading this clean power as it's generated and selling it back to the
grid on demand.

Cars spend on average more than 90 percent of their time parked,
explained Reicher, so this would give them a function when they are
otherwise useless. The collective potential is sizable: Tim Vail, a former
engineer at General Motors, has calculated that "the power-generation
capacity trapped under the hoods of the new cars sold in America each
year is greater than that of all the country's nuclear, coal, and gas power
plants combined."

It's a thrilling concept. Reicher is talking about meshing together the power utilities and the automobile sector, two of the biggest—and historically the dirtiest—industries in the nation. The result would be a seamless link that effectively eliminates the need for oil (and ultimately coal) in transportation altogether. Google sees its potential role, appropriately, as a networker—helping to link together cars, homes, and the clean power sources of the future.

To make such a system work, the devil is in the details. Let's say you go to your friend's house for dinner and want to charge up your car in her garage—how do you put that charge on your own account? If you want to charge up your car when electricity is cheapest, how will you know what the real-time pricing is? What if you want to buy green energy for your car, not dirty coal-generated power—how do you tell that to the utility? And if you want to sell your energy back to the grid, how do you broker that deal?

The answers to these questions are still in the works, but they are precisely what Google is exploring. No doubt, the company could play a natural and pivotal role in this smart future: the grid must evolve into a multidimensional, complex, and dynamic web of information, much like the Internet. "The car will no longer be an autonomous machine that gets filled up with gasoline once a week," Reicher told me. "It'll be a component of a larger intelligent network—a traveling battery, essentially, that uploads and downloads electricity to and from a grid that's constantly growing smarter, greener, more decentralized. When I imagine this kind of power network, the whole world of software and the digital age gets incredibly exciting—and immeasurably hopeful."

The vision is there, but the practical plan of action is not. Where to begin such a vast undertaking? There is a crucial role for the federal government to play—creating mandates and incentives, helping to orchestrate and propel the many different players along an ambitious timeline. But there's also a lot of adjusting, engaging, and information sharing that the automobile and power industries will have to do among themselves. "Sociologically it is a very interesting moment where these two corporate behemoths are having to talk to each other. The auto manufacturers grew up with the oil industry—they're old friends. But the electricity industry is a whole new world to them," explained Reicher. So now Detroit's leaders are wondering: if we're really going to build millions of electric cars, are utilities going to have the juice available for our customers to run them? Meanwhile the utilities are wondering about the timetable of Detroit's shift to plug-in vehicles, and about the practical challenges of integrating millions of Detroit's products into their systems.

"Three years ago," said Reicher, "the utility industry was not talking to the automobile industry. Today there are regular meetings, regular conference calls. The alliance is well under way." If only Thomas Edison—builder of the first electric car and the first power grid—could sit in on some of these meetings and see his inventions evolve and fuse together a hundred years later. Here again, the ingenuity that drove us into America's energy problems over the last century is precisely what could propel us beyond them in the next.

MAIDEN CHINA

The most impressive display at Detroit's 2009 International Auto Show was not presented by America's biggest automakers, or even Japan's. The star of the show was a little-known car company from China, BYD Auto. Founded as an offshoot of the world's biggest mobile phone battery maker, BYD got significant backing from billionaire investor Warren Buffett. In Detroit's Cobo Center, BYD debuted the world's first mass-produced plug-in car, which had appeared in Chinese dealerships in December 2008, priced at around $22,000. "We are confident of exporting our electric cars to the U.S. market in 2011," said Li Zhuhang, general manager of BYD's auto export trade division.

The pressure on Detroit to develop electric-car technology comes not just from our need to cut America's carbon emissions, ease our dependence on foreign oil, grow our grid, and compete with Europe and Japan—it comes from the meteoric growth of the developing world, primarily China. The emerging Chinese middle class wants cars, lots of them—just the way America's rising middle class in the twentieth century celebrated its newfound wealth with a great rush of investment in tail-finned, chrome-plated, high-octane status symbols. Nowhere is the potential economic opportunity—and environmental peril—of automobile technology greater than in this nation of 1.3 billion inhabitants. We'd be foolish to think about our automotive future without thinking about theirs.

As both a producer and a consumer of cars, China will have a greater impact on global auto markets than any other nation. China is second only to America in terms of its purchasing power, and will handily surpass us on this front within a decade. In 2000, China had 16 million cars; in the following seven years that number more than tripled. Car-purchasing trends will continue to accelerate for the foreseeable future: government studies predict that the number of cars on China's roads will soar to 120 million by 2020.

Meanwhile, China is in the midst of the greatest road-building boom since Eisenhower embarked on his interstate highway project in the 1950s. "The United States built 41,000 miles of new highway from 1957 to 1969," *USA Today* recently reported. "China plans 30,262 miles this decade." Paved highways will stretch between all of its provinces, from the Tibetan Himalayas to the Gobi Desert. This is a nation, in short, that is just now embarking on its own power trip.

China represents not only a huge new potential market for American automakers, but also a huge new burden from the standpoint of global warming emissions. Already, China is producing greenhouse gas emissions per unit of GDP at six times the rate of the United States. The nation also has twenty of the thirty most polluted cities in the world. The consequences are economic as well as environmental: pollution-related illnesses are restricting the country's GDP growth by as much as 15 percent.

Meanwhile, China is rapidly depleting its domestic petroleum reserves, having already become the world's second-largest oil consumer. As recently as 1992, China was exporting its oil; today the country is importing nearly half of its petroleum from suppliers in Russia, the Middle

East, and Africa (and in the process sending a new slew of petrodollars to regimes often hostile to the United States). A recent McKinsey & Company report found that at China's current rate of growth, its oil imports will roughly double by 2030.

In response to these worrisome trends, the Chinese government has begun taking some big steps toward a clean energy future. Beijing recently imposed tough fuel-efficiency requirements—so tough, in fact, that many of the SUV and light-truck models produced in Detroit will soon fail to meet the standards on the roads of China. The country is also experimenting with programs that offer subsidies for electric and hybrid cars.

This presents a wake-up call to the American auto industry for two reasons. Detroit will be missing out on a massive market if it doesn't aggressively begin to produce ultraefficient cars in the coming years. And China's automakers are beginning to pose even stiffer competition than Japan's. In the outskirts of Shanghai, the country is quickly building an auto-manufacturing hotbed known as Jaiding to rival Detroit. "We've built this in three years," a Jaiding official told the *Economist*. "It took Detroit a hundred years."

China is a country, after all, that excels in producing high-tech gadgetry—it's where most of the world's major electronics companies manufacture their cell phones, laptops, flat-screen TVs, and other IT products. Moreover, the nation's state-backed education system is producing science-minded and tech-savvy students by the millions, creating a hyperintelligent workforce to shape the world's automotive future. In February 2009, just months after BYD began selling its 62-miles-per-charge plug-in hybrid, China's top domestic carmaker, Chery Automobile, released its own model—the S18—which it claimed could get 93 miles on a single charge and be fully recharged in four to six hours.

China has several huge advantages in moving over to sustainable transportation. First, it has the heft of a centralized (and nonelected) government that can impose strict gas taxes and product standards with minimal pushback from lobbyists or consumers, and steamroll new technologies into the mainstream. Second, China's emerging auto manufacturers can appeal to the built-in loyalty of their own exploding domestic consumer base. Third, the country doesn't have a major existing investment in infrastructure for gasoline-powered cars. America's infrastructure, by comparison, is fully built around combustion engines and gasoline stations, and will have to be scrapped and replaced at great cost.

China also has a relatively clean slate when it comes to its grid infrastructure. Leaving aside the fact that the country is building coal plants at a harrowing rate, it also has ambitious plans to construct a new smart grid. The superconducting cables and intelligent software of this system will be able to accommodate a new era of clean power sources, plug-in cars, and smart homes.

China has a psychological advantage as well: its consumers don't struggle with the kind of nostalgic attachment to gas-guzzling cars that has for so long provoked irrational consumer behavior in the United States; nor do Chinese consumers flinch at high gas prices, for they never got used to cheap gas in the first place. With a fresh perspective on the world of automobiles, they can move into an era of efficiency without feeling that they're letting something go.

PILOT PLANT

One of the world's smallest nations could also hold an essential key to the future of automobiles. The government of Israel recently sanctioned a joint venture between the Better Place corporation and the automakers Renault and Nissan to build a nationwide electric-car network. Better Place is installing and managing the grid, and plans to offer an innovative "battery swapping program" that will enable a driver to stop at one of 1,000 stations nationwide when the car's battery is running low and exchange the spent unit for a freshly charged one. (The station will simply pop out the dead battery and charge it up for a future customer.) The carmakers, meanwhile, will lease basic, standard car models to consumers at prices below the cost of conventional cars, thanks to substantial government subsidies.

Better Place founder and CEO Shai Agassi had the groundbreaking concept of basing his business model on that of the mobile phone industry: Think of the electric cars as cell phones manufactured by Nokia, and think of the car batteries and the infrastructure of wires, charging ports, and battery-changing stations that will service them as the AT&T network. Better Place customers can choose to subscribe to either a monthly plan based on estimated miles they intend to travel in driving, or a pay-as-you-go plan.

Smart digital chips in the vehicles and in the specially designed charging ports located in garages and on city streets will read the car owner's identity and register the use of electricity every time the car is charged (just as the cell phone network reads your phone's identity and charges minutes to your account accordingly). Customers can also sell their unused electricity back to the Better Place network.

The smart software that the company is developing to enable these power exchanges with the grid is not unlike the software now being tested by Google and its corporate partners. The competition doesn't bother Agassi, forty, who previously worked for software giant SAP—in fact he welcomes it. Though he fully intends to earn a profit for his private investors—who have committed $200 million to his plan—Agassi has insisted on building "an open, standards-based network. So that we can't lock any company out, and competitors can't lock us out when they show up." He believes that this kind of open competition offers the best hope for the rapid adoption of his model worldwide: "The mission is to end oil," he recently told *Wired*, "not create a company."

Better Place projects that one hundred thousand electric cars will be on the road in Israel by the end of 2011. President Shimon Peres has stated to his constituents that he wants his country to serve as "a daring world laboratory and a pilot plant" to demonstrate the benefits of electric cars. The global business community is taking notice: Deutsche Bank analysts have forecast that the Better Place model could cause "massive disruption" to the auto industry and has "the potential to eliminate the gasoline engine altogether."

The success of this project would serve a clear political purpose for Israel, a nation that is in direct conflict with many of its oil-rich neighbors in the Middle East. Better Place has pledged to purchase only homegrown, renewable energy sources to power its network. When voicing his support for a major investment in solar power, Peres said "the Saudis don't control the sun."

Agassi's plan to break Israel's oil addiction is catching on: the governments of Denmark and Australia have recently signed agreements with Better Place, and the company is discussing a partnership with the Chinese automaker Chery. In the United States, the mayors of San Francisco, Oakland, and San Jose, California, recently appeared with Agassi to announce plans to offer incentives for electric car infrastructure in their cities; and the state of Hawaii announced a partnership with Better Place to build a network and subsidize the mass-market availability of electric cars for its residents by 2012.

So far, Agassi has tried without success to pitch his idea to members of the U.S. Congress. Certainly his concept would strike many Americans as counterintuitive—forcing us to rethink our relationship to cars in a fundamental way. In this kind of subscriber-based automotive system, cars serve as functional, efficient hardware to be leased, not owned— not the identity-defining accessories we grew to love over a century of joyriding.

BIG WHEELS KEEP ON TURNING

Cars, above all else, inhabit a particularly sacred place in the hearts of American consumers. In the twentieth century, cars came to signify freedom and mobility for all in a way that no other invention has. Few of us will have much emotional difficulty in letting go of the sprawling old coal plants that clutter the Midwest, or seeing our ancient electricity cables replaced with new superconducting wires. But many of us will have a hard time laying our old gas guzzlers to rest.

My father had described his first car, a secondhand Alpha Romeo that he bought for $800, as "my first real taste of freedom . . . my passage into adulthood," and I felt the same about my first car, an 18-mile-per-gallon rumbling SUV that he handed down to me. I have particularly fond memories of the behemoth 1979 Oldsmobile Cutlass Cruiser station wagon my grandfather drove when I was a kid, with its two-tone navy and faux wood color scheme, its plush bench seating in back that felt to me like a moonbounce, the throaty cough of its engine when he fired it up, and the strangely comforting smell of gasoline and motor oil it carried. It felt wonderfully heavy as it rolled around town, rocking and bouncing on its shock absorbers like a ship on the open seas.

I voiced these laments to a Tesla Motors executive I met at the Washington, D.C., plug-in car conference when he took me for a test ride in his ultraglamorous all-electric sports car. I'm all for eco-friendly cars, I told him, but it's a sad day in the lives of car aficionados when you can take a vehicle from 0 to 60 in four seconds flat but barely hear a whisper. "What would a NASCAR race be without the sound of roaring engines?" I challenged. Saying good-bye to the combustion engine, I tried to explain, is almost like seeing lions or some other big-game species go extinct. America will mourn the loss of a mighty creature when it ushers out the era of gas-powered cars and welcomes in the smart, clean, whisper-quiet electric age.

Not to worry, he told me. Tesla engineers can address this problem. The car could be outfitted with a high-end speaker system and its engine could be programmed with a broad range of different sounds that would broadcast as the car accelerates, simulating the rumbling, roaring, groaning engines of yore. "The driver could select his favorite sound," he told me, "just like we choose ringtones on our cell phones."

INTERSTATE 2.0

The automobile was a revolution in transportation that brought with it much of what we know as the American dream today: suburbia and its goal of home ownership for all, the freedom of independent movement from anywhere to anywhere at any time of day or night. Furthermore, through the model of Henry Ford's assembly line, the automobile brought the mass production that elevated America to its status as the world's leading industrial power, paving the way for the cheap, plentiful consumer goods that we all now buy and use. The freedom made possible by cars helped give rise to the sense of pride, rebellion, and optimism that has come to represent the American spirit—a nation of "fast, proud, fearless go-getters with rebel hearts," as one NASCAR fan had described it to me.

But today, the combustion engine automobiles that did so much to build and establish this American dream could threaten its survival.

Strong government action will be needed to secure America's post-petroleum future, both to support the development of a smart grid infrastructure and to help bring new car technology into middle-class driveways. "The extra battery and software costs of a plug-in add a minimum of $8,000 per car to the manufacturing cost of a standard hybrid," according to clean energy analyst Joseph Romm, and that needs to be defrayed for consumers and producers alike before these innovations can be widely adopted. Even if the Chevrolet Volt proves commercially successful, it will still reach only a narrow slice of the market because of its $40,000 price tag. The industry needs federal leadership. As Romm put it: "No country has ever delivered a mass-market alternative fuel vehicle without government mandates."

Lawmakers have introduced measures to offer an $8,000 federal consumer tax credit for purchasers of the first million plug-in vehicles, and a $4,000 rebate for purchasers of the second million, to help spur on economies of scale. Other policy proposals include increasing fuel economy standards to roughly 35 miles per gallon by 2016 (a big leap for America, but still significantly behind standards set by Europe and Japan). These cars will provide huge cost savings in the long run to American consumers. Given today's average electricity prices, battery-powered hybrid vehicles offer consumers the equivalent of 75¢ per gallon when cruising on electricity. If electricity prices rise, this calculation would change, but even a doubling of current electricity prices would yield the equivalent of $1.50-a-gallon gasoline—dirt cheap by global standards.

Still, long-term policy measures to encourage this industry are struggling to get passed into law. The 2009 Recovery and Reinvestment Act offered a substantial plug-in tax rebate, but it applies only to the first 250,000 sold—a small number compared to the 300 million combustion-engine vehicles currently on American roads. While Detroit welcomes tax rebates for its cars, it is still resisting tough fuel economy targets. The petroleum industry, meanwhile, is throwing its lobbying weight against policies that would accelerate the shift away from oil. Political will to support clean energy innovation has never been more urgently needed.

Above all, if an all-electric automotive future is going to come of age in America, we need to build not a system of highways but a sophisticated and dynamic network of wires. We need a smart grid that crisscrosses the country, carrying energy created by renewable sources to propel a new fleet of clean vehicles.

We can borrow a lesson from our automotive past as we shape Detroit's twenty-first-century future. In 1954, when President Eisenhower pushed hard for the passage of his 40,000-plus-mile interstate highway plan, he argued down critics who claimed it was unnecessary, stressing national security above all else. He spoke of the road system's "appalling inadequacies . . . to meet the demands of catastrophe or defense." The same criticisms, in recent years, have been made about our electrical grid.

Like Eisenhower, America's leaders will need to tackle vast logistical challenges and commit significant resources over many decades to see a national smart grid through to completion. They may, in fact, be able to draw from Eisenhower's model in practical ways as well as symbolic: In recent years, engineers have been puzzling over where to locate new superconducting wires and smart transportation networks—a task that requires complex and lengthy permitting processes. Some grid and transportation experts have pointed out that the interstate highway system offers an already-existing nationwide "right of way" in which new high-speed railways, new power lines, and smart charging ports for electric vehicles could be installed.

Futurists have also envisioned "smart highways" embedded with software that can communicate with cars—automatically guiding them along roads, selecting the most efficient routes, and avoiding congested areas, thereby saving significant amounts of fuel. It is a hopeful paradox that the massive road network Eisenhower designed a half-century ago—which has been so widely criticized for its environmental burdens—could serve as a backbone for the clean-energy infrastructure of tomorrow.

For a hundred years, we've been designing buildings that work against nature. Now we're designing buildings that work with it. We're transforming them from structures that consume energy into structures that produce it, from air polluters into air purifiers, from resource hogs into resource savers.

—Richard Cook, green architect

10

City, Slicker

BUILDING ENERGY-SMART HOMES
AND THE CITIES OF TOMORROW

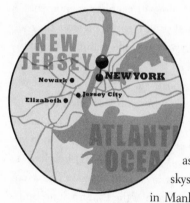

"We shape our buildings; thereafter they shape us." I had heard that Winston Churchill quote many times, but it never struck a chord with me until one hot August afternoon when I made a white-knuckled ascent to the top of America's greenest skyscraper. The second-tallest building in Manhattan, it serves as headquarters for Bank of America—an angular, all-glass high-rise jutting up like a colossal piece of quartz from midtown's sooty streets. I was touring the structure in its final phase of construction with one of its architects, Rick Cook, an award-winning Manhattan-based designer who heads up Cook+Fox Architects with his partner Bob Fox. The team has worked on projects ranging from a holistic wellness clinic to the rebuilding of the World Trade Center. The central elevators weren't operating yet, so we could only ascend to the roof via the open-air construction elevator—a chain-link cage shakily rigged to a lift that climbed the exterior of the tower, halting and lurching up story after story as we dropped workers off along the way.

At the Department of Energy's living lab in Lenoir City, I had seen the kinds of affordable green homes that might one day shelter millions of Americans—tapping into renewable energy sources and warming, cooling, and illuminating our lives with utmost efficiency. I wanted to look even further ahead, so I had come to the Bank of America Tower to see a demonstration of the most sophisticated green design concepts and technologies in development. Like all-electric cars, these technologies may take some time to reach the mass market, but many will shape the cities and communities of tomorrow.

While Cook had the austere appearance that's typical of architects—lean, meticulously groomed, dressed head to toe in black—his demeanor was anything but somber. He palpably loves his work, and discussed the possibilities emerging in the world of architecture—from solar panels to waterless urinals—with almost breathless excitement. "A skyscraper can be optimistic," he mused as we stepped out onto the roof. "It can rise above the texture of the city and generate optimism about the future."

Nine hundred and forty-five feet above Times Square, the roof was fiercely windy. It still had no walls or railings and was piled high with heaps of construction equipment. I followed Cook through a haphazard assortment of I-beams, steel cabling, and huge pieces of machinery set against the distant backdrop of Manhattan, its streets and buildings so far down below they looked like pieces on a board game. The scene reminded me of the famous 1931 photograph of the Empire State Building under construction in which workers tinker high up on the building's exposed steel ribs, casually dangling their boots hundreds of feet above firm ground. It's an image that captures the last revolution in architecture, nearly a century ago, when the invention of steel beams and electric elevators first enabled architects to build skyward.

Now Cook's construction workers were perched above the Manhattan skyline installing a new generation of technologies: an on-site clean-burning power plant that will supply nearly 70 percent of the building's energy needs annually; an air filtration system that removes 95 percent of indoor pollutants; and huge cisterns that will capture rain and funnel it into a series of filters and gravity-fed pipes supplying the building's toilets and sinks.

Beginning at the roof and descending fifty-five stories into the cellar, Cook showed me dozens of green features within his tower. Just as the technologies behind electric cars, wind turbines, and solar panels are at once archaic and futuristic, the novel design features of Cook's building were both simple and state-of-the-art—or, as Cook put it, "as much a result of breakthrough ingenuity as plain old common sense."

Common-sense innovation is sorely needed, I had learned, when it comes to America's buildings. Motionless, soundless, and unassuming, buildings are collectively one of the single largest contributors to global warming in the United States. We harvest and manufacture vast amounts of materials to construct and renovate them. In the process of warming, cooling, ventilating, lighting, powering, and plumbing them, we devour energy and create pollution.

Globally, buildings account for nearly 40 percent of all energy use and contribute nearly 40 percent of the world's annual greenhouse gas emissions—a greater net impact on the environment than all transportation systems combined. In the United States, buildings consume roughly three quarters of the electricity we produce. One recent government report found that yearly carbon emissions from U.S. buildings alone are "greater than total carbon dioxide emissions of any country in the world, except China." America's buildings also consume 30 percent of our nation's domestic raw materials and produce 30 percent of the waste in our landfills. They are, generally speaking, as inefficient and outdated as our oversized combustion engine cars and our polluting coal plants. Just as most of the cars on the road today have the same basic technology as the Ford Model T, the design of America's buildings and the cities that house them has scarcely improved over the last century.

Going forward, we face an enormous amount of new construction in the next two decades: "In 2030, about half of the buildings in which Americans live, work, and shop will have been built after 2000," according to a Brookings Institution report. Which means our built landscape will change dramatically in the next few decades. Will it change for the better? An architectural movement is under way that aims to transform buildings from burdens into environmental benefits—"from structures that consume energy into structures that produce it," as Cook told me, "from air polluters into air purifiers, from resource hogs into resource savers."

Pioneers in green architecture, including Cook, are attempting to make high-performance buildings that function less like inanimate objects and more like intelligent organisms, capable of harnessing natural elements such as sunlight and rainwater, automatically shutting down appliances when they're not in use, and dimming electrical lights when the sun provides enough natural illumination. These smart homes and offices will be able to communicate with the grid using digital technologies like those being developed by Google. They will monitor the price of electricity as demand fluctuates throughout the day, and automatically run the dishwasher and laundry machines when power is cheapest. Eventually, many green architects claim, intelligent buildings could be networked into communities that are as beneficial to the environment as forests and as dynamic as ecosystems.

Alongside this architectural trend, local officials are drafting plans for cleaner, greener American cities. They are purchasing clean cars and trucks for municipal fleets, designing futuristic public transit systems, expanding carpool lanes and bike paths, converting traffic lights and streetlamps to ultraefficient bulbs, implementing strict green building codes, and ramping up recycling.

These visions for green living in America have a long way to go before they're fully realized, but talk to your city council member or a local architect or real estate agent and they'll likely tell you a green building boom is quickly gaining ground in America. Smart, efficient technologies have been applied everywhere from the White House to your local Walmart superstore. Shea Homes, the nation's biggest supplier of prefabricated houses, is now offering popular green models up to 50 percent more efficient than conventional homes, with solar equipment as a standard feature. Dozens of solar-powered planned communities—modern-day Levittowns—have sprung up in California and throughout the Southwest. Habitat for Humanity is building on the success of the Lenoir City project with a plan to construct five thousand new low-income green homes over the next five years in partnership with Home Depot.

At the other end of the spectrum, high-end developers are designing trophy buildings with billion-dollar budgets and green design features befitting the science fiction fantasies of Philip K. Dick. The new Bank of America Tower designed by Cook and Fox was the world's first skyscraper to nab the highest rating, platinum, from the U.S. Green Building Council, the standard setter for sustainable design.

TOWER OF POWER

It's a precedent-setting move to install a 4.6-megawatt power plant inside a building. Edison attempted a similar feat on a much smaller scale a century ago when he built a generator in J. P. Morgan's house, but fumes and noise foiled the plan. Now small-scale, clean-burning on-site power plants are looking like increasingly practical energy solutions for commercial buildings. The first of its size and kind in Manhattan, the Bank of America's natural-gas-powered "microplant" captures and recycles its own waste heat and eliminates energy loss over long-distance transmission lines, producing power three times more efficiently than the grid.

What was more innovative than the energy that this building produced was the energy that it saved. Take, for instance, the efficient air-conditioning system. It was designed to take advantage of an often overlooked law of nature: heat rises. The average cooling system in buildings today is shockingly inefficient—blowing chilled air through a long, winding network of ducts and forcing it downward into rooms from vents in the ceiling. "The cool air loses much of its chill along its journey through the ducts," Cook explained, "then has to mix with the layer of hot air that naturally collects at the top of each room, becoming warmer still, before it floats down to where you're sitting." So even if you set your thermostat to 75 degrees, the old system actually has to cool the air down to about 60 degrees to offset the heat gains. The Cook-Fox team worked with engineers to implement an entirely new approach, dropping ducts altogether and creating a hollow 14-inch-high chamber beneath the floors where cool air could freely circulate and move upward through floor vents. "Hot air naturally rises, creating a current that naturally pulls cool air up with it. This system is responding to the natural flow of things, rather than working against them."

Cook described this seemingly simple feat of engineering as just one example of his bigger-picture philosophy: architects can look to nature to solve design problems. "For a hundred years, we've been designing buildings that work against nature. Now we're designing buildings that work with it."

The architects also installed "ice batteries" in the building's cellar: forty-four steel tanks each holding thousands of gallons of water. At night, when the building's electricity demands plummet and its on-site power plant is still humming, the water is frozen into blocks of ice by electricity that would otherwise go unused. During the day, as it slowly melts, that ice chills the air that wafts upward through the ventilation system. "Sometimes it's the simplest solutions," said Cook, "that have the most profound benefits."

The architects found a new way to conserve energy while addressing the number one complaint in every office building in America: "I'm too hot" or "I'm too cold." We've all been there—wearing turtlenecks at the height of summer because the office is overly air-conditioned, or stripped down to our T-shirts in the dead of winter because it's so overheated. "It's one of many examples of a terrible flaw in conventional building design," Cook said, "that's grossly inefficient, unhealthy, and uncomfortable." Problem solved: Cook and Fox installed individual vents in the floor of each office so that occupants can control the flows of warm and cool air into their spaces, and the entire building isn't kept at one broiling or bone-chilling temperature.

Electric lighting is the second biggest energy demand on a building, and here, too, the architects implemented both common-sense and high-tech strategies. First, they flooded the building with natural light. "That's the beauty of an all-glass building exterior," Cook told me. "You get floor-to-ceiling windows and gorgeous views, but also maximum sunlight means minimum use of electric bulbs." And since electric bulbs generate heat, using them sparingly takes a burden off the air-conditioning system. The architects used two panes of glass with a layer of air between for superior insulation. And to keep the building from getting too hot in the summer sun, they coated the glass with tiny ceramic dots that reflect the sun's warmth while letting in the light. The ultraefficient electric lighting is controlled by a computer sensor that tracks the arc of the sun throughout the day, automatically dimming the lights when the sun is bright and illuminating them as it sets.

Other sensors throughout the building monitor the level of carbon dioxide that humans naturally exhale. The ventilation system pumps fresh air into the rooms with high CO_2 levels (high CO_2 tends to make people drowsy) but automatically turns off the air going to rooms with low CO_2 levels (indicating that people aren't present). "Why pump fresh air into an empty room?" Cook said. Strange as it may seem that a building could "know" which of its rooms are occupied, this will eventually offer big advantages from an energy standpoint: as soon as occupants leave a room, a smart system like this could automatically cut the lights and power down Xerox machines, computers, televisions, and other electronic equipment into hibernate mode.

Equally clever is the smart system for elevators, an energy-scrimping feature that has been widely used in recent years and saves another coveted resource—time. Here's how it works. Say you enter a building in a hurry to get to the fourteenth floor. You head over to the bank of elevators and punch in 14 at a central switchboard. The switchboard does a rapid-fire internal computation and tells you which elevator will get you there quickest—grouping you with other visitors who need to get to 14 and neighboring floors so you won't have to make stops along the way. Likewise, people who need to get to higher and lower floors will all be grouped according the most direct and efficient routes. Eventually, this kind of centralized system will oversee all of the electricity-consuming devices throughout the building, controlling them via a centralized switchboard that can cut back unnecessary demands (coffee percolators, decorative lighting, vending machines) during peak-load times when electricity is most expensive and the grid is under stress.

Cook's team also managed to reap energy savings from less obvious sources—construction materials, for instance. Nearly half of the tower's building materials—glass, steel, concrete, and drywall—were recycled, which cut down on the energy used and pollution emitted when making these products from scratch. (For every ton of cement that's mixed, for instance, roughly a ton of CO_2 is produced; cement production alone is responsible for about 4 percent of total U.S. greenhouse gas emissions.) Moreover, much of the construction materials were acquired locally—within 500 miles of Manhattan—which limited the huge volume of transportation fuel that's typically used to haul supplies to a building site.

Cook showed me dozens of efficiency features that get as minutely detailed and technical as the specific aerodynamic design of vent openings. Many features pertained to water conservation: the Bank of America building is as miserly with water as it is with electricity. The rooftop cisterns are designed to capture virtually every raindrop that falls on the 2-acre building site. The fresh water is cleaned and then funneled by gravity (instead of electric pumps) into bathrooms that are equipped with low-flow water fixtures. The men's urinals are waterless, kept clean with nontoxic chemicals instead of H_2O. The building's wastewater from sinks is recaptured, filtered, and reused. Together these features save more than 7.7 million gallons of water annually.

For all its efficiency measures, the Bank of America building was not designed to leave visitors with the impression that it is "miserly" or "scrimping," as I've described it. The majority of these features are invisible. What I quickly learned on my tour with Cook is that conserving resources is only one aspect of sustainable design. The rest is not about cutting back—it's about enhancing the experience of workers and residents within the built environment.

ART IMITATES NATURE

"How does this make you feel?" It's a question commonly posed by psychiatrists and self-help books, not one I expected to hear from an architect. But Cook posed it halfway through our tour as we faced a wall of windows on the fifty-first floor looking out over the Hudson River. He held his arms out and tilted his head back, relishing the vista: "There's no reason why being in a building can't be as pleasing to the human senses as being in a forest, on a mountain, at the edge of the sea." Cook explained that a key tenet of green design is that buildings should connect people to nature rather than remove them from it—excite the senses rather than deaden them.

"Humans crave nature," Cook continued. "We evolved over thousands of years in natural environments rich with sensory experience, with texture and dimension." Only in the past century have we cut ourselves off from the outdoors and trapped ourselves inside buildings with narrow cubicles, stark walls, low ceilings, harsh light, and poor ventilation—an aesthetic that spread through our suburbs and cities via the prefab, assembly-line construction process popularized by William Levitt in the 1950s. Most contemporary American office and apartment buildings, Cook said, are unnatural, two-dimensional environments that can be, quite literally, depressing. He cited studies showing that modern city dwellers suffer from a "nature deficit disorder" that has contributed to the rise in clinical depression over recent decades. Green buildings with ample sunlight, views of the outdoors, fresh air, open space, and natural materials, on the other hand, energize their inhabitants.

Cook faults conventional buildings for promoting not just depression and poor health but also wasteful and negligent behavior. He cited a comment from David Orr, head of the environmental studies program at Oberlin College:

Most buildings reflect no understanding of ecology or ecological processes. Most tell their users that knowing where they are is unimportant. Most tell their users that energy is cheap and abundant and can be squandered. Most are provisioned with materials and water and dispose of their wastes in ways that tell their occupants that we are not a part of a larger web of life.

In other words, architects like Cook believe that most of the nation's buildings cut people off from the consequences of their actions. These buildings don't communicate when the electricity we're using is in high demand or oversupply, when it is costly or cheap. They don't tell us whether the reservoir providing the water to our faucets is drying up or full. They seal us off from the dumps that receive our waste and from the hidden consequences of the materials mined for their construction. Cook holds that buildings can subtly remind their inhabitants—with features like electrical meters that show time-of-day pricing, water-conserving fixtures, and recycling centers—that the resources we use are scarce and precious.

Cook is part of a growing subset of architects who espouse the theory of "biomimicry"—the notion that nature holds the answer to all design challenges. In addition to working with airflow instead of against it, his firm designed the exterior of the building to resemble a quartz crystal, ceiling lights to resemble cloud structures, and a lobby awning patterned after a tree canopy.

Other examples of nature-inspired design from contemporary architecture include exterior paints and tiles that self-clean, just as plant leaves do, by causing rainwater to bead up and rinse away grime. German manufacturer Ziehl-Abegg took inspiration from an owl's wing when it designed a fan blade with serrated edges for a quieter, more efficient air-conditioning system. "The goal, in essence," explained environmental reporter Marc Gunther, "is to distill the wisdom of 3.8 billion years of evolution" into practical design tools.

William McDonough, the architect and green-plastics innovator known to many as the godfather of sustainable design, says that in the coming decades buildings will increasingly function like trees—organisms that don't deplete resources but rather generate and replenish them. A tree's single "purpose" is to grow and produce seeds so that it can regenerate itself, but this process has the unintended benefits of feeding other organisms, conditioning the soil, and purifying the air.

Likewise, buildings primarily provide shelter for people, but they have their own unintended drawbacks—devouring resources and spewing pollutants. Going forward, says McDonough, buildings can transform sunlight into energy from rooftop solar panels, capture and purify their own water, and create oxygen through rooftop gardens that absorb carbon dioxide, cleaning the air of greenhouse gases. They can be designed for easy disassembly, and their construction materials can be infinitely reused and recycled. They can produce more clean power than they need to operate.

The Bank of America building doesn't achieve all of these goals, but it's an important step toward that sustainable future. The building's air-filtration system, for instance, removes 95 percent of particulate emissions from smog as well as ozone and volatile organic compounds, all of which cause asthma and other respiratory problems. "The air we exhaust from the building is considerably cleaner than the air we draw in," says Cook.

Douglas Durst, the developer who helped bankroll the building and leases it to Bank of America, put his money behind the project because of the benefits it offers for its inhabitants, and for New York City at large. "Growing up we were always taught to leave a place better than we found it," he said. "That's what we're trying to do."

GREEN DIVIDENDS

What's most promising about the Cook+Fox building from an environmental standpoint are not these do-gooder goals or holistic design theories, however admirable they may be. It's the bottom line. Bank of America's executives chose to go green chiefly for dollar-sign reasons. They stood to save millions of dollars annually through the building's efficiency measures. But what appealed to them most was something subtler, though considerably more valuable: green design, they believed, would boost the productivity of their employees.

Cook+Fox made a convincing case that a commodious building with ample sunlight and filtered fresh air would keep workers energized and focused, extend the amount of time they spend in the office, and reduce the total number of sick days taken. The architects presented data showing that students tested considerably better in classrooms with good air quality and natural daylight than they did than in windowless rooms. The team argued that greening office buildings will yield benefits that go well beyond saving on direct costs and protecting the environment—it will transform the way we feel and work.

Ever pragmatic, the bank crunched some numbers and found that it could reap at least twice the value it would save on energy bills through increased worker productivity. The total cost of the building was reportedly more than $1 billion; and of that, tens of millions went to its additional green components (those not otherwise necessary for the building to function). With savings of $4 million per year in energy bills (compared to a conventional building its size), the efficiency measures would pay back the up-front costs of the green technology in just a few years. Not bad. But if a clean, healthy workplace increased productivity by just 1 percent—five minutes a day per worker in an average workweek—it would translate into a value of $10 million annually. A 10 percent productivity increase—fifty minutes a day—could accrue an annual value of $100 million.

Bank leaders also noted that building green brings with it a certain cachet, particularly among young recruits in the banking industry. Cook told me that, in the final analysis, the bank concluded that building a high-performance structure would help attract and retain the best talent: "People would feel good working here."

Cook believes that his skyscraper will be seen as a tipping point in the green building revolution not so much because it employs breakthrough technologies as because Bank of America agreed to come onboard: "It proved that doing a green building at the very highest levels was good business." Without direct financial incentives, green design theories, however impressive, tend to stay on the drawing board.

Dozens of other titans of finance and commerce have opted to build green office complexes and retail spaces in recent years, embracing the dollars-and-cents logic and the prestige of sustainable design. Goldman Sachs, Nike, Ford, the Gap, the New York Times Company, Starbucks, Herman Miller, and of course Google are among the many to go this route, each adopting radical efficiency measures while taking its own innovative approach to green design.

The Ford Motor complex where trucks are manufactured in Dearborn, Michigan, for example, has the world's largest "living roof": the 10-acre rooftop of the sprawling assembly plant is crowned with a thick layer of sedum (a drought-resistant groundcover) that serves as a blanket of insulation, helping to keep the complex warm in the winter and cool in the summer. The green roof cuts down on energy bills while absorbing CO_2 and beautifying the building. The Herman Miller headquarters in Michigan, to take another example, has huge windows and glass skylights that enable the building to house full-grown, leafy trees, freshening the air and energizing the space. The well-heeled employees of Goldman Sachs tread on carpets made from 100 percent recycled plastic. Nike built six wind turbines at its distribution center in Belgium, providing enough clean energy to power the 2-million-square-foot facility.

Perhaps no major company is embracing the green building trend more aggressively than Walmart. The retail giant has historically been a leading driver of suburban sprawl and our car-addicted, strip-mall-centered lifestyles. Nevertheless, as previously mentioned, Walmart chairman H. Lee Scott and CEO Michael Duke have pledged to eliminate fossil fuel use entirely in their stores—eventually running them exclusively on renewable energy. The pilot green store that Walmart built in Aurora, Colorado, for instance, has rooftop skylights that flood the store with sunshine, LED lights illuminating its display cases, and a hot water supply for its bathroom sinks that's warmed by waste heat captured from the grocery section's refrigeration units.

These measures make good business sense. Walmart is among the biggest private consumers of electricity in the world, so the company has a vested interest in cutting its energy bills. And just as Bank of America sees green building as a way to recruit talented employees, Walmart sees it as a way to draw customers. "In retail you want shoppers to come in, be comfortable, and stick around for as long as possible," said Matt Kissler, director of Walmart's sustainability effort. "In green buildings you have a sense of well-being."

To the U.S. military, green buildings are less a feel-good marketing strategy than a tactical necessity. On domestic bases and at its D.C. headquarters, the Pentagon has installed some of the world's biggest renewable energy installations so that in the event of a terrorist attack or grid blackout its troops and civilian employees can continue operations. But efficient, energy-independent structures are even more valuable in the combat zones of Iraq and Afghanistan, where there's no grid to plug into, and transporting fuel becomes very costly (as any sizable network of cables or pipelines would be vulnerable to sabotage in these cases). In addition to other green initiatives I'd discussed with Al Shaffer, director of alternative energy programs at the Pentagon, Defense Department engineers have recently been working to develop ultraefficient "living pods" that self-cool. Already the military has begun using tents in Iraq that have a thick layer of spray-foam insulation, substantially reducing energy demands; eventually, the igloo-shaped pods under development will have an exterior membrane that deflects the sun's heat while capturing its energy to power interior lights and air-conditioning.

Shaffer describes these living spaces as "little microcosms of energy independence." There's no reason why America's homes and offices can't function the same way, enabling the nation to move toward energy independence one building at a time.

Several recent studies by McGraw-Hill Construction found that trends are heading in this direction. Some of the world's top construction professionals reported that they expect more than 60 percent of their commercial projects to incorporate green building practices within the next few years. The market for green building is expected to fully double by 2012. Surprisingly, this shift was strengthened—not hampered—as the recession and housing crisis accelerated in 2008. More than 70 percent of home buyers said that they were "more" or "much more" inclined to buy a green home over a conventional home in a depressed housing market. Why? Because they saw it as a more prudent, penny-wise, and better-performing investment.

HOUSE BEAUTIFUL

When and how will these innovations in green construction become available to and affordable for mainstream American consumers and communities? One Silicon Valley entrepreneur, Steve Glenn, who founded the company LivingHomes, has built a prototype of a zero-energy home in San Francisco that resembles a one-story version of the Cook+Fox tower—with an all-glass exterior, elegant modern lines, sustainable building materials, rainwater capture, and a massive ceiling fan that exhausts hot air, cutting the need for air-conditioning. Glenn is now offering a line of prefabricated houses based on this prototype, with built-in solar panels, moderately sized at 2,500 square feet—but they cost more than $500,000 apiece, not counting delivery and the land they're built upon. This is an elite product marketed to "people who buy organic food, do yoga, and shop at Design Within Reach," said Glenn, not to the majority of American consumers. There are other, more affordable "near zero" homes coming on the market with top-notch efficiency measures and solar roofs that are part of green communities in the Southwest, but most of these are in the mid-six-figure price range—which excludes many lower- and middle-income home buyers.

That's why the DoE's experiment in Lenoir City, Tennessee, is so intriguing. It's aimed at bringing ZEH technology to new homes that average around $100,000. Designed top to bottom for a combination of money and energy savings—from its modest rooftop solar panels and

special insulated walls to its geothermal heating and cooling system—
this prototype could go a long way toward shifting the energy-use patterns of American home owners.

But what about the nation's older buildings? I wondered about the
more immediate, intermediate steps that can be taken as a bridge between this new green construction and the inefficient homes and offices
that many of us still live and work in today. My own home, for instance,
was built in the 1920s, when oil and power were cheap and efficiency
was a distant concern. I love its creaky floorboards, rippled window-
panes, and time-softened wood, but it's also a drafty place with shoddy
insulation, outdated appliances, and a boiler that racks up huge energy
bills month after month. My husband and I have taken some measures
to curb our costs: we replaced our burnt-out incandescent bulbs with
curlicue compact fluorescents, and we wrapped our old boiler in an insulated sheath. Those actions alone cut our energy bills by more than 10
percent, covering the cost of the bulbs and the insulation within a matter
of months. Impressed with the savings, we began to investigate what it
would cost to do a soup-to-nuts green makeover of our house—sealing
the wall cracks, stopping air leaks, replacing every bulb with efficient
lighting, and installing double-paned glass, a new boiler, better insulation in the attic, superefficient kitchen appliances, and an Energy Star
entertainment system and home office equipment—maybe even solar
panels on the roof.

The good news was that most of the products we would need for
such a makeover were available off the shelf at our local Middle America
strip mall—appliances, insulation, and glass at Home Depot, lightbulbs
at Walmart, efficient TVs, stereos, and computers at Best Buy. The bad
news was cost—we'd have to pull together up to $25,000 to cover all
the new materials and devices. In the long run, the green retrofits would
more than pay off in energy savings, cutting our monthly energy bills by
up to three-quarters. But in the short run, we didn't have the funds to
cover even a few of the upgrades.

For one thing, I was still paying $500-a-month in financing charges on
my Toyota Prius (three years after the initial purchase). I realized that if

I'd put my total $24,000 Prius investment into efficiencies for my home, I would almost certainly have gotten more environmental bang for my buck. Retrofitting my home, however, was a laborious, complex, piecemeal process that couldn't be financed, at least not by my Nashville bank.

I wondered why not. This would seem to be one simple and obvious way in which banks and nonprofit organizations alike could support and encourage energy-efficiency improvements that would generate dollar value and eco-value over time.

Retrofitting (or weatherizing) old buildings for improved energy efficiency is in fact rapidly gaining support at the local and federal levels as a way of helping save taxpayers money, reducing greenhouse gas emissions, and generating jobs. The 2009 Recovery and Reinvestment Act included funding to go toward retrofitting 1 million American homes, along with significant funding to help federal and local governments make their existing buildings more energy efficient.

Just as Bank of America realized it could save money both directly (in utility bills) and indirectly (through increased cachet and worker productivity), many municipalities and smaller businesses throughout the country are realizing they can do the same on an immediately affordable scale—not by building from scratch but by improving their existing structures.

A recent U.S. Green Building Council study reported that building owners can save an average of 90¢ per square foot per year by retrofitting, recovering the up-front costs of the process in around two years. The software company Adobe, for example, invested $1.4 million to retrofit its headquarters in San Jose, California, from 2001 to 2006—and estimated that it had recouped those up-front costs in less than a year.

Most clean energy construction today is happening in new buildings such as the Bank of America tower—but they represent only 2 percent of the total number of commercial buildings in the United States. Retrofitting the nation's 4.9 million older commercial structures would offer the most direct, immediate prospect for energy savings and greenhouse gas emissions reduction. The Green Building Council's Leadership in Energy and Environmental Design (LEED) program created a standard to encourage the retrofitting of existing structures in 2004, called LEED-EB (for "existing buildings"), that includes a detailed manual of guidelines to help building owners and property managers save big on energy and money.

Chicago's Merchandise Mart is the best-known of American buildings to have received LEED-EB certification. Occupying 4.2 million square feet on two entire city blocks, big enough for its own zip code, Merchandise Mart is the largest commercial building in the world. In 2007, this structure—built in 1930—also officially became the world's largest green building. Renovations implemented at the Mart between 2005 and 2007 included instituting a dynamic recycling program, installing meters to track energy consumption and identify particular areas of waste, fixing leaky pipes, and reusing nonpotable water. No detail of energy consumption was too small: according to *BusinessWeek*, even matters such as the color of carpets were considered, with bright carpet colors being more light-reflective and reducing the need for artificial lighting.

The savings at the Mart so far have been impressive: total utility bills immediately dropped 10 percent. Water use alone has been cut 35 percent, saving $100,000 per year. The utility bill savings, improved working conditions, and prestige offered by the revamped interior proved to be a big draw for new occupants: occupancy rates were 77 percent ten years ago; they've since jumped to 96 percent.

Following on this success, the Empire State Building, built in 1931, recently launched a $100 million green retrofit that includes 6,500 window replacements and a new air-conditioning system, with projected energy-bill savings of about $5 million a year.

Hundreds of companies both large and small across the country have signed on to receive LEED guidelines and efficiency ratings. Inspired in part by the success of its Merchandise Mart, the city of Chicago currently has more than 250 LEED-EB projects in process, more than any other city in the United States. Nonprofit programs including the Clinton Foundation's Climate Initiative are now offering financing and other assistance to make retrofitting more practical and affordable for businesses around the world.

That's great for existing commercial buildings. But what about existing homes? Even as the economic recession has slowed sales of technologies such as household geothermal systems and solar panels, other trends offer reason for hope. A Seattle-based start-up company, G2B Ventures, is in the process of raising $50 million to buy up homes that have been foreclosed or are currently priced below their potential market value. The company will retrofit these properties with green features and then rent or sell them for a profit. Aaron Fairchild, one of the company's managing partners, said, "In the midst of chaos there's opportunity. We're trying to direct the opportunity in a socially and environmentally positive direction that has enhanced rates of return for our investors."

Some states, including Pennsylvania and Kansas, have begun offering programs designed to encourage energy-efficiency improvements by individual home owners. New York State now offers reduced-interest-rate loans for home energy improvements, and has a pilot program in the works to subsidize mortgages for energy-efficient homes.

In cities and towns across the country, banks now offer mortgages to help home owners pay for green retrofits. One such type of loan is known as an energy-efficient mortgage, or EEM. Here's how it works: the lender requires a licensed energy consultant using a home energy rating system (HERS) to evaluate the cost of the improvements and estimate the dollar savings per month. The cost per month must be lower than the amount saved.

For instance, Pat and Mynette Theard, first-time home buyers in California, got an FHA mortgage of $144,480 to pay for their house and energy-efficiency improvements: $142,500 for the cost of the home and $2,300 set in escrow for the retrofit. The extra $2,300 raised their monthly mortgage bill by $17—but the improvements saved them $45 per month in reduced utility bills. "This is our first home," Pat said, "and the EEM saved us a lot of headaches because we knew what we needed to do to the house. It's nice and comfortable now." Given the $28 difference each month between what they save and what they're paying, the family is essentially getting paid to go green. Homeowners often save much more than the amount tacked onto the mortgage per month, so it's a win on all fronts—one that can also increase the resale value of the home.

How much would it take to make an old home into a fully green home? Retrofitting can reduce energy use between 20 to 50 percent in most buildings. Part two of the process would be exchanging the conventional energy from the grid for renewable sources of energy—either through on-site solar or wind technologies, or through purchasing electricity from utilities that use clean sources for their power. EEMs typically allow for a total of 5 percent to 15 percent of the property's value to be financed (a maximum of $30,000, say, on a home that costs $200,000). The Australian-based green architecture firm Davis Langdon calls such investments "future-proofing your asset."

ALL TOGETHER NOW

America can march toward energy independence one home and one building at a time, but it can also hurdle in that direction one community, one county, one whole city at a time. A new trend in community-supported energy is spreading across the country, driven by residents who are banding together to apply wind, solar, and efficiency measures to their homes and farms collectively. Working as a community is more affordable and rewarding than working individually—it gives residents purchasing power to buy the technology at cheaper group prices.

Consider, for instance, the Kennecott Daybreak land development outside of Salt Lake City, Utah, a modern-day Levittown gone green that will have 200,000 homes when it's completed. Designed for efficiency, those homes will emit 55,000 fewer metric tons of greenhouse gases annually than a standard community of about the same size. "One home owner doesn't make a big difference," as one of its residents put it, "but 200,000 home owners do—I feel like a part of something meaningful."

The Windustry organization, to take another example, is an advocacy group that promotes community-owned wind projects among rural land owners nationwide.

Based in Minnesota, which has the highest concentration of farmer-owned wind projects in the country, Windustry works with rural communities in twenty-seven states that collectively produce more than 1,000 MW of renewable power—enough for one million homes. Instead of partnering with utilities or big investors, the farmers themselves own, install, maintain, and operate the turbines. Working together, they share risk and are better positioned to get financing for their projects.

There's a hopeful message rising out of these grassroots efforts. The era of increasingly scarce and expensive fossil fuels is driving Americans—home owners, farmers, small businesses—not just to embrace green technologies but also to reconnect to their communities. Local politicians, too, are taking action. In the absence of federal-level leadership on climate solutions over the last decade, hundreds of American mayors have been mobilizing citywide energy-efficiency programs and developing smart growth strategies to reverse urban sprawl and reduce car usage. They are designing residential and commercial developments around public transit, pedestrians, and bike lanes. It's a U-turn in the trend toward suburban living that has been gaining ground for half a century.

Today, more than half of the residents of America's 100 biggest metropolitan areas live in suburban communities. Removed from commercial centers, they must travel by car to get to the office, the library, the dry cleaner, or the doctor's office. The New York–based developer Jonathan Rose estimates that, when you factor in car travel, suburban households use four times more energy than urban homes near mass transit.

Suburban families not only travel longer distances by car, they occupy significantly more square footage of living space. Whereas single-family suburban homes generally have attics, basements, and four external walls exposed to outside temperatures, urban residences in mid- and high-rise buildings have walls adjoining those of other residences, keeping them inherently insulated. On average, a typical suburban family uses 20 percent more energy to heat and cool its home than a family living in a compact urban area.

Demographic trends have been moving away from suburban (low-density) living and toward high-density urban living in recent years.

Thanks in part to the upward trend in gasoline prices, "U.S. cities have been making a comeback," said Reid Ewing, an urban planning professor at the University of Utah. He predicts that the density of residential development in urban areas will increase from seven units per acre today to eleven units per acre by 2050.

In light of these trends, state and city governments have been allocating funds and rezoning downtown areas to spur the development of residential complexes clustered around shopping centers and office buildings. Their hope is to bring people closer to their workplaces, shrink commutes, and create hubs for public transit. But for all the benefits of urban living, cities are still tremendously resource-intensive. They are twenty-four-hour commercial hubs in which products, people, airplanes, truck fleets, buses, cars, construction equipment, and the utilities powering office, retail, and residential buildings are constantly in motion. (A medium-sized city of 350,000 people can spend about $4 million per year simply to power its streetlights.) All told, cities are responsible for more than three-quarters of the world's greenhouse gases.

It stands to reason, then, that mayors have a critical role to play in transforming America's energy landscape. "We're the ones building roads, designing mass transit, constructing schools and libraries, buying police cars and dump trucks and earthmovers," Patrick McCrory, the Republican mayor of Charlotte, North Carolina, told me. "We're the ones lighting up the earth when you look at those maps from space." Mayors have significant purchasing power, so when they decide to build green buildings, buy green auto fleets, expand mass transit, or mandate recycling programs, their decisions can have ripple effects that move entire markets.

Between 2005 and 2009, the mayors of more than nine hundred American cities ranging from Eugene, Oregon, to Dallas, Texas, signed the U.S. Mayors' Climate Protection Agreement to adopt aggressive targets for reducing greenhouse gas emissions 7 percent below 1990 levels by 2012 (targets consistent with the Kyoto Protocol). The results have been remarkable: dozens have met their targets well ahead of time, and collectively they've saved billions annually in total energy expenditures.

A quick snapshot of the mayors' green strategies shows the breadth of what's possible. Seattle instituted mandatory recycling and imposed a pay-as-you-throw plan for garbage, reducing the amount of the city's waste and the fuel used to haul it to landfills. Houston developed a "Flex in the City" program that encourages employees of the city's largest companies to telecommute and cut down on traffic and emissions. Chicago imposed green building guidelines for new development and offered tax breaks to new businesses that develop clean energy technologies. San Francisco installed tidal turbines in its bay and banned plastic bags. New York City is trying to convert its massive taxicab fleet to hybrid engines by 2012. Portland, Oregon, instituted an "urban growth boundary" that prohibits development outside designated city lines, preventing sprawl. Tucson, Arizona, imposed restrictions on landscaping practices, allowing only native plants that can live off rainfall, limiting the amount of water that has to be brought in by electric pumps.

SPRAWL REVERSAL

These actions are encouraging, but no such progressive programs seemed to be gaining traction in my own neck of the woods. Important as it is to improve efficiency and environmental performance in places such as New York, San Francisco, and Portland, these cities are already densely developed and have world-class public transit systems. It seems a far more daunting—and potentially rewarding—challenge to green a place such as Nashville, Tennessee, which encompasses a large amount of suburban sprawl and has a meager transit system—no subway, bus stops that are few and far between, and buses that run infrequently. Even sidewalks and bike paths are hard to come by. What would a visionary green strategy look like in Nashville, I wondered, and how would you go about implementing it?

I contacted Rick Bernhardt, director of Nashville's metro planning department, to find out. Bernhardt, a tall, affable man in his mid-fifties, with wire-framed glasses and a jaunty yellow-patterned tie, spent nearly two decades applying sustainable-planning principles in Orlando before returning to his home state, Tennessee. Nashville and other sprawl cities, Bernhardt told me, are not a lost cause: "You can build a green future here if you put the right guidelines in place."

He rattled off a litany of smart growth strategies that he's advancing in Nashville: density targets to encourage infill in downtown areas so that the city grows upward, not outward, and residential and commercial areas become increasingly integrated; zoning changes so that every neighborhood can offer housing, transportation, and access to services and employment. Bernhardt is also channeling government spending into areas where development is most strategic—improving infrastructure, roads, electricity, water lines, and sewer lines around designated population "hot spots." He is implementing "complete streets" programs that remove and narrow car lanes, add bike lanes, and widen sidewalks for pedestrians. Once Nashville has established more concentrated population nodes, said Bernhardt, it can then consider expanding public transit by increasing service and adding more bus stops or even light rail stations.

Halfway through the interview, Bernhardt pulled out a map of Nashville from the 1920s. "You're not going to be happy about this," he said. There were blue lines running like octopus tentacles throughout the streets of Nashville—a map of the old streetcar network. It looked nearly as intricate and far-reaching as the New York subway system. "The streetcar lines were bought up by Detroit; the tracks were ripped up and covered in asphalt," he continued—part of that larger trend beginning in the 1930s that eliminated railways all across the country. Without streetcars, the city's development quickly began to sprawl outward around highway networks, not transit lines.

Reversing this trend won't be easy. Bernhardt has had to fight bureaucratic inertia and swim against the tide of opposition from city public works and from commercial developers who bridle at political red tape, preferring to throw up buildings and shopping centers at maximum speed and minimum cost, with little regulatory control. But the resistance to smart growth strategies has been waning somewhat in recent years as the environmental, economic, and national security arguments for curbing energy consumption have gained ground.

The recession has also had one positive and unanticipated effect for sprawl cities: *Time* magazine recently reported that as shopping malls and large retail stores have failed across the country (roughly 148,000 stores closed in 2008 alone), urban planners in places ranging from Long Beach, California, to Lakewood, Colorado, have successfully used these vacant spaces—and their already-existing steel and concrete infrastructure—as central hubs for new mixed-use developments combining apartments, office space, and smaller retail stores.

"It's clear," Bernhardt told me, "that demographic patterns are trending toward urban centers that provide an enhanced quality of life. Any city that wants to remain livable will have to respond to and accommodate those trends." Referring back to that old map of lost streetcar lines running through clustered neighborhoods, he added, "In a sense, the challenge now is about making things more like the way they used to be."

DO THE LOCOMOTIVE

Travel to Western Europe and you'll find that more than half of the population lives within two blocks of a mass transit station. That percentage is even higher in Japan, which boasts the largest ridership numbers on mass transit systems of any country in the world. In Singapore, public transport is used for nearly 65 percent of commuter trips during peak hours; in Hong Kong, light rail accounts for over 30 percent of daily domestic travel.

In the United States, by comparison, public transit accounts for only 5 percent of all commutes to work, and 2 percent of local trips in general. Only about half of U.S. households are located within even a mile of a bus line, and only about 40 percent are within a quarter mile. The situation is worse for rail: only about 10 percent of the U.S. population lives within a mile of a light rail or subway stop, and only about one-quarter lives within 5 miles.

Even the best efforts of urban planners such as Bernhardt aren't likely to put public transportation usage in the United States on par with that of its European and Asian counterparts anytime soon. Robert Cervero, a transit expert at the University of California, Berkeley, estimates that based on current planning trends, public transit could comprise up to 15 percent of daily local trips in the United States by 2050. "Given the settlement patterns we have in the United States, it's difficult to make a large shift in a short time on public transport," said Amory Lovins of the Rocky Mountain Institute. "It's much easier to make the technology behind cars, trucks, and planes three times more efficient—and that will have greater immediate payback." Smart highway systems will also play a role.

But aggressively developing mass transit is a crucial opportunity on the horizon to improve energy efficiency and clean up emissions. Simply increasing the use of buses and trains from 5 to 15 percent of daily trips in the next few decades would be equivalent to taking millions of cars off the road and would make a huge dent in our daily greenhouse gas emissions. Government at the local, state, and federal levels could work to rehabilitate those old rail lines that once crisscrossed the United States and have been retired into greenways, and use the existing right-of-ways here and alongside highways in the effort to develop a national network of interstate light rail and bullet trains. Federal leadership has recently begun to encourage a shift in this direction, committing nearly $8 billion to the development of mass transit and high-speed rail, nearly a 70 percent increase over present spending levels.

That's a fraction of what a comprehensive public transit plan will cost, but it will move the country forward toward a green future—in part, as Rick Bernhardt said, by "making things more like the way they used to be." Many of the most promising urban-planning strategies for the future are the ones that apply new technologies to the transportation patterns of the past. American cities and towns can narrow their streets and allocate car lanes for above-ground electric trams and buses. They can reduce traffic flows with congestion pricing mechanisms whereby drivers pay a fee when they enter high-traffic areas, which in turn encourages carpooling and the use of mass transit.

Citizens of Europe and Japan are less wedded to their cars in part because they routinely pay two to three times more than U.S. consumers per gallon of gas. That also helps explain why these nations are taking the lead in developing futuristic transit networks. Virtually all of these networks are aboveground—digging new underground subways is so energy-intensive and the underground facilities are so expensive to maintain that, going forward, they may well be viewed as an obsolete technology. The streets of Strasbourg, France, for instance, now have exhaust-free, space-age-looking trams with bubble-glass exteriors and whisper-quiet electric propulsion. They operate on greenways in downtown zones where cars are restricted, sharing the streets with pedestrians.

In Lyon, France, the "civic buses" being used today are also cutting-edge, and virtually indistinguishable from trams. These electrically powered buses have optical guidance components that follow a path indicated by coded markings embedded in roads; an image processor detects and corrects deviations by activating a motor on the steering wheel. This enables the buses to snake through tight streets, safely avoiding collisions, and pulling up to passenger platforms like trains. Pneumatic tires produce a silent, smooth ride. Such bus networks are less expensive and more versatile than trams because they don't require the installation of actual tracks, and they could therefore be relatively easy systems to implement in cities such as Nashville with little existing public transit infrastructure.

The maglev bullet trains of Tokyo, Japan, and Shanghai, China, don't actually glide on tracks—they are suspended on a cushion of air above the tracks and propelled by magnetic forces. The frictionless trains travel at an average speed of roughly 125 miles per hour but are capable of reaching speeds of over 350 miles per hour—the cruising speed of a commercial airplane.

These trains could conceivably replace not just the old Amtrak dinosaurs that travel throughout the Northeast corridor but also eventually the gas-guzzling flights between major U.S. cities. Ancillary light rail networks—akin to those people-mover systems in large airports—could then emanate from major rail hubs into the outlying employment centers and shopping plazas of surrounding towns. Beyond that, public car-sharing systems and neighborhood car co-ops could be established to service even lower-density areas.

TOMORROWLAND

Just as intriguing as the technologies that already exist to improve our cities are the ones that have yet to be invented. To catch a glimpse of the future that beckons further out—a future that's still in its earliest wild-idea stages—I talked to urban planner and futurist Mitchell Joachim. Joachim has proposed ideas as far-flung as building homes inside trees that have been "trained" to grow in the shape of dwellings using high-tech software (the concept, called Tree Fab Hab, has been exhibited at the Museum of Modern Art).

Joachim also envisions building cities out of garbage. Like Robert Malloy, the UMass plastics engineer I'd interviewed about recycling, Joachim sees today's garbage dumps as future stockpiles of usable materials. He notes that if you could somehow convert waste into construction material, you could make another Empire State Building out of what New Yorkers throw away every two weeks. "We will have robots to mine our landfills, programmed to sort through the trash and tell a toilet from a Big Wheel—think *Wall-E*," he told me, referring to the recent Disney movie in which human civilization is overrun with trash. There's enough trash in New York City's landfill, he said, to "rebuild all the buildings in Manhattan at full scale seven times over."

The most practical and compelling of Joachim's concepts—one named among *Time* magazine's 2007 inventions of the year—is his City Car, which combines the comfort of a private vehicle with the efficiency of public transportation. "For almost a century we have been designing cities around cars," he said. "Why not design a car around a city?" Joachim's concept borrows from the public car-sharing model of ZipCars, in which vehicles are leased hourly by a network of continual users, but takes it a step further: his shared vehicle would be a foldable, stackable electric minicar that could be rented and then returned to any station around the city, almost like an airport luggage cart. Electric motors would be embedded in each wheel (cutting out a cumbersome mechanical drive train), and these 5-foot-long two-seaters could speed along at up to 55 miles per hour. "Made from neoprene and other soft materials, cars would no longer suffer traffic-fouling fender benders, merely what [Joachim] calls 'gentle congestion,'" reported *Wired* magazine. "Picture a flock of urban sheep grazing against one another."

Joachim described his City Car system as a logical successor to the public bike-sharing systems of Paris and Montreal—nearer-term, motor-free solutions to urban transit that have been extremely successful. In 2007, Paris mayor Bertrand Delanoë launched a system of more than twenty thousand public bicycles known as Vélib' (the name combines *vélo*, "bicycle," and *liberté*, "freedom"). The city installed 1,450 rental stations throughout its streets—roughly one every 900 feet. The bikes are rented on credit cards for the equivalent of about $1.50 a day, and a $200 penalty is charged if a bike is not returned. In spring 2009, Montreal installed a slightly higher-tech version of the Vélib' system, known as Bixi, in which solar-powered bike-rack stations are linked to the Internet so that potential renters can monitor online whether bikes are available. The bikes are embedded with RFID tags so they can be remotely tracked to prevent theft. It's perfectly conceivable, says Joachim, that we could soon have comfortable miniature cars folded up in public rental slots where those bikes once were.

As Joachim sees it, humanity is progressing from the Age of Industry (the twentieth century) to the Age of Recovery, our current era, in which we struggle to redress the environmental damages of our past, to the Age of Positive Ecology, in which our imprint on the earth is not just clean and sustainable but also nourishing and ecologically productive.

THE RIGHT TRACK

For now, for most Americans, the movement toward this potential future is a change that is happening one lightbulb, one retrofitted house, and one green community at a time, in places such as the Lenoir City homes in Tennessee and the Kennecott housing development outside of Salt Lake City, in the California labs of renewable technology innovators and on the drawing boards of Detroit's automakers, in the fields of farmers leasing land for wind turbines in Texas and the Windustry farming organization in Minnesota, in the city planning offices of places like San Francisco and Nashville.

The changes that are being proposed and put in place today in America's homes and offices, transportation systems, city planning departments, and electric grid will help us rise to the twin challenges of improving national security through freedom from oil and combating global warming. To accelerate, support, and fund this clean, green infrastructure, a tax could be placed on gasoline and parking—an expenditure that would be akin to the sacrifices widely accepted by Americans during World War II, enabling the United States to meet that global challenge.

Americans could lay a regulatory groundwork for the modern metropolis by supporting the establishment of guidelines for dense, mixed-use developments that bring people and jobs closer to each other. We could encourage federal leaders to develop codes for zero-energy buildings and require that every new home and office built within the United States be green—I can't think of one good reason why they shouldn't.

The late Jane Jacobs, who wrote the 1961 bestseller *The Death and Life of Great American Cities*, promoted a vision for urban design that is today more relevant than ever—mixed-use zoning, diverse architecture, robust public transit, dense downtowns, and pedestrian-friendly street life. Jacobs's philosophy followed one central rule: "Design is people." Good design, she argued, fosters vibrant, productive communities, while poor design burdens communities—harming their health, productivity, and potential to succeed.

The U.S. Census Bureau expects America's population to grow from 310 million today to roughly 400 million by 2050—a jump of nearly 30 percent. The nation as a whole will pay a toll if the cities and buildings created for these future Americans are not designed to be smart, clean, and sustainable.

"Whenever and wherever societies have flourished and prospered rather than stagnated and decayed," Jacobs wrote, "creative and workable cities have been at the core of the phenomenon; they have pulled their weight and more." Her comment is particularly resonant now, as America faces both the threat of decay and the opportunity to flourish. In the quest to secure America's future, Jacobs added, "The humble, vital services performed by the grace of good city streets and neighborhoods are . . . as good a starting point as any."

If Dr. King were with us today, he would
be working to build a green economy—
an economy strong enough to lift people
out of poverty and restore hope to America.

—Van Jones, activist

11

Fresh Greens

NOT YOUR GRANDMA'S ECO-MOVEMENT—
MEET THE NEW PIONEERS

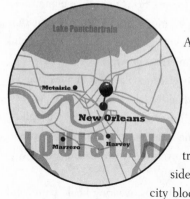

A woman picks daffodils from a small garden behind a concrete platform—all that remains of her house, which was torn from its foundation after the levees broke on August 29, 2005 in the wake of Hurricane Katrina. Weeds sprawl out of cracks in sidewalks still tinged with mold. Entire city blocks are vacant, grassy lots occupied only by "stairs to nowhere"—freestanding flights of cement steps that once led to front stoops. Scattered across this flat landscape, which sits below sea level, are purple bursts of thistle and gnarled, tenacious oak trees that somehow survived for days under 20 feet of water.

This is New Orleans' Lower Ninth Ward, where the floodwaters flattened buildings, leached chemicals into the soil, and left an entire community homeless. An estimated sixteen hundred people in the city died as a result of Katrina; well over half were from this neighborhood. Many residents drowned in their attics or succumbed to dehydration and heat exhaustion stranded on their rooftops. As of 2009, about three-quarters of the families that lived in the Ninth Ward before Katrina were still displaced.

But here in this blighted community—more perhaps than in any other place nationwide—the future of America is taking root. Two blocks away from the 5-mile-long Industrial Canal that links Lake Pontchartrain to the Mississippi River, a cluster of new, candy-colored homes—lemon yellow, tangerine, berry blue—on 8-foot stilts rises up from narrow grassy lots, adorning the scarred landscape like jewels in ash. The homes all have sleek, modern angles, expansive windows, lofty ceilings, rooftop solar panels, and front porches with rocking chairs and hanging wooden swings—a mixture of southern tradition and cutting-edge technology. These eight homes are the models for many more to come. The Make It Right Foundation, a nonprofit founded by actor Brad Pitt, has commissioned a group of international architecture firms to build a community of 150 affordable green houses in the Ninth Ward by 2010. Its mission is to make the neighborhood that has suffered the worst environmental damage in U.S. history a model for America's green future.

"We started out just wanting to give people their homes back—rebuilding the hardest-hit area with affordable, sustainable houses," the foundation's executive director, Tom Darden, told me as we walked through the neighborhood one spring afternoon. "What we've come to realize is that we have the opportunity to help revolutionize the way buildings are built, the way people live."

Darden, a fresh-faced man in his thirties, wearing khaki pants and wraparound sunglasses, first joined the project as a volunteer after working at a North Carolina real estate firm that developed new properties on cleaned-up brownfields. He signed on to work for Make It Right full-time after a meeting in 2007 at which he and the foundation's leaders pitched their concept to Ninth Ward residents: "There was one elderly man who was initially very suspicious of our proposal. He came up to me after the meeting and shook my hand. He wouldn't let go. He had tears welling up in his eyes and he just said 'I believe you—I believe you are going to do this.'"

Make It Right has since raised more than $30 million to build and help finance the first 150 structures, all designed in accordance with rigorous environmental standards. Geothermal heating and cooling and efficient appliances and lighting systems radically cut the houses' energy demands. Solar panels provide about three-quarters of their electricity, channeling power back to the grid on sunny days. Architects also designed the homes to be "zero-runoff," capturing 100 percent of the rainwater that falls on each lot to be used for indoor plumbing and native landscaping—an important feature in a city that routinely struggles with storm water management. They incorporated "zero waste" principles, using framing techniques that enable building components to be easily disassembled and reused. All of their construction materials are nontoxic, sustainably produced, and mostly recycled (roughly 1 ton of waste is produced, per house—a tenth the amount of waste generated in conventional construction).

The homes have been engineered to withstand Category 4 hurricanes—some were even designed to float in the event of another catastrophic flood. As tropical storms intensify with global warming trends, the Ninth Ward is a neighborhood that some skeptics argued should never be rebuilt. But instead of fleeing a future fraught with environmental risk, this settlement is facing it head on.

Designed to cost less than $150,000—the average price of a house here pre-Katrina—the homes are being purchased by residents with the help of insurance money, government disaster relief funds, and loans subsidized by Make It Right. (The foundation has a homeowner counseling arm to guide residents through the financing process.) Pitt launched the foundation in December 2007, just over a year after he'd visited the city and found the devastation from Katrina largely unrepaired due to political neglect. The group of visionaries he assembled included green building's "godfather," William McDonough, and award-winning green architect Lars Krückeberg; the coalition eventually grew to include Frank Gehry, Kieran Timberlake, and other notable members of twenty-one architecture firms from the United States, Europe, Africa, and Asia.

"I can't think of another example in history," said Darden, "where you have world-class architects driven by a common passion for social justice and a shared devotion to the environment coming together to design low-income housing."

So far, residents are overwhelmingly pleased with the results. "I feel good as hell every time I walk in here," fifty-four-year-old Melba Leggett told me as we sat in her sunny living room. "I used to keep the TV on. Now I like to turn it off and sit in this room. You can see the birds and sometimes you can see the dirt divers, you can see the butterflies. It is so peaceful." Leggett, who works in a school cafeteria and has five children and six grandchildren, described the details of her house with the intuitive ease of a woman describing the traits of an old friend. "My house is designed like a walk-in refrigerator," she explained, referring to the exterior walls which are made of steel and polystyrene, much like a thermos, keeping the interior cool in the summer and warm in the winter. Leggett said that her energy bills have gone from $500 a month down to well under $100: "I don't really pay much attention to them because they're so low."

Leggett praised her rooftop solar panels for their simplicity and low maintenance. "You just have to mop them sometimes so the sun gets through." She also noted her natural-fiber carpets: "They're biodegradable like everything else—my fans, my paint, my cupboards." Leggett, who has struggled with asthma for decades, said the house she lost to Katrina had been through several hurricanes—"Betsy, Camille, I can't even count. It had mold in it that made us sick." She pointed to an intake system in her new home that conditions and filters air. "This is my fresh air machine. Since the house is sealed so tight, this keeps the oxygen cycling through. It keeps my asthma down. What's good for this house is good for me."

Leggett's neighbor, seventy-four-year-old Collins Foots, a retired bus driver, was also fluent in his home's every working detail. "We've got this type of heater that only warms up the water when you need it." He turned on a hot water faucet to demonstrate. "Hear it working in the next room? You don't need a 35-gallon tank of boiling hot water all the time." He took me outside to see the guts of his geothermal system, pointing to the visible ends of the pipes that were buried some 10 feet below his front yard: "These here tubes keep the house the same temperature as the earth down underground. It means my heating and my air conditioning don't have much work to do." Foots then displayed his electricity meter, pointing to its needle which was wagging leftward—the sign that his solar panels were producing a surplus. "We should've been building houses like this all the while," he said. "The energy people probably don't like it much—taking money out of their pockets and putting it in ours."

Going forward, Tom Darden hopes that the Ninth-Ward settlement will serve as a proof-of-concept for other low-income green building settlements—nationally, and perhaps worldwide. "We're working to crack the code on affordable green living. There's a long way to go to fully get to zero-energy, zero-waste, zero-runoff, but we're moving in that direction as fast as we can." For applications beyond New Orleans, the homes will have to be modified for different climates, he explained: what works in a warm, humid climate might not work in Chicago, for instance—or for that matter in Shanghai or New Delhi. Make It Right offers contractors incentives to find ever-cheaper green building methods and supplies—giving them half of every dollar they save on construction costs. "With every home we build, the environmental performance gets a little better and the cost of building gets a little cheaper."

At a time when green living is widely perceived to be a privilege available only to the moneyed few, this Ninth-Ward settlement, like the DoE's zero-energy experiment in Lenoir City, Tennessee, makes the case that every single American, including—and especially—those on the lower economic rungs of society, can live sustainably. If all goes well, Make It Right could help do for green homes what Walmart is doing for organic and local foods, or what the Smart Car is doing for fuel-efficient vehicles—making sustainable lifestyles ever cheaper and more available to the masses. But what sets the New Orleans homes apart from other efforts to mainstream green innovation is its conspicuous link to social justice.

As I walked down Deslonde Street, speaking to other families who had only weeks earlier moved into their new green homes—sixty-seven-year-old Gloria Guy, who was living with her daughter and six grandchildren; Gertrude and Marvin LeBlanc, who had lived in the Ninth Ward for more than sixty years before Katrina hit—I began to see in this place where rooftops once bore frantic hand-scrawled SOS signs an unexpected marriage of environmentalism and civil rights, of technology and social justice: issues that are fusing, I was to learn, in communities across the nation.

WE THE PEOPLE

The green homes of the Ninth Ward are being built against an unlikely industrial backdrop. Apart from the threat of hurricanes, this community and those that adjoin it carry multiple environmental burdens in the form of sewage treatment plants and waste transfer stations (the very types of facilities that released their pollutants into the floodwaters after Katrina hit). They also shelter refineries that service the Gulf of Mexico's oil drilling operations, including rigs like the one I'd visited, Chevron's Cajun Express.

As the solar roof of Melba Leggett's home glinted against the shadow of nearby smokestacks, I began to consider one last facet of America's fossil-fuel-based past: The vast majority of America's dirty industrial facilities—our power plants, factories, chemical refineries, waste transfer stations, airports, truck fleets, bus hubs, and toxic dumps—are heavily concentrated in low-income and mostly minority communities. The Ninth Ward is just one example of this larger social and environmental trend.

New York City's South Bronx region is another. This predominantly African-American and Latino community is one of the poorest congressional districts in the United States. Spanning just 15 square miles, the neighborhood accommodates four power plants, two prisons, the world's largest food distribution center, and other facilities that collectively draw more than sixty thousand diesel trucks weekly. It also has a waste-transfer station that processes nearly half of New York City's commercial garbage and a sewage-treatment plant that handles over three-quarters of Manhattan's sludge. South Bronx residents suffer from 25 percent higher rates of asthma, on average, than do residents of other communities in America, and children are particularly affected, with one in four in the neighborhood diagnosed with the respiratory ailment—that's several times the national average.

Low-income neighborhoods bear most of the environmental burdens of virtually all cities in the United States. Government statistics show that roughly half of the low-income housing units in America are located within a mile of industrial facilities reporting toxic emissions.

Majora Carter, a graduate of Wesleyan University who won a Mac-Arthur Fellowship for organizing her South Bronx community around environmental solutions, shared some alarming statistics: African-Americans are 80 percent more likely than whites to live in areas where air pollution levels pose health risks, according to a recent analysis of EPA data. Lead-poisoning rates among Latino and African-American children are roughly double those among white children. More than 30 percent of America's public schools are located within a quarter mile of highways that serve as main truck and traffic routes, which makes them air pollution danger zones.

I was astonished, however, to find the most hopeful stirrings of America's energy future in places that were among the most burdened by this environmental legacy. A new kind of green activism—a broader, more diverse, and more inclusive green consciousness—is surfacing in New Orleans, in the South Bronx, and in thousands of other urban and rural communities that have traditionally shouldered most of the hidden costs of our antiquated, fossil-fuel-based industries.

These activists don't fit the tree-hugger stereotype of Prius drivers who vacation in Yosemite, shun plastic water bottles, and shop at Whole Foods. But the residents of these pollution-laden neighborhoods are deeply connected to and concerned about the health and sustainability of their communities. They are protective of the land they live on. And their lives are poised to be radically and intimately transformed by the benefits of cleaner, more efficient technologies.

The Make It Right project leaders in New Orleans and Majora Carter in the South Bronx are among of hundreds of leaders nationwide helping to grow the burgeoning environmental justice (EJ) movement. In its philosophy and organization tactics, EJ brings a new dimension to environmentalism. It is by definition a grassroots, community-based movement that creates change by knocking on doors, distributing petitions, and marching in the streets; it's not a national movement that exercises its power via Capitol Hill lobbyists and circuit-court litigators. More fundamentally, the EJ movement is not one that considers "the environment" to be only exotic wildernesses; rather, it includes the industrial landscapes where people live, work, and play—their sidewalks, schoolyards, homes, factories, and offices.

I did not expect to end my power trip examining these emerging new dimensions of the environmental movement. Two years before, I'd begun my journey through America only a few miles from where I ended it—at a heliport in New Orleans climbing into a Chevron chopper, consumed with an interest in industry and innovation. I wanted to learn about extreme drilling technology, military machinery, cars, plastics, fertilizers, supply chain logistics, and power grid design. But in exploring these fields I'd come to realize that it's not so much technology that has shaped America's past and present as it is *people*—the actions and convictions of individuals ranging from larger-than-life visionaries such as Thomas Edison, Henry Ford, and Franklin Roosevelt to unsung administrators like Chevron's Paul Siegele, the Defense Department's Al Shaffer, and Con Edison's Lou Rana, as well as ordinary consumers with oil- and coal-based households, diets, medical histories, and lifestyles like my own. It wasn't the innovations themselves that shaped our cultural identity; it was the ways in which we chose to wield them—to popularize, covet, or condemn them.

An energy landscape and history that had seemed unemotional, monolithic, and inevitable to me when I first started my journey had by its end come to seem more surprising, inspired, and ultimately *human*. "I never perfected an invention that I did not think about in terms of the service it might give to others," Edison said when asked about the source of his inspiration. "I find out what the world needs, then I proceed to invent." Technological innovation is an act that rises, like music or story-telling, from a combination of necessity and creative human impulse.

I saw that creative impulse throughout my travels and research—in the work of innovators making solar panels and batteries from nanoparticles, digital engineers transforming a one-way grid of cables and wires into a nimble and multidirectional network, and green architects designing buildings to function like trees. I also saw it in the work of the Chevron engineers digging for treasure in bedrock located under 7,200 feet of water, NASCAR mechanics making oil burning into a kind of ballet, the plastic surgeon in California resculpting the human body with man-made materials, and the Kansas farmer fertilizing exhausted soil using advanced GPS technology. Technological innovation is, in its way, an art form—a creative act, an emotive act, a human act.

In this there is cause for hope. The errors of our industrial past were human errors, and our technological triumphs were human triumphs. As President Barack Obama has phrased it, "We are the change we have been waiting for." While our technological innovations can have destructive human impacts, they can also sow justice.

DREAM REBORN

On the fortieth anniversary of Martin Luther King Jr.'s death, April 4, 2008, a crowd of thousands gathered at the Cook Convention Center in downtown Memphis, Tennessee, for a three-day symposium. They had come to celebrate King's life and to explore strategies for advancing his dream. This was the first national event organized by Green For All, a group founded by green jobs advocate Van Jones, and the largest assembly the environmental justice movement had yet convened. The venue was low-ceilinged, with drab gray carpeting and polyester drapes, but this setting did nothing to dampen the crowd's enthusiasm; cheering activists periodically jumped up from folding chairs and threw their arms in the air, roller coaster style. Behind the podium hung a floor-to-ceiling poster with a graphic of King's face set against a backdrop of workers in hard hats erecting windmills and affixing solar panels to rooftops. The caption: "A Dream Reborn."

"If Dr. King were with us today," Jones told the crowd, "he would be working to build a green economy. . . . He would be standing with those communities that have been locked out of the last century's pollution-based economy. And he would be working to ensure that all our people, the entire beloved community, are included in the emerging clean and renewable economy."

The EJ movement, as Majora Carter told me, holds that "no community should be saddled with more environmental burdens"—like power plants and waste facilities—"and fewer environmental benefits"—like parks and green innovations—"than any other." The movement's emphasis is as much on *promoting* environmental benefits for everyone as it is on *preventing* environmental burdens. "We are not antidevelopment," said Carter. "Our goal is not to block industrial facilities—we welcome them, and the jobs they bring. Our goal is to ensure that they are clean, healthy, high-performance facilities, just as they would be if housed in a wealthy neighborhood." If power plants, refineries, and waste dumps had been located in upper-income neighborhoods throughout the twentieth century, she argued, "we would have had a green economy long ago."

In recent years, members of the Sustainable South Bronx organization Carter founded have teamed up with other local activists and city officials to plant hundreds of saplings throughout a landscape of mostly bald concrete. They've replaced junkyards with waterfront parks and miles of tree-fringed bike paths. They've sprouted acres of rooftop gardens on top of old warehouses. Carter herself has drawn up proposals for clean-energy facilities ranging from a recycling park to a biofuels factory.

Perhaps most notably, these activists have begun training workers to clean up brownfields, install green roofs, weatherize buildings, and tend the acres of new parklands, creating much-needed jobs in a community struggling with a 25 percent rate of unemployment. Likewise, the Make It Right project in New Orleans is cultivating a workforce to construct its homes and install clean energy technology, training local residents for skilled jobs in green construction. And Van Jones's Green For All has partnered with Al Gore's organization the Alliance For Climate Protection to develop a job training academy, laying the foundations for a "green-collar" economy. As Jones told me, "We want to connect the people who most need work with the work that most needs doing."

Jones started Green for All in September 2007. After graduating from the University of Tennessee at Martin and Yale Law School, he worked as a community leader in California and eventually created the organization to combine his concerns for racial, social, and environmental justice. He invoked John F. Kennedy's aphorism "A rising tide lifts all boats" when he explained, "The green wave must lift *all* boats."

Jones and Carter's shared goal—to expand the production of clean technologies in a way that spreads their benefits widely throughout society—represents a complex and compelling vision. I had already come to understand that the shift away from fossil fuels could improve America's diplomacy, protect our national security, buoy the economy, bolster public health, and revolutionize the design of our buildings, cars, and infrastructure. What I was coming to realize, as my trip neared its end, had even broader and deeper implications.

When I asked Jones how he hoped to join environmental and economic progress with social justice, he outlined a vision for a green New Deal (many components of which would be woven months later into the Obama Administration's economic stimulus plan). His federal economic aid package would devote billions of taxpayer dollars to retrofitting buildings, redesigning cities, rehabbing and expanding public transit systems, building a new superconducting electricity grid, and installing corridors of solar, geothermal, and wind power. In turn, this federal investment would create "millions of skilled jobs," he told me—jobs targeted at communities beset with both pollution and unemployment, jobs that are inherently local and can't be shipped overseas. Jones additionally argued for clean technology educational centers in every public high school and green vocational training centers in every community college. He advocated a system that "doesn't give your daughter asthma and cost you

$10,000 a year for inhalers," a system in which the energy companies "pay you rather than you paying them." Jones was laying the groundwork for a new green politics—an "eco-populism," as he calls it.

The Dream Reborn convention had drawn a diverse group that demonstrated this green movement's populist appeal. I spoke with activists from Los Angeles, Milwaukee, Chicago, Detroit, Newark, and Boston battling a glut of industrial facilities similar those in New Orleans and the South Bronx. I spoke with rural leaders, too: Appalachian activists from West Virginia and Kentucky, where mountaintop coal mining routinely leaches chemicals into local water sources; denizens of the Deep South, where poultry and livestock farms pollute the air and water with millions of tons of animal manure; Alaskans from coastal communities who are losing their homes and livelihoods as the permafrost melts with rising global temperatures.

Some common themes resonated in these conversations—truths that are, in a way, obvious, but also strikingly absent from conventional wisdom. First, a clean environment is a civil right. It is, moreover, a universal human right—one that has often been miscast as a privilege. "You can't have equality without environmental equality," one young organizer told me. And while there are countless laws protecting far-flung wildernesses that most Americans rarely visit and endangered species that we scarcely see, no legislation to date has been tailored to protect the communities that contain the vast majority of America's polluting facilities. Just as the civil rights movement culminated in the passage of the Civil Rights Act of 1964, which prohibited discrimination based on race, religion, gender, ethnicity, or national origin, EJ leaders want to see a cornerstone law passed in the coming decade protecting communities from the unequal impacts of pollution.

EJ leaders tend to emphasize solutions rather than grievances. The best way to change the order of things, most believe, is not to complain that the old system doesn't work but to demonstrate the virtues of a new system. "Dr. King didn't build a movement by saying 'I have a complaint,'" said Van Jones. "King had a dream. He built monuments to hope and possibility." Four decades after King's death, Jones, Carter, and the hundreds of other assembled activists were working to rebuild those monuments.

GREEN OLD PARTY

It was clear to me in New Orleans, the South Bronx, and Memphis that the green movement is undergoing a huge transformation, and that the changes happening within traditional green groups are being paralleled by even more sweeping and startling changes without. Environmentalism is shifting from a fringe political issue endorsed by a narrow slice of American voters—predominately white, wealthy, liberal, and middle-aged—to a more universal movement. Diverse new political coalitions have been forming to advance a shift to clean energy, not only among the EJ leaders I'd encountered, but also among the business leaders I'd met earlier in my travels, including Walmart's H. Lee Scott, GE's Jeffrey Immelt, and Jim Rogers of Duke Energy, who had all praised the financial upsides of green innovation. These and other surprising new coalitions are forming independently, each for its own reason, but they are slowly beginning to strip away the traditional barriers—of race, class, politics, religion, and age—that for decades have separated their members.

Why did contemporary green activism come to be seen as the narrow domain of the liberal elite, when it was formed through events such as the Santa Barbara oil spill and the Cuyahoga River fire in the 1960s, which stirred widespread public concern? This is a question both of perception and reality. The narrow focus of environmental groups has been shaped in large part by the financial realities of their fund-raising process. Recently, in a single year, green groups collectively received about $1.7 billion (a huge amount by nonprofit standards) from foundation grants and generous contributions from a typically well-heeled constituency. To keep their donors happy, these groups have tended to fight high-profile legislative battles via lobbying and litigation expenditures, and to sink sizable investments (to the tune of $400 million a year) into land trusts to protect pristine wilderness areas. Less than about 10 percent of those green-group dollars has been annually allocated to organizing and outreach at the local level, and a small fraction of that has gone to battles in low-income neighborhoods.

Contemporary green activism has also tended to be perceived as the implacable opponent of a force with formidable lobbying and PR heft—big business. Trillions of dollars were made throughout the twentieth century by the defense, auto, plastic, agricultural, and supply chain industries operating as though they existed in a limitless and invincible environment. (We have all been complicit, readily consuming their products, from the plastic fibers interwoven in our clothing to our cell phones and the cheap year-round produce grown by agribusiness.) The green movement evolved in response to a genuine need to rein in these industries—to regulate and restrict their impacts on the environment. But this, in turn, gave activists a general reputation for being hostile to economic progress.

It is only in the past few years that the environmental movement has come to be seen as a potential catalyst—not a hindrance—to economic growth. Global warming has played a large role in this altered perspective, both through the growing consumer consciousness of its perils (which mainstream business has had to allay) and through the stiff greenhouse gas regulations abroad with which America's multinational companies have had to comply. The general upward trend in oil prices—occurring as resources become more remote and riddled with political conflict—has also played a role in motivating companies to take measures to reduce their oil dependence. In recent years, when green groups began to partner with big business in alliances such as the U.S. Climate Action Partnership—a coalition of energy corporations and green groups pushing for federal greenhouse gas limits that includes ConocoPhillips, Shell, and Pacific Gas and Electric—it signaled a reconciliation between longtime foes.

Traditional political barriers are also being crossed in surprising ways. The fissure between environmentalists and Republicans has recently become a lose-lose scenario for both sides, with greens lacking a much-needed constituency and conservatives running the risk of appearing unresponsive in the face of escalating climate change. It is, in point of historical fact, a relatively new political fissure, dating back to the antiregulatory attitudes prevalent in the 1980s. Republican President Theodore Roosevelt, who served from 1901 to 1909, is widely regarded as one of the first environmentalists. "True conservatives," Jim DiPeso of the nonprofit Republicans for Environmental Protection told me, "protect the health and well-being of current and future generations." Conservative philosophy has always embraced "moral obligations to posterity," said DiPeso, and condemned "waste, greed, and materialistic hedonism."

In recent years, a number of moderate conservatives have been trying to reclaim the green movement. Leading the pack is Arnold Schwarzenegger of California, who has vowed to put his state on track to get as much as 33 percent of its electricity from renewable sources by 2020, and has supported landmark legislation to reduce the global warming impact of automobiles through improved fuel mileage standards and the development of cleaner fuels. Florida's Republican governor Charlie Crist signed an executive order calling on Sunshine State utilities to produce 20 percent of their electricity from renewables. Republican Senator Olympia Snowe of Maine has led the charge to curb government subsidies for oil and gas companies, and Republican-turned-Independent mayor Michael Bloomberg has vowed to make New York City one of the world's greenest metropolises by 2030.

Core groups within the Republican constituency, meanwhile, have been independently pushing renewables. Heartland farmers have become impassioned supporters not only of ethanol but also of wind energy. They see a future in which they can earn more revenue from harvesting wind than they can from farming the soil. Ranchers have been protesting the increased domestic natural gas and oil drilling that is encroaching on their lands, arguing for less invasive alternatives. Military hawks such as James Woolsey, meanwhile, see renewable energy as the sharpest weapon in their arsenal against terrorism—a weapon that could eventually bleed Al Qaeda and others of the oil money that funds their acts of violence.

These various interest groups are increasingly cooperating in pursuit of common goals, and as they do so, they are gaining political currency.

HIGHER POWER

One of the most surprising new bastions of green advocacy is currently arising in conservative religious communities. Since the Evangelical Climate Initiative was launched in 2006, it has garnered signatures from more than 240 evangelical leaders calling on all Christians to join the fight against climate change and pressuring lawmakers to implement mandatory restrictions on planet-warming emissions. More and more evangelical churches have been preaching a gospel of "creation care," an ethic inspired by biblical passages in which God calls on humanity to "to watch over and care for" the earth and its creatures (Genesis 2:15). The Book of Revelation states that "God will destroy those who destroy the earth."

Reverend Richard Cizik, former vice president for governmental affairs of the National Association of Evangelicals, is at the helm of the creation-care movement. NAE boasts a membership of 30 million that includes the members of forty-five thousand churches and seven thousand megachurches, some with billion-dollar budgets. "We represent 40 percent of the Republican Party," Cizik told me during a phone interview from Washington as he dashed between meetings with members of Congress. Though NAE is not an ally in most respects to the traditional, left-leaning environmental base—among other hot-button political issues, it stands against abortion, gay marriage, and embryonic stem cell research—there are some points of congruence. "We are commissioned by God the Almighty," Cizik said of the organization's environmental beliefs, "to be stewards of the earth. It is rooted not in politics or ideology, but in the scriptures. Genesis 2:15 specifically calls us 'to watch over and care for' the bounty of the earth and its creatures. Scripture warns that the earth is not ours to abuse, own, or dominate."

In addition to throwing its political heft into lobbying for federal climate change legislation, NAE takes a more personal approach to climate activism, promoting lifestyle changes among its members. It asks believers to engage in "creation-friendly" living habits by recycling and conserving energy. "There are still plenty who wonder: 'Does advocating this agenda mean we have to become liberal weirdos?'" Cizik mused. "And I say to them, 'Certainly not. It's in the scripture. Read the Bible.'" Reverend Jim Ball, president and CEO of the Evangelical Environmental Network, has described energy conservation as a sacred act in line with the Golden Rule: "Energy efficiency is a new way to love your neighbor; it's a new way to love your kids."

Other religious communities have also taken up the mantle of climate activism. Interfaith Power and Light is a coalition of four thousand congregations in twenty-eight states that collaborate on the Regeneration Project, which organizes the aggregate purchasing of renewable energy in an effort to help scale up demand and bring down cost. "We focus on tangible results in congregations—putting our faith into action," said the coalition's founder, Reverend Sally Bingham. This work includes educating congregations, helping them buy energy-efficient lights and appliances, providing energy audits of members' homes, sending "Green Teams" to help wrap boilers and change out bulbs, and developing carpool networks so people drive less. Rabbi Julian Sinclair, cofounder of the global Jewish Climate Initiative, is working to "inspire a vision for the future that matches the magnitude of climate change" and exhorts his organization's members to "live better and walk more wisely and responsibly on this earth." Emily Askew, a professor of theology at Lexington Theological Seminary, said of a recent Jewish, Christian, and Muslim interfaith dialogue on climate change, "This may be the issue that hurtles us over our differences."

To get a feel for the kind of movement building that's happening in faith-based communities, I accompanied a friend to a recent Sunday service at a Nashville church actively involved in climate outreach. The pastor opened the service by placing four plastic water bottles on the altar, one of them filled with oil. "Producing the [water] bottles for American consumption requires the annual equivalent of more than 17 million barrels of oil," he told his congregation. "That's enough to fuel more than 1 million cars for a year." He condemned the waste that accompanies America's water habit: 85 percent of all PET bottles for noncarbonated drinks end up in landfills or as litter in parks. That's 24 billion empty water bottles, 66 million every day. "Can you imagine the size of that pile? When I look at that mountain [I think,] Lord God, what are human beings that you put up with us? . . . I often wonder if the planet wouldn't be better off without us, kings and queens who trample all over the things put under our feet."

At the end of the communion line—after the church members ate bread and sipped grape juice—each was handed a four-pack of energy-efficient lightbulbs along with the reminder to "spread light throughout the world—but not pollution!" As he offered my friend a packet of bulbs, the pastor added: "You are blessed. Be a blessing on this earth."

COMMON PURPOSE

The growing green activity in low-income neighborhoods, in the business arena, in the military, in farming, in faith communities, and throughout the Republican Party bodes well for a robust and expansive environmental movement in the coming years. It is this emerging consensus, rather than the agenda of any one specific cause or interest group, that has the potential to effect large-scale change at a federal level. But this change will face entrenched resistance from many members of Congress and industry lobbyists; and there will be much debate about the exact form it can and should take, and over what timeline. "Profound and powerful forces are shaping and remaking the world," as President Bill Clinton put it in 1993. "And the urgent question of our time is whether we can make change our friend and not our enemy."

President Obama has promised a swift and aggressive push toward climate solutions and energy independence. In February 2009, the new administration began the process of implementing its sweeping economic stimulus plan aimed at providing relief to middle- and lower-income Americans through measures that included significant funding for green jobs. "We can create millions of jobs," Obama said, "starting with a twenty-first-century recovery plan that puts Americans to work building wind farms and solar panels and fuel efficient cars. We can spark the dynamism of our economy with long-term investment in renewable energy that will give life to new businesses and industries. . . . We will make public buildings more efficient, modernize our electricity grid, [and] reduce greenhouse gas emissions."

To help shape his administration's energy, economic, and environmental policies, the president selected advisors and cabinet members who have strong backgrounds in climate science, renewable energy development, and environmental justice activism. "The green talent in this administration is staggering and without precedent," said Bracken Hendricks of the Center for American Progress. These officials have described their roles as having momentous historical significance. Obama's secretary of energy, Steven Chu, a Nobel Prize–winning physicist and renewable energy proponent, said when accepting his post that what America does to transform the energy sector "in the next decade will have consequences that will last for centuries." Obama's top White House environment advisor, Nancy Sutley, a longtime renewable energy advocate in California, has pledged to "do my part to restore the American dream to the fullest measure by strengthening our economy and creating a more sustainable future." Obama's EPA chief, Lisa Jackson, who worked to cut greenhouse gas emissions as a state-level official in New Jersey, has vowed to tackle a unifying concern relevant to all Americans: "Worried about the economy? I've got a green answer for you. . . . Worried about security? I've got an answer for you: the clean energy economy."

Obama's green team includes unconventional assignments and roles, too. Hilda Solis, Obama's labor secretary, is a longtime environmental justice activist who ardently pushed green jobs legislation when she was in the House of Representatives. Van Jones was named the administration's Special Advisor for Green Jobs, Enterprise and Innovation, working inside the White House Council on Environmental Quality. Obama also created the new position of energy czar at the White House level to coordinate activities across the many departments that deal with energy, economic, and climate policy. This title went to Carol Browner, the former EPA director for the Clinton administration and a zealous supporter of green jobs and clean energy innovation. "It is a tremendous responsibility," she said when accepting the appointment. "It is a pivotal moment in our nation's history. The challenges are immense but the opportunities are even greater."

The League of Conservation Voters president Gene Karpinski characterized the Obama adminstration's first 100 days as "the most environmentally important of any in history." In addition to allocating tens of billions of federal dollars for clean-energy industries, President Obama directed the Department of Transportation to increase fuel economy requirements for Detroit's vehicles. He toured factories producing wind turbines, solar panels, and biofuels, and aggressively promoted legislation that would cut greenhouse gases 83 percent below 2005 levels by 2050. His EPA reclassified "carbon dioxide"—formerly considered a benign compound—as a pollutant that poses a danger to public health and welfare.

President Obama has also introduced a new urgency to the public discourse: "Each day brings further evidence that the ways we use energy strengthen our adversaries and threaten our planet," he said at the opening of his inauguration speech, and went on to invoke the issue no fewer than six times. It was unprecedented: the issue of energy had never been mentioned in a previous inaugural address. The new president brushed off skepticism about his plans to transform America's energy landscape, improve its economy, and protect the environment at the same time. "There are some who question the scale of our ambitions—who suggest that our system cannot tolerate too many big plans," he said. "Their memories are short. For they have forgotten what this country has already done; what free men and women can achieve when imagination is joined to common purpose, and necessity to courage." As I listened to Obama's words standing among the euphoric throngs on the National

Mall, I thought through the succession of big plans I'd encountered in my research of America's energy past: FDR's New Deal, Truman's Marshall Plan, Eisenhower's interstate highway system, Nixon's Environmental Protection Agency, Carter's bid for energy efficiency, Clinton's NAFTA.

These are encouraging precedents for sweeping federal initiatives. But history also shows that truly transformative change—the kind brought about by cultural revolutions, including the women's suffrage movement of the 1910s, the civil rights movement of the 1960s, the environmental movement of the 1970s, and the overthrow of Communism in the 1980s—doesn't happen from the top down, it happens from the ground up. The solutions to our energy crisis cannot be found and implemented through federal leadership alone; they must be understood and embraced by all Americans. "We've got to stop crying and start sweating, stop talking and start walking, stop cursing and start praying," as Jimmy Carter put it in a 1979 address to the nation, quoting an activist friend. "The strength we need will not come from the White House, but every house in America."

REGENERATION

A vibrant popular movement appears to be rising out of the businesses, churches, and low-income neighborhoods I visited—one that spans the partisan divide. But none of these efforts would have any chance of enduring if they didn't resonate with America's youth.

"When people talk about the planetary impacts that we'll be facing in 2050, it's not hypothetical to us," twenty-six-year-old Jared Duval told me. "That's just about the time I'll be retiring." Duval has taught economics in Dar es Salaam, Tanzania, served as Howard Dean's youngest policy advisor during the 2004 presidential primary season, and directed the Sierra Student Coalition, the national student chapter of the Sierra Club and the largest student environmental organization in America. A graduate of Wheaton College in Massachusetts, Duval wants to devote his life "to enhancing democracy," as he explained it to me, "in ways that can realize historic opportunities for sustainable human progress."

Duval's rhetoric and résumé are not uncommon among his peers—members of the millennial generation who showed up in record-shattering numbers to vote in the 2008 election (a 65 percent turnout among youth voters, more than ever before in American history). These young people don't exhibit the "slacker" mentality that became the media stereotype for the generation that preceded them; that's a good thing, given how much work they have ahead of them and how much they have at stake.

Duval is part of a burgeoning climate movement involving hundreds of thousands of youth activists nationwide. "I'm a grizzled veteran at this point," he noted. "The real action is among the students—college and younger." Closely allied with environmental justice activists such as Van Jones and Majora Carter, these young leaders are harnessing the Internet and other modern communication tools to spread information, forge new grassroots organizing tactics, mobilize voters, green their campuses, and pressure lawmakers.

One of the principal engines of this youth movement is the Energy Action Coalition, a student-led campaign that aggregates hundreds of local student organizations nationwide under the motto "Youth united for clean and just energy." It launched a Campus Climate Challenge in 2006 that has since garnered commitments from hundreds of U.S. institutions of higher learning—ranging from Ivy League schools such as Princeton and Brown to small private colleges such as Bowdoin and Reed to agricultural and technical universities such as Western Kentucky and Appalachian State—to go carbon-neutral through investments in renewable energy, efficient dorms and facilities, green roofs, and hybrid vehicle fleets.

Today there are more college student organizations devoted to fighting climate change than to any other issue. Student climate blogs now number in the hundreds, and many universities are expanding their climate-related curricula.

Like the college students who opposed segregation at lunch counter sit-ins throughout the 1960s South, these activists are thinking outside the campus quad. In 2008, they orchestrated a Power Vote campaign that collected nearly a million signatures from students pledging to vote and endorsing a shift to clean energy. In March 2009, youth activists staged a "Power Shift" conference in Washington, D.C., pressuring the new administration and the new Congress to adopt radical shifts in policy. "You can't solve this problem with incremental change," one student organizer told me. "It's worse than doing nothing at all. We need an avalanche."

Power Shift's organizers have supported a complete freeze on all coal plant construction, mandates for plug-in cars and bold renewable energy targets, a federal cap that would cut greenhouse gas emissions not just over the long term, but immediately (25 percent by 2015), and reforming trade policies to protect environmental conditions overseas. In daily planning sessions, twelve thousand students from all fifty states convened according to their geographical regions to discuss strategies for pressuring their mayors, governors, and congressional representatives to promote their goals. Many in the crowd wore emerald-colored hard hats, symbolizing an era of green jobs. The green hard hat is their answer to the panda bear that signified the protection of remote endangered species in the past—a symbol for the new populist green politics.

For all their idealism, the youth activists I've met are eager to ground their big-picture goals in pragmatic outreach programs. Timothy DenHerder-Thomas, a twenty-two-year-old from Newark, New Jersey, is a senior at Macalester University in St. Paul, Minnesota. As a sophomore, he established the Macalester Conservation and Renewable Energy Society (MacCARES), in which he partnered with Minnesota farmers developing rural wind farms to buy offsets for the university. He assembled an advisory panel of tax lawyers, investment experts, and university administrators to help him establish a Clean Energy Revolving Loan Fund for investments in on-campus technologies ranging from low-flow showerheads and waterless urinals to efficient lighting and electric shuttle buses—innovations that reduced the university's utility and fuel costs, and over time generated savings that could be reinvested in more green projects. "The idea is simply to fund environmental projects through the savings that they generate," DenHerder-Thomas told me.

The fund has been so successful within Macalester that DenHerder-Thomas applied the framework outside of campus. He founded Cooperative Energy Futures, an organization that assembles coalitions of neighbors to collectively invest in efficiency measures and on-site solar panels for their homes. The organization educates middle-income communities about the financial logic of environmental upgrades—demonstrating how the savings they achieve are far greater over time than up-front costs—and creates mechanisms for loans that families and communities can gradually repay. The bulk purchasing of materials for green home retrofits both lowers cost and expands the market for green contracting.

DenHerder-Thomas sees a lucrative industry emerging through these kinds of community-led energy and efficiency investments. "Right now we're seeing relatively modest investments of about $4,000 in efficiency measures per home and savings of about $1,000 a year on energy costs, so the payback adds up quickly. If you were to do this on a nationwide scale, you get a multi-hundred-billion-dollar industry annually. It's just not being tapped right."

While DenHerder-Thomas welcomes the role of digital-era tools—e-mail, blogs, listservs, YouTube, Twitter, Skype—in youth climate outreach, he tends to avoid technology in his personal life. He does not own a cell phone or a car. He commutes exclusively by bike, even to rural meetings and through the Minnesota winters, cycling on average "probably thirty miles a day." He insists that no amount of online organizing can ever take the place of engaging people face to face: "Facebook won't convince people to transform their lives and communities, human connection will." He believes that collective investments in next-generation technologies can foster a growing collective consciousness. "By getting neighbors to pool resources and work in partnership," he told me, "you create social norms. You remove the barriers that often inhibit lifestyle changes." He considers himself a realist, not an idealist. "Maybe it's ridiculous to imagine that a bunch of students and farmers and community members could lead a societal revolution that fundamentally transforms the foundation of our entire global economy from the ground up," he said. "But it happens."

Other youth activists I encountered take the same view—that their local efforts are part of a historic shift. Marcie Smith, a recent graduate of Transylvania University, has helped to organize a Clean Energy Corps to improve energy efficiency in a hundred low-income homes in Kentucky. Jesse Jenkins, a graduate of the University of Oregon, founded the Cascade Climate Network, an organization of students at universities throughout Washington and Oregon who are promoting community-scale renewable energy projects throughout both states. Nineteen-year-old Marisol Becerra developed an award-winning interactive map of industrial pollutants in her inner-city neighborhood in Chicago, which houses two major coal plants. She has worked with another young Chicagoan, Martin Macias Jr., to develop an after-school program for local public schools that teaches principles of sustainability. Macias has also raised funds from the state of Illinois that will help retrofit all facilities in the Chicago public school system with green improvements. And Chelsea Chee, a recent graduate of the University of Arizona, works with the Black Mesa Water Coalition, a nonprofit that fights coal mining and promotes green jobs on Navajo lands.

Other student activists are currently engaged in forging international partnerships with youth organizers. Jared Duval told me, "When we began building the youth movement in 2005 the question was: how do you make campuses a model of sustainability? Then we added another layer: how do you push for national solutions? Now it's: how do you build a global movement equal to the task of solving a global problem?"

Morgan Goodwin, who attended Williams College in Massachusetts, has worked with student organizers in China to seed the movement in their nation. May Boeve, a graduate of Middlebury, helped found the organization 350.org, based on the number that scientists have identified as the safe upper-limit of CO_2 in the atmosphere (350 parts per million). "We want to make 350 the most important number in the world," Boeve explained; the group has raised millions to fund demonstrations worldwide advocating policies that reduce global emissions below this thresh-

old. Caroline Howe, a recent graduate of Yale University, is developing
and advocating climate solutions in India via the Indian Youth Climate
Network. In early 2009, Howe spearheaded an electric-car caravan powered by solar energy that traveled 2,100 miles throughout India, from
Chennai to New Delhi, training climate organizers and educating local
residents about homegrown climate solutions.

For many of us, the combination of global warming, war in the Middle
East, and economic recession seems a more daunting, complex, and cata-
clysmic challenge than any America has faced in the past. But these young
people are so rigorously solution-oriented that they don't seem to have
the time or energy to fear the problems themselves. When I pressed the
student activists to describe their thoughts about this triple threat, they
were unflappable. Every generation faces mortal threats, they reasoned,
ticking off the challenges of their forebears: slavery, civil war, world war,
genocide, nuclear proliferation, bigotry, AIDS. And while today's crises
are gargantuan and harrowing, they all have connections to a common
source—energy—and can therefore be managed with a common strategy.

The young activists' impulse toward optimism in the face of adversity
is perhaps best captured in this statement by Martin Luther King Jr.:
"When our days become dreary with low-hovering clouds and our nights
become darker than a thousand midnights," he said when accepting the
Nobel Peace Prize in 1964, "we will know that we are living in the creative turmoil of a genuine civilization struggling to be born."

MORNING IN AMERICA

It is often hard to believe, I'll admit, that our civilization is really struggling to be reborn in a genuine way at a time when we are constantly,
blithely assured that we can "green our lifestyles with ten easy steps," that
we can be "lazy environmentalists," that we can "lighten our footprints"
by driving hybrid-engine luxury SUVs that get 25 miles per gallon. Environmental lip service abounds these days. Every company from Gerber
to Peabody Coal is putting a shiny green gloss on its products. It is dan-

gerous to assume that there will be an easy fix to the energy and environmental problems we face—that just by buying a high-performance car or wrapping our boilers or taking communion at a green church we are doing our part to solve America's escalating crises. It belittles the problems we face to promote the myth that half-measures will suffice.

But I don't think that what we are witnessing today is merely a half-hearted trend of greenwashing. As individual nodes of activism grow and expand—within churches, on college campuses, in EJ communities, throughout America's workforce, and in the halls of diplomacy—they are breaking down the barriers of exclusivity. We are witnessing the growth of a postpartisan, cross-industrial, interfaith, multiracial movement that spans class lines, city limits, state borders, and nationality. Eventually, I believe, it will unify the globe. We will see the growth of a global green economy, strong enough to completely redesign our industries. Strong enough to reverse prodigious trends in global warming. Strong enough to end forever the need for wars tied to oil.

What we are witnessing isn't just the birth of new technologies or a new economy, it's the birth of a new politics—and potentially the birth of a new consciousness. "The solution of our energy crisis can also help us to conquer the crisis of the spirit in our country," President Jimmy Carter said in a 1979 address to the nation as America continued to grapple with the effects of the Arab oil embargo. "It can rekindle our sense of unity, our confidence in the future, and give our nation and all of us individually a new sense of purpose." These sentiments were largely overlooked in Carter's era, but their time may finally have come. As new groups and individuals are drawn into the green movement, they are joining a collective consciousness with the power to create a society more mindful, more respectful, more reverent of the natural world.

As I was walking through the Ninth Ward watching families move into their freshly painted houses, I came across an elderly man who was pushing a small lawn mower along a communal patch of grass by the sidewalk in front of the growing settlement. He was still living in a FEMA trailer, but he would soon be moving with his wife and grandchildren into a green home near Collins Foots's and Melba Leggett's. Tied to his belt loop was a plastic bag that he was filling with candy wrappers, beer cans, plastic soda bottles, and other debris from his street. "I used to throw my trash on the ground," he told me. "Now the ground is something sacred."

It reminded me of the comment Kim Charles made in her Habitat for Humanity home in Tennessee, likening her household chores to gardening—mindful of "what the sun is offering" that day. I thought of the neighborhoods in Minnesota that are building wind cooperatives and those in California whose residents are linking together their homes in solar networks. I thought of T. Boone Pickens transforming himself from an oil tycoon to a wind-energy evangelist. I thought of the youth climate activists building a global outreach network. These American stories represent not just technological progress but a shift in our collective consciousness.

The United States in its early days represented a shift in consciousness of some magnitude to the people of other nations around the world. French social historian Alexis de Tocqueville, who traveled throughout United States in the 1830s, wrote about this country's extraordinary resilience in his classic work *Democracy in America*. As we look forward to the unknown energy future, these words that offered consolation and encouragement nearly two centuries ago still have resonance today: "The greatness of America lies not in being more enlightened than any other nation, but rather in her ability to repair her faults."

Acknowledgments

I am grateful to the following travel companions who took me in, fueled me up, let me get lost, and steered me back on track during the three years I spent researching and writing this book. Richard Abate, my agent, had the patience of a Holland Tunnel tollbooth operator, fielding idea after idea as I pieced the concept together and labored to get it on paper. He launched the trip and saw it through with clarity and conviction.

My editor, Gail Winston, was as wise and agile a guide as I could have hoped for. Editing a book that skips from the Pentagon to pantyhose in the span of a few pages is, I'd imagine, like herding cats; Gail's focus, tenacity, and faith in this project enabled the straying pieces to come together. She kept her eye on the big picture as I got lost in the details, gave the story room to shift and evolve, and cheerfully pushed it, chapter by chapter, along its way. She was gentle and tough, practical and inspired—a pleasure and a privilege to work with. Thank you to Jason Sack for expertly leading me through the production process, to Heather Drucker for her marketing savvy, to Richard Ljoenes for his provocative cover design, and to Jonathan Burnham for a life-changing opportunity.

It would have been well worth the trip just to work with Deborah Murphy. She came on as an "active reader" and ended up my Ambassador of Quan. Deborah understands storytelling on every level—technically, strategically, poetically—like a mechanic, a dealer, and a NASCAR driver understand cars. She immersed herself in research, crafted transitions, coaxed out characters, restructured chapters, and championed the nubby details. I am immensely grateful for—and forever changed by—our collaboration. Thank you also to Anthony Walton for his prodigious talent, his sage counsel—and his wife.

Thank you to the hosts of my journey: Paul Siegele of Chevron, Al Shaffer of the Pentagon, Dr. Grant Stevens of Marina Plastic Surgery Associates, Ken McCauley of McCauley Farms, Lou Rana and Dennis Romano of Con Edison, T. Boone Pickens of The Pickens Plan, Dan Reicher of Google.org, Richard Cook and Bob Fox of Cook + Fox Architects, Tom Darden of Make It Right, Van Jones of Green For All and the Obama administration, and Majora Carter of the Majora Carter Group. They are builders and pioneers of our energy landscape, and they shared generous hospitality and insight as I explored their turf.

Equal thanks to the scholars and luminaries who educated me on all things energy: David Painter, James Woolsey, Dan Nolan, Daniel Schmidt and his colleagues at University of Massachusetts Lowell, Paul Hesse and his team at the Energy Information Administration, Richard Munson, Malia Mills, Tom Philpott, Fred Below, Edgar Blanco, Marc Levinson, Kurt Yeager, Jeff Christian, Ashok Gupta, Joseph Romm, Bracken Henricks, Felix Kramer, Reid Ewing, Richard Bernhardt, and Jared Duval.

Daniel Yergin was both an educator and a trailblazer throughout my research. His Pulitzer Prize-winning book *The Prize: The Epic Quest for Oil, Money and Power,* which I reference more than a dozen times in these pages, was a key inspiration for *Power Trip,* and should be read by anyone who wants a deeper knowledge of this topic.

Brian Sides—researcher, sounding board, and fellow adventurer—debated and shaped many of the ideas in this book. He contributed his knowledge and opinions on topics as varied as race cars, economics, farm policy, and war tactics. Michael Hill provided excellent archival research, plumbing the stacks of the Library of Congress and digging up century-old periodicals to bring alive the stories of the past. Angela Watercutter, *Power Trip*'s fact-checker, met implausible deadlines and still made her fact-finding into a kind of poetry. I am immensely grateful for her rigor and her good judgment.

Heidi Ross gave the book pictures. She created the circular map-pin images, supervised stock photo research, and shot the portraits. Clark Williams-Derry is the wry wordsmith who helped with section headings and chapter titles. My brother-in-law Courtney Little, a NASCAR savant, taught me about checkered flags, carburetors, and the need for speed. Emily Volz offered her knowledge and insight on the solar industry (to say nothing of her expertise on poop power). My industrious intern, Mary Cooper, organized heaps of research, exhibiting what I can only describe as a human form of renewable energy.

I greatly benefited from the encouragement of other writers. Kathryn Schulz was my compass and cotraveler. Alix "Sue Ellen" Barzelay, Alice Randall, Olivia Barker, Amely Greeven, Jeff Howe, and Warren St. John offered me invaluable advice and feedback. Thank you to my colleagues who encouraged my reporting as I researched this book—especially Chip Giller and Lisa Hymas at Grist.org, Chris Keyes and Dianna Delling at *Outside,* and Nick Thompson at *Wired.*

The women of WNWC served as a spirited focus group for the stories in these pages; special thanks to Mary Brooke Bonadies, who inspired the subtitle. Thank you to my officemate Barbara Rhodes, and to Robin Porter, who was both a family member and a colleague—offering instrumental ideas and research while caring for my daughter when I couldn't be at home.

My brothers Rufus and Bronson are my closest advisers and dearest friends; they read through these pages with expertise and candor. I never would have tried my hand at journalism or found my way to the subjects of energy and the environment without their influence. I am deeply grateful to my mother Nancy Florance and her husband Colden for helping me see this project through with their tireless support and a steady supply of newspaper clippings. I am deeply grateful to my history-loving father Rufus Griscom and his wife Hopie, who encouraged me to consider the topic of energy from many perspectives. Heartfelt thanks to Lady Lyn Little, and to the memory of Frank Little, which motivated me every day. Thank you especially to my feisty Grammy, who taught me to love to read, to have an opinion—and to voice it.

Above all, to my husband Carter, who gave immeasurable time, patience, and inspiration to this project: Thank you for feeding me, cheering me, talking me down, lifting me up, scouring the news, braving five seasons of *Dallas*, and doing the best damn impression of Jock Ewing this side of Fort Worth. On this road trip—as he does on all our adventures—Carter kicked the tires and lit the fires. Last, to our daughter Aria, who accompanied me up skyscrapers, across farmlands, into the Pentagon, and onto private jets, who pushed me along and pulled me home, this journey is yours as much as it is mine.

The writing of *Power Trip* was as much a journey through books, government data, academic studies, magazines, newspapers, Web sites, and the occasional TV show as it was a voyage through the American landscape. This is a book that skims across the surface of many topics rather than diving deeply into a single field of study. Which is another way of saying it only scratches the surface. I benefited greatly from the expertise and research of many other authors, historians, and journalists. I hope these notes will be a gateway for readers to further explore the characters, places, and events that I only briefly described. A version of these notes with links to online materials can be found at www.amandalittle.com/powertrip. Please contact me at info@amandalittle.com if you would like further information about sources.

1. Over a Barrel

3   *The oilfield known as "Jack"*: Some of the material from this section first appeared in a feature I wrote in *Wired* magazine, "Going Deeper," September 2007. Also see: Jesse Bogan, "Shell's Radical Rig," *Forbes*, October 30, 2008 for an article on more recent developments in the Gulf of Mexico's ultradeep drilling frontier.

3   *largest discovery in the U.S. since 1968*: Steven Mufson, "U.S. Oil Reserves Get a Big Boost," *Washington Post*, September 15, 2006.

4   *director of Chevron's offshore drilling*: Since the time of our voyage in 2007, Siegele has been promoted from vice president of Offshore Drilling to a higher executive post within the company.

5   *Gulf yields 25 percent*: From the Energy Information Administration (EIA), a research arm of the Department of Energy. Its data on annual crude oil production can be found here (Gulf of Mexico data is listed as Federal Offshore [PADD 3]: http://tonto.eia.doe.gov/dnav/pet/pet_crd_crpdn_adc_mbbl_a.htm.

5   *Three out of four exploration wells*: From my interviews with Siegele. All of the subsequent material in the chapter on Chevron and its deep-sea rigs came from my interviews with Siegele and other employees of the oil company.

6   *Chukchi Sea of the Arctic Ocean*: Details on the conditions in this region from an email exchange with Jonathan Gupton, industry economist at Energy Information Administration, www.eia.doe.gov.

6   *workers have to carry gas masks*: "Developing Kashagan is Painfully Slow," *Financial Times*, January 2, 2007.

6   *On Sakhalin Island*: See Stanley Reed, "Sakhalin Island: Journey to Extreme Oil," *BusinessWeek*, May 15, 2006.

6   *five workers died*: From Sakhalin Energy company's 2005 Annual Review, www.sakhalinenergy.com/en/documents/sakhalin05_Eng.pdf.

6   *fewer than eight hundred fields*: The statistics on supergiants came from an interview with Trevor Morgan, senior economist at the International Energy Agency.

7   *three-quarters of the world's oil*: I took into consideration the following countries when compiling this statistic: Sudan (5 billion barrels), Iran (136 bbl), Iraq (115 bbl), Venezuela (99 bbl), Congo (1.9 bbl), China (16 bbl), Kuwait (104 bbl), Libya (43.7 bbl), Nigeria (36.2 bbl), Saudi Arabia (266.7 bbl), Russia (60 bbl), and the United Arab Emirates (97.8 bbl). That amounts to roughly 981 billion, or roughly 73 percent of the world's reserves. The Energy Information Association has a good comparative breakdown of known reserves in oil-producing nations: http://www.eia.doe.gov/emeu/international/reserves.html.

9   *9,500-horsepower engines*: From Guy Cantwell, head of Communications at Transocean, the company that manufactures the rigs, supplied the information on the thruster's power requirements. www.deepwater.com.

9   *40,000 gallons of diesel*: From my interviews with Chevron engineers.

9   *13,300 Hummers*: The average daily commute is about 32 miles. A Hummer gets roughly 11 mpg for actual mileage, consuming roughly 3 gallons per day

on a standard commute. The Cajun Express, by comparison, consumes a total of roughly 40,000 gallons a day, which could therefore power 13,333 Hummers on a typical day of driving.

10   *burn fossil fuels to harvest them:* The numbers on the diminishing ratios of energy invested to oil produced come from an article by Charles A. S. Hall and John W. Day Jr., "Revisiting the Limits to Growth After Peak Oil," *American Scientist*, May–June 2009, p. 230.

13   *$250 million blow:* David Greising, "Troubles Run Deep on Gulf Oil Platform," *Chicago Tribune*, May 28, 2007.

14   *hold steady—not decline:* "New EIA Energy Outlook Projects Flat Oil Comsumption in 2030," Department of Energy, December 17, 2008, www.eia.doc .gov/neic/press/press312.html.

14   *global demand for energy could triple:* Andrew Revkin, "The Energy Gap and the Climate Challenge," *New York Times*, July 10, 2008, http://dotearth.blogs .nytimes.com/2008/07/10/the-energy-gap-and-the-climate-challenge/. In this statistic, Revkin was referring to demand for all energy sources including fossil fuels and new renewables. Demand for oil alone is also expected to rise. The CNBC documentary *The Hunt for Black Gold* reports that global oil consumption is expected to jump 50 percent by 2030.

15   *4 percent shortfall . . . 177 percent increase:* From Oil Shock Wave, a computer program that simulates oil crises—terrorist attacks on pipelines in Alaska, an Iranian oil embargo—launched at Harvard's Kennedy School of Government and commissioned by Belfer Center for Science and International Affairs. www.oilshockwave.com.

15   *less than 10 percent . . . roughly two thirds:* "Petroleum Basic Statistics," EIA, February 2009.

16   *90 percent of the world's oil:* A Citizen's Guide to Nationalized Oil Companies, Part A: Technical Report, published by the World Bank and the Center For Energy Economics at University of Texas, Austin, October 2008.

17   *U.S. forces . . . costs U.S. taxpayers some $100 billion a year:* Anita Dancs, "The Military Costs of Securing Energy," National Priorities Project, October 2008, p. 35.

17   *"$2.5 trillion . . . at risk":* Fredrich Kahrl and David Roland-Holst, "California Climate Risk and Response," University of California, Berkeley, November 2008, p. 10.

19   *"It is the light of the age":* Quoted in Daniel Yergin, *The Prize* (New York: Simon & Schuster, 2008), p. 12.

19   *"the gift of 'new light' ":* Ibid., p. 38.

19   *two rogue wildcatters disagreed:* James Anthony Clark and Michel Halbouty,

Spindletop: The True Story of the Oil Discovery That Changed the World (New York: Random House, 1952, Chaps. 1–3).

19– *"tiny bodies of animals buried in the sediments"*: Mikhail Lomonosov, "Slovo o
20 reshdenii metallov ot tryaseniya zemli," Proceedings of the Imperial Academy of Science, St. Petersburg, 1757.

21 *"people of Beaumont . . . are in a feverish state"*: Jason Grumble, "A Big Oil Geyser," *Dallas Morning News*, January 11, 1901.

21 *"There is wild excitement"*: "Big Oil Strike In Texas," *New York Times*, January 13, 1901.

21 *"weary and oil-saturated Hamills"*: Halbouty, *Spindletop*, p. 12.

22 *"Texas oil was burning in Germany"*: Quoted in "Texas Became Texas," *American Heritage*, April 1977.

23 *"Now every bubble of gas"*: "Albert Phenis' Story," *Galveston Daily News*, February 3, 1901.

23 *One farmer sold his land for $20,000*: David Uhler, "The Gusher That Changed the World," *San Antonio Express-News*, January 10, 2001.

23 *Oil was regarded . . . like a wild animal*: Anthony Sampson, *The Seven Sisters: The Great Oil Companies and the World They Made* (New York: Viking Press, 1975), p. 76.

23– *"Oil City, with its one long crooked"*: John Herbert Aloysius Bone, *Petroleum
24 and Petroleum Wells: A Complete Guidebook* (Philadelphia: J.P. Lippincott & Co, 1865), p. 80.

24 *3.5 million barrels of oil in 1901*: Numbers on Spindletop's declining production from E. DeGolyer, "Anthony F. Lucas and Spindletop," *The Southwest Review*, vol. 31, Fall 1945.

24 *"The cow was milked too hard"*: Quoted in ibid.

25 *"It is the most fortunate thing"*: Quoted in "Spouts Fuel Oil: Product of the Big Beaumont Geyser is too Light for Illuminating Purposes," *Dallas Morning News*, January 17, 1901.

26 *Texas Company . . . Joseph Cullinan*: The story of Buckskin Joe came from Sampson, *Seven Sisters*, p. 40.

26 *$235 million had been invested*: Halbouty, *Spindletop*, p. 112.

27 *Rockefeller's . . . instincts took hold*: Most of the Rockefeller biographical material was culled from a combination of the following: My interview with Daniel Yergin and *The Prize*, ch. 2; H. M. Briggs, "An Intimate View of John D. Rockefeller," *American Magazine*, vol. 71, November 1910; Ron Chernow, *Titan: The Life of John D. Rockefeller, Sr.* (New York: Random House, 1998); Ida M. Tarbell, *The History of the Standard Oil Company* (New York: McClure, W. W. Norton & Co., 1966); The PBS *American Experience* documentary *The*

Rockefellers, and from a supporting essay for PBS by Keith Poole, a professor of political science at the University of Houston: www.pbs.org/wgbh/amex/rockefellers/peopleevents/p_rock_jsr.html.

28   *"Many of us who fail . . . achieve"*: Quoted in Chernow, *Titan*, p. 174.

29   *"A man of words and not deeds"*: Quoted ibid., p. 174.

29   *"A wise old owl"*: Quoted in Yergin, *The Prize*, p. 47.

29   *"The most unemotional man"* . . . *"I guess he's 140"*: Quoted in ibid., p. 31.

29   *"I am eating celery"*: Quoted in Chernow, *Titan*, p. 218.

30   *"Almighty had buried the oil"* . . . *"The Lord will provide"*: Ibid., p. 283.

31   *"To know every detail of the oil trade"*: Tarbell, *The History of the Standard Oil Company*, p. 110.

32   *"Like an epic card game"*: Quoted in Yergin, *The Prize*, p. 88.

32   *"Standard Oil people . . . system of morality"*: Theodore Roosevelt and Willis Fletcher Johnson, ed., *Addresses and Papers* (New York: The Unit Book Publishing Co., 1909), p. 423.

33   *"[It] became a kind of morality play"*: Sampson, *The Seven Sisters*, p. 42.

33   *"It must in good time be perceived"*: Quoted in "Rockefeller Sees No Portent of Disaster," *New York Times*, October 20, 1907.

33   *"Well, gentlemen, shall we proceed?"*: Quoted in Yergin, *The Prize*, p. 109.

36   *"produce all possible oil"* ": Carl Solberg, *Oil Power* (New York: Mason/Charter, 1976), p.73.

36   *"inducements to produce the stuff"*: Ibid, p. 75.

36–   *"Travel but a little in the country,"*: E. H. Davenport, *The Oil Trusts and Anglo-*
37   *American Relations* (New York: Macmillan Company, 1924), p. 2.

38   *"110 million population, 90 percent . . . lawyers"*: Quoted in Carl Solberg, *Oil Power* (New York: Mason/Charter, 1976), p. 6.

38   *"spotted with this flood of oil"*: Quoted ibid.

39   *"Eisenhower's smashing personal victory"*: Robert Engler's, *The Politics of Oil: A Study of Private Power and Democratic Directions* (New York: MacMillan, 1961) p. 358.

38–   *tens of millions of acres:* From an interview with David Alberswerth, senior
39   policy advisor of the Wilderness Society.

39   *in 1970 when the country hit peak oil:* Information on peak oil in America can be found at www.theoildrum.com or in *Hubbert's Peak: Impending World Oil Shortages* by Kenneth Deffeyes (Princeton, New Jersey: Princeton University Press, 2008).

40   *more than fifty Oval Office staff members:* Lita Epstein, C. D. Jaco, and Julianne Neiman, *The Complete Idiot's Guide to the Politics of Oil* (Indianapolis, Indiana: Alpha Books, 2003), p. 197.

40 *"Oil drilling isn't for the pessimist"*: "Michel Halbouty, 95; Oilman Was Friend of Two Presidents," *Los Angeles Times*, November 18, 2004.

42 *21 billion barrels*: EIA, "World Proved Reserves of Oil and Natural Gas, Most Recent Estimates," Energy Information Administration, March 3, 2009, www .eia.doe.gov/emeu/international/reserves.html.

42 *our current levels of consumption*: EIA, "Petroleum Basic Statistics."

42 *Strategic Petroleum Reserve*: Ibid.

42 *outer Continental shelf . . . 18 billion barrels*: EIA, "Impacts of Increased Access to Oil and Natural Gas Resources in the Lower 48 Federal Outer Continental Shelf," www.eia.doe.gov/oiaf/aeo/otheranalysis/ongr.html.

2. War and Grease

45 *"Priority 1" request for emergency battlefield supplies*: Good detail on the request can be found in David Sandalow, *Freedom from Oil: How the Next President Can End the United States' Oil Addiction* (New York: McGraw-Hill, 2008), pp. 24–27. For the original text of the request, see the Marine Corps Systems Command Web site: www.marcorsyscom.usmc.mil/.

46 *one particularly brutal incident*: T. Christian Miller, "Families Want Answers in Iraq Convoy Disaster," *Los Angeles Times*, March 27, 2005.

46 *"trucks slid through like hogs"*: Thomas Hamill, *Escape in Iraq: The Thomas Hamill Story* (Accokeek, Maryland: Stoeger Publishing Company, 2005) p. 40.

46 *suspend convoys around Baghdad*: See T. Christian Miller, "Halliburton Suspends Some Iraq Supply Convoys," *Los Angeles Times*, April 3, 2004.

46 *bigger losses*: See Joseph R. Chenelly, "Then All Hell Broke Loose," *Army Times*, May 2, 2005. See also www.iht.com/articles/ap/2008/05/29/america/ KBR-Truckers-Killed.php.

46 *"We had a duty to deliver fuel"*: Hamill, *Escape in Iraq*, p. 29.

46– *reported attacks on supply convoys*: Jim Michaels, "Attacks Rise on Supply
47 Convoys," *USA Today*, July 13, 2007. The article reads: "Attacks on supply convoys protected by private security companies in Iraq have more than tripled, as the U.S. government depends more on armed civilian guards to secure reconstruction and other missions. There were 869 such attacks from the beginning of June 2006 to the end of May this year. For the preceding 12 months, there were 281 attacks."

47 *1.5 million gallons*: From my interview with Colonel Shawn Walsh, but higher numbers of daily fuel use in Iraq have been published. Robert Bruce in his article "Gas Pain" (*Atlantic Monthly*, May 2005) said that "the U.S. military

now uses about 1.7 million gallons of fuel a day in Iraq." An American Forces Press Service article estimated that the military was using between 10 million and 11 million barrels of fuel each month to sustain operations in Afghanistan, Iraq, and elsewhere, which translates to about 15 million gallons a day. (Gerry J. Gilmore, "DoD Has Enough Petroleum Products for Anti-Terror," August 11, 2005), www.defenselink.mil/news/newsarticle.aspx?id=16915.

See also this Energy Bulletin article by Dr. Sohbet Karbuz, a former statistician at the International Energy Agency: www.energybulletin.net/node/13199.

47 *first time . . . formally requested renewable:* From Chris Isleib in the Pentagon communications office. See also: "Commanders in Iraq Urgently Request Renewable Power Options," *Defense Industry Daily*, June 27, 2006.

47 *largest consumer of petroleum:* Defense Science Board Task Force, *More Capable Warfighting Through Reduced Fuel Burden*, January 2001, p. ES-1. See: www.acq.osd.mil/dsb/reports/fuel.pdf. According to this report: "Military fuel consumption for aircraft, ships, ground vehicles and facilities makes the DoD the single largest consumer of petroleum in America, perhaps in the world," www.acq.osd.mil/dsb/reports/fuel.pdf. This Energy Bulletin article by Sohbet Karbuz, "U.S. Military Oil Pains," February 17, 2007, says the Pentagon is the largest consumer in the world: www.energybulletin.net/node/26194.

47– *between 130 and 145 million barrels:* Defense Energy Support Center Fact
48 book 2007, www.desc.dla.mil.

48 *consumption of the United Arab Emirates:* The UAE consumes about 399,000 barrels each day; the Pentagon consumes 400,000 barrels per day. An overview of energy use by country can be found at Energy Information Administration's Country Energy Data center: http://tonto.eia.doe.gov/country/index.cfm.

48 *"And herein lies the dilemma":* Michael Klare, *Blood and Oil* (New York: Henry Holt, 2004), p. 11.

48 *Zilmer's memo in . . . USA Today:* Mark Clayton, "In the Iraqi War Zone, U.S. Army Calls for 'Green' Power," *Christian Science Monitor* (syndicated by *USA Today*), September 7, 2006.

48 *world's third-largest oil reserves:* See EIA chart of proven reserves by country: www.eia.doe.gov/emeu/international/reserves.html.

49 *oil had become . . . a tactical necessity:* The material in this section was mostly gathered from my interviews with David Painter and from his book *Oil and the American Century: The Political Economy of U.S. Foreign Oil Policy, 1941–1953.* (Baltimore: Johns Hopkins University Press, 1986). Daniel Yergin also comprehensively covers this topic in his book, *The Prize: The Epic Quest for Oil, Money, and Power* (New York: Simon & Schuster, 2008).

50 *two thirds of the world's petroleum*: From my interview with Yergin.

50 *development of synthetic fuels from coal*: Thomas P. Hughes, "Technological Momentum in History: Hydrogenation in Germany, 1898–1933," *Past and Present*, 44 (August 1969).

51 *"Ships were . . . burning everywhere"*: "Interview with Pearl Harbor Eyewitnesses," Scholastic, December 4, 1996, http://teacher.scholastic.com/pearl/transcript.htm.

52 *"Two hundred pine roots"*: Quoted in *The Prize*, p. 345.

53 *"In peace prepare for war,"* Sun Tzu, *The Art of War*, Ralph D. Sawyer, ed. (Boulder: Westview Press, 1994), p. 17.

54 *"defense of Saudi Arabia is vital"*: Quoted in Anthony Sampson, *The Seven Sisters: The Great Oil Companies and the World They Made* (New York: Viking Press, 1975), p. 95.

54 *"the center of gravity"*: Quoted in Edward J. Marolda, *Shield and Sword: The United States Navy and the Persian Gulf War* (New York: Crown, 2001), p. 7.

55 *King Ibn Saud was a descendant*: Biographical material drawn in part from Robert Lacey, *The Kingdom Arabia and the House of Saud* (New York: Avon Books, 1981); and Rachel Bronson, *Thicker Than Oil: America's Uneasy Partnership with Saudi Arabia* (Oxford: Oxford University Press, 2006).

55 *"bargainers worthy of anyone's steel"*: Wallace Stegner, *Discovery!: The Search for Arabian Oil* (San Francisco: Selwa Press, 2007), p. 16.

55 *1945 . . . producing 21 million barrels*: From my interview with David Painter.

56 *Saud arrived for his voyage*: Account drawn from transcripts and memoirs of Colonel William Eddy, *F.D.R. Meets Ibn Saud* (Vista, California: Selwa Press, 2005, reprinted from original in 1954); "Arabian King Brings Live Sheep, Own Food for Talk With Roosevelt," *Washington Post*, February 21, 1945; "U.S. Warship Becomes Arab Court in Miniature for Ibn Saud's Voyage," *New York Times*, February 21, 1945; and "Arabian Ruler At Home on One of Our Warships," *New York Times*, March 14, 1945; and Ensign W. Barry, "Ibn Saud's Voyage," *Life*, March 19, 1945.

These books also included accounts of the Ibn Saud meeting: Michael F. Reilly, *Reilly of the White House* (New York: Simon & Schuster, 1947); Aaron David Miller, *Search for Security: Saudia Arabian Oil and American Foreign Policy* (Chapel Hill: University of North Carolina Press, 1980); Charles E. Bohlen, *Witness to History, 1929–1969* (New York: Norton, 1973); Elliott Roosevelt, *As He Saw It* (New York: Duell, Sloan, and Pearce, 1946); and William D. Leahy, *I Was There: The Personal Story of the Chief of Staff to Presidents Roosevelt and Truman, Based on His Notes and Diaries Made at the Time* (New York: McGraw Hill, 1950).

57   *to keep the royal dynasty in power*: Michael Klare comprehensively makes the case for this thesis in his book *Blood and Oil: The Dangers and Consequences of America's Growing Petroleum Dependency* (New York: Henry Holt and Company, 2004). Rachel Bronson also addresses it in *Thicker than Oil*, as does Arron David Miller's *Search for Security: Saudia Arabian Oil and American Foreign Policy* (Chapel Hill: University of North Carolina Press, 1980).

58   *never made any attempt to colonize*: Eddy, *F.D.R. meets Ibn Saud*, p. 33.

58   *"Gentlemen, the Japanese offered me twice"*: "Saudi Arabia: Fish to Jidda," *Time*, February 19, 1940.

58   *not interfere in his domestic affairs*: Raymond Daniell, "By 'Flying Carpet' To Arabia's Oil Fields: Ibn Saud's Primitive Kingdom Already Feels the Impact of American Enterprise and Ideas," *New York Times Magazine*, January 18, 1948.

58   *"I wish to renew to Your Majesty"*: Yergin, *The Prize*, p. 409.

59   *"We do, of course, have historic ties"*: This was part of Dick Cheney's testimony to the Senate Armed Services Committee on September 11, 1990. Quote in Klare, *Blood and Oil*, p. 47.

60   *"Although significant warfighting"*: Defense Science Board Task Force, *More Capable Warfighting Through Reduced Fuel Burden*, p. 65.

60   *"the true cost of the fuel"*: Ibid., p. 67.

62   *seven hundred supply trucks circulate the roads of Iraq*: From the Pentagon's Chris Isleib.

63   *multibillion-dollar contracts*: "Defense Contract Management: DOD's Lack of Adherence to Key Contracting Principles on Iraq Oil Contract Put Government Interests at Risk," U.S. Government Accountability Office, July 2007, www.gao.gov/new.items/d07839.pdf.

63   *overcharged . . . $61 million*: See "Bush Warns 'Oil Overcharge' Firm," BBC News, December 13, 2003. http://news.bbc.co.uk/1/hi/business/3312015.stm. See also this commentary from U.S. Senator Patrick Leahy (D-VT): http://leahy.senate.gov/press/200603/030206d.html.

63   *Pentagon reportedly terminated the company's fuel deliveries*: James Glanz, "G.O.P. Donor Is Accused of Overcharging Pentagon," *New York Times*, October 17, 2008.

63   *could get paid between $80,000*: James Glanz, "Truckers of Iraq's Pony Express Are Risking It All for a Paycheck," *New York Times*, September 27, 2004.

63   *"They'll find new meat"*: Cathy Booth Thomas, "Fear and Loathing on Iraqi Roads," *Time*, June 7, 2004.

64   *"They don't no more care"*: from Wheeler's September 28, 2006, appearance

on CNN's *Anderson Cooper 360,* http://transcripts.cnn.com/TRANSCRIPTS/0609/28/acd.02.html.

64 *"We've got trucks":* David Wood, "Convoys Feed Ravenous Force," *Baltimore Sun,* June 4, 2007.

65 *Saudi Arabia and some of its partners:* As David Painter explained to me: "The 1973–74 oil embargo was not an OPEC embargo, but rather an OAPEC (Organization of Arab Petroleum Exporting Countries) embargo; Iran and Venezuela, two very important members of OPEC, did not participate."

65 *"To many Americans":* Quoted in Sonia Shah, *Crude: The Story of Oil* (New York: Seven Stories Press, 2004), p. 30. Originally from Brian Truumbore, editor of Stocksandnews.com.

65 *"send everything that can fly":* Walter J. Boyne, *The Yom Kippur War: And the Airlift Strike That Saved Israel* (New York: MacMillan, 2003), p. 119.

66 *"America's complete support for Zionism":* Yergin, *The Prize,* p. 579.

66 *"a household word—not just an obscure acronym":* Quoted in Shah, *Crude,* p. 29. The comment was originally published in *National Petroleum News.*

67– *reduce its daily petroleum consumption by 18 percent:* EIA, "Table 5.1 Petro-
68 leum Overview, 1949–2007," www.eia.doe.gov/emeu/aer/txt/ptb0501.html.

68 *"that basic fundamental doctrine":* Remarks of Secretary of Defense Dick Cheney at the American Defense Preparedness Association conference on December 10, 1990, Sheraton Hotel in Washington, D.C.

69 *set fire to nearly six hundred Kuwaiti oil wells:* Thomas Hayes, "The Job of Fighting Kuwait's Infernos," *New York Times,* February 28, 1991.

69 *"reading maps by flashlight":* Thomas W. Lippman, "Gulf War Leaves Environment Severely Wounded," *Washington Post,* March 2, 1991.

70 *"$36 trillion from Muslims":* Steve Coll, "Young Osama: How He Learned Radicalism, and May Have Seen America," *New Yorker,* December 12, 2005.

70 *"attackers struck at the umbilical cord":* Martin C. Libicki, Peter Chalk, Melanie Sisson, "Exploring Terrorist Targeting Preferences," *RAND Homeland Security Report,* 2007, www.rand.org/pubs/monographs/2007/RAND_MG483.pdf.

73 *F-15 fighter jet burns about 1,580 gallons:* Steven Komarow, "Military's Fuel Costs Spur Look at Gas Guzzlers," *USA Today,* March 8, 2006. See also statistics on burn rates from *Fuels Management Pocket Guide,* Airforce Logistics Management Agency, May 2007, www.af.mil/factsheets/factsheet.asp?fsID=83.

75 *"there will be no fundamental limits":* Defense Science Board Task Force, *More Capable Warfighting,* January 2001, p. 110.

76 *$10 a gallon . . . "more than $400 a gallon":* Ibid., p. 36. The report reads: "a reasonable estimate of the total cost of fuel when delivered to Army combat

platforms over even modest distances is in the $10's/gallon range. Over large distances the total cost would range from at least $40–$50 per gallon for overland transport up to more than $400 per gallon for air delivery using platforms with today's capability."

77   *"unleash us from the tether of fuel"*: *More Fight—Less Fuel*, Report of the Defense Science Board Task Force, February 2008, p. 4. http://www.acq.osd .mil/dsb/reports/2008-02-ESTF.pdf.

77   *Woolsey . . . lives in a rambling farmhouse:* The reporting for this section was first published in *Men's Journal* in 2006 in my article titled "Ending Our Addiction." See also my interview with Woolsey in Grist.org: http://www.grist.org/ article/little-woolsey/.

3. Road Hogs

82   *NASCAR's XXL Big Gulp–sized speedway:* Talladega facts including "12,000 pounds of Ballpark Franks" can be found at www.talladegasuperspeed way.com.

83   *"motherless, dirt-poor southern teens"*: Neal Thompson, *Driving with the Devil: Southern Moonshine, Detroit Wheels, and the Birth of NASCAR* (New York: Three Rivers Press, 2006), p. 10.

87   *four hundred companies . . . $1.5 billion:* Susanna Hamner, "NASCAR Sponsors, Hit by Sticker Shock," *New York Times*, December 14, 2008.

86   *fuel to keep . . . pageant in motion:* Fuel-mileage estimates from my interviews with crew members. Coverage of NASCAR fuel consumption also here: Jenna Fryer, "Even NASCAR Isn't Immune to Skyrocketing Gas Prices," Associated Press, May 7, 2008.

86   *forty to eighty tires per race:* Interview with NASCAR crewmembers.

86   *fully loaded . . . 4.5 miles per gallon:* Fryer, "Even NASCAR Isn't Immune to Skyrocketing Gas Prices," Associated Press.

86   *earliest days of the sport:* For more on the early decades of NASCAR is Tom Wolfe's "The Last American Hero Is Junior Johnson. Yes!" *Esquire*, March 1965.

87   *"Win on Sunday, sell on Monday"*: Thompson, *Driving with the Devil*, p. 89.

87   *ardent brand-loyalists:* Karl Greenberg, "Nascar: So Many Brands, So Many Opportunities," *Marketing Daily*, November 29, 2007.

89   *1.5 gallons of gasoline:* EIA's figure for the gasoline consumption of a one-person household is 550 gallons per year, www.eia.doe.gov/emeu/rtecs/ contents.html. Divide that by 365 and you get 1.5 gallons per day.

89   *quadruple that of the average European:* Based on consumption rates in *Foreign Policy* magazine (Gerhard Metschies, "Prime Numbers: Pain at the Pump," July/August 2007) and population numbers from CIA World Factbook.

90   *Henry Ford, the father:* Ford biographical material drawn from "Who IS Henry Ford," *Time,* May 19, 1923; Douglas Brinkley, *Wheels for the World: Henry Ford, His Company, and a Century of Progress* (New York: Penguin, 2003); Daniel Gross, *Greatest Business Stories of All Time* (New York: Wiley and Sons, 1996); and David Halberstam, "Citizen Ford," *American Heritage* 37, (6): 49–64, 1986.

90   *hundreds of units in the first year:* According to Bob Kreipke at Ford Motor Company, there were 308 Model T's built between October 1908 and January 1909.

90   *$850 ($19,300 in today's dollars):* All numbers adjusted for inflation were calculated using data from the Consumer Price Index, www.bls.gov/cpi/.

90   *By 1912 . . . selling for $575:* Most of the production and cost statistics on the Model T in this passage are from James J. Flink, *The Automobile Age* (Cambridge, Massachusetts: MIT Press, 1988), pp. 37–38.

90   *"assembling . . . took only ninety minutes":* Ibid., p. 50.

91   *two-thirds of America's cars:* Thompson, *Driving with the Devil,* p. 66.

91   *declined even a taste:* Ibid., p. 32.

91   *"Chop your own wood":* Quoted in "Who IS Henry Ford," *Time,* May 19, 1923.

91   *contemplated a bid for president:* R. A. Haughton, "FORD'S WHITE HOUSE BEE; Dearborn Club Is Pushing His Presidential Candidacy—Raising $50,000," *New York Times,* June 25, 1922.

91   *believed in reincarnation:* "Reincarnationist," *Time,* Sept 3, 1928.

91   *anti-Semitic comments:* Ford owned the *Dearborn Independent* from 1920 until 1927, during which time the newspaper published commentary that the American Jewish Historical Society described as "anti-immigrant, anti-labor, anti-liquor, and anti-Semitic." See also Charles Y. Glock and Harold E. Quinley, *Anti-Semitism in America* (New Brunswick: Transaction Publishers, 1983), p. 168.

91   *"The way to make automobiles":* Flink, *The Automobile Age,* p. 43.

92   *"like a Colossus":* Robert Payne, *Report On America* (New York: John Day Company, 1949), p. 3.

93   *"The primary object of the corporation":* Quoted in Flink, *Automobile Age,* p. 233.

93   *just over 150 million citizens:* Census statistics can be found at www.census.gov/ipc/www/idb/; go to "Country Rankings" and enter the year 1950).

93   *50 million cars on the road:* From the Department of Energy Vehicle Facts

Sourcebook, www1.eere.energy.gov/vehiclesandfuels/facts/2007_fcvt_fotw474 .html.

93   *more than one for every household*: The number of households in 1950 was about 43.5 million, according to Census data: U.S. Census Bureau, "No. HS-12. Households by Type and Size: 1900 to 2002," *Statistical Abstract of the United States, 2003: Mini-Historical Statistics*, www.census.gov/statab/hist/ HS-12.pdf.

93   *"The Chevy was for blue-collar people"*: David Halberstam, *The Fifties* (New York: Villard Books, 1993), p. 120.

94   *"economic benchmark on life's journey"*: Ibid., p. 120.

94   *"the Cellini of chrome"*: Ibid., p. 124–126.

94   *drew from offbeat inspirations*: Karal Ann Marling, "America's Love Affair with the Automobile," essay in *Autopia: Cars and Culture*, eds. Peter Woolen and Joe Kerr, (London: Reaktion Books, 2003) pp. 356–57.

95   *age of "too-muchness"*: Ibid., p. 358.

95   *$1,270 at the beginning . . . to $1,822*: Flink, *The Automobile Age*, p. 287.

95   *50 million to about 74 million*: According to U.S. Census data, there were 180,671,000 people in the U.S.; in 1960 and 408.8 cars per 1,000; that comes out to 73,858,304 cars.

95–   *"Complacency carried a high cost"*: Flink, *Automobile Age*, p. 293. Further in-
96   formation on the waning of the American auto industry can be found in Brock Yates, *The Decline and Fall of the American Automobile Industry* (New York: Houghton-Mifflin, 1983).

96   *a large Caddy . . . 13.5 miles per gallon*: Ibid., p. 287.

96   *the birth of Big Inch*: This 2000 study by the Cultural Resource Group offers a good overview of the birth of the Big Inch and Little Inch pipelines, www.culturalresourcegroup.com/pdf/inchlines.pdf. See also Yergin, *The Prize*, pp. 375–376.

96   *the first drive-in . . . the first motel*: The first drive-in was In-N-Out Burger opened in Los Angeles in 1921. The first shopping mall, Country Club Plaza, opened near Kansas City, Missouri, in 1922. The first motel was built in San Luis Obispo, California, in 1925, and the first drive-in movie theater was built in Camden, New Jersey, in 1933.

96   *"determined couples found ways"*: Flink, *The Automobile Age*, p. 160.

97   *Along came Dwight D. Eisenhower*: Eisenhower bio materials and anecdotes drawn from his memoir *At Ease: Stories I Tell to Friends* (Garden City, New York: Doubleday, 1967); Stephen Ambrose, *Eisenhower: The President* (New York: Touchstone, 1990); and Carlo D'Este, *Eisenhower: A Soldier's Life* (New York: Henry Holt and Company, 2002).

97   *"After seeing the autobahns"*: Quoted in Dan McNichol. *The Roads That Built America: The Incredible Story of the U.S. Interstate System* (New York: Sterling Co., Inc., 2006), p. 100.

98   *"Some days when we had counted"*: Quoted in Yergin, *The Prize*, p. 190.

98   *half of the nation's 3 million*: McNichol, *The Roads That Built America*, p. 103.

98   *"greater convenience, greater happiness"*: Quoted in Chester J. Pach, Jr., and Elmo Richardson, *The Presidency of Dwight D. Eisenhower* (Wichita: University Press of Kansas, 1991), p. 123.

99   *"Our highway net is inadequate"*: The full text of Eisenhower's speech, delivered by Nixon to the Governor's Conference in Lake George, New York, on July 12, 1954, can be found here: www.fhwa.dot.gov/infrastructure/rw96m.cfm.

99   *signal the decay of cities*: One of the most outspoken opponents of highway development was Jane Jacobs, who wrote a withering critique of 1950s urban planning in her book *The Death and Life of Great American Cities* (New York: Random House, 1961).

100   *"Our unity as a nation is sustained"*: from President Eisenhower's address to Congress, delivered at noon on February 22, 1955; transcript obtained through the Dwight D. Eisenhower Library.

100   *the "highway-motor lobby"*: Stephen B. Goddard, *Getting There: The Epic Struggle Between Road and Rail in the American Century* (Chicago: University of Chicago Press, 1996), pp. 58–64. See also: Flink, *The Automobile Age*, pp. 368–373; and Richard F. Weingroff, "Creating the Interstate System," *Public Roads*, U.S. Department of Transportation Federal Highway Administration 60, no. 1 (1996):10, www.tfhrc.gov/pubrds/index.htm.

100   *"Obviously we have a selfish interest"*: Quoted in Flink, *The Automobile Age*, p. 368.

101   *William Levitt*: See: "How William Levitt Helped to Fulfill the American Dream," *New York Times*, February 6, 1994. See also Richard Lacayo, "William Levitt," *Time*, December 7, 1998.

101   *Levitt's portrait appeared*: "Up from the Potato Fields," *Time*, July 3, 1950.

101   *living in makeshift dwellings*: Halberstam, *The Fifties*, p. 134.

101   *"got itchy"*: Quoted in "Up from the Potato Fields."

102   *"No man who owns his own house"*: Quoted in Halberstam, *The Fifties*, p. 132.

102   *"Houses were updated annually"*: McNichol, *The Roads that Built America*, p. 110.

103   *"community has . . . antiseptic air"*: "Up from the Potato Fields."

103   *"one house to a Negro family"*: Quoted in Halberstam, *The Fifties,* p 141.

104   *"universally and exclusively"*: Quoted in McNichol, *The Roads That Built America*, p. 108.

104 *"developments conceived in error":* Quoted in Joseph Weber, "William J. Levitt: A Social Architect," *BusinessWeek*, May 31, 2004.

104 *60 million Americans flocked:* Halberstam, *The Fifties*, p. 142.

104 *triumph of the automobiles:* Flink, *The Automobile Age,* chapter 19.

104 *eighty-five thousand U.S. towns and cities:* This and other statistics in this passage came from interview with National Railway Historical Society librarian L. J. Dean, citing *The Official Guide of the Railways* of December 1918.

105 *"Streetcar companies were bought up":* This and *"abruptly curtailed"* quoted from Flink, *The Automobile Age*, p. 365.

105 *control of rail systems in forty-five cities:* From the January 3, 1951, antitrust hearing *United States v. National City Lines, Inc.* A transcript of the court proceedings can be found here: http://bulk.resource.org/courts.gov/c/F2/186/186 .F2d.562.9943-9953.html.

105 *By the 1960s, most . . . scrapped:* Interview with L. J. Dean.

105 *"I cannot accept the argument . . .":* Mayor Joseph Alito's quote taken from archival footage included in the 1996 documentary *Taken for a Ride*, directed by Jim Klein.

106 *50 percent in . . . seven years:* According to the DoE's Energy Information Administration, when the CAFÉ standards were enacted in reaction to the oil crisis, "the standards set a corporate sales-fleet average of 18 miles per gallon beginning with the 1978 model year, and established a schedule for attaining a fleet goal of 27.5 miles per gallon by 1985."

106 *Nine out of ten Americans:* U.S. Census Bureau. "Most of Us Still Drive to Work—Alone: Public Transportation Commuters Concentrated in a Handful of Large Cities," Washington, D.C., June 13, 2007.

106 *Midas actually gave an award:* Nick Paumgarten, "There and Back Again: The Soul of a Commuter," *New Yorker*, April 16, 2007.

106 *4 million miles of . . . roadways:* Bureau of Transportation Statistics. "System Mileage Within the United States," www.bts.gov/publications/national _transportation_statistics/html/table_01_01.html.

107 *three-quarters of Americans vacationed:* Travel Industry Wire Survey, Summer 2008, www.travelindustrywire.com/article32637.html.

107 *14,000 miles a year:* Pat S. Hu, "Summary of Travel Trends: 2001 National Travel Household Survey," Center for Transportation Analysis, Oak Ridge National Laboratory. nhts.ornl.gov/2001/pub/STT.pdf. The equator about 24,900 miles around, 1.8 times the annual distance of miles traveled by the average American driver.

107 *"113-pound housewife":* Introduction to *Classic Crews* anthology, Harry Crews, *Class Crews* (New York: Simon & Schuster, 1993).

108   *SUV . . . sales plunged 40 percent:* Interview with environmentalist and fuel-economy watchdog Dan Becker, director of the Safe Climate Campaign.

4. Plastic Explosive

111   *snub-nosed, two-seater:* Smart car information from Ken Kettenbeil of Smart USA.

112   *downsizing and down-weighting:* Matthew L. Wald, "Study Says Minicar Buyers Sacrifice Safety," *New York Times,* April 14, 2009.

112   *"You're gonna see steel phasing out":* The materials engineer I spoke with, employed by a major auto company in Detroit, asked to remain anonymous given the sensitivity of the fuel economy issue in his industry. Information on the auto industry's use of plastics can be found here: www.plasticsnews.com/automotive/.

113   *eighty thousand different types of commercial plastics:* Based on the number of commercial resins listed in IDES Plastics Prospector, the leading search engine for materials information in the plastics industry: http://prospector.ides.com/.

114   *5 percent of . . . U.S. energy consumption:* The Energy Information Administration offers a chart of energy consumption by sector here: www.eia.doe.gov/emeu/aer/txt/ptb0201a.html. My source at the EIA explained the data this way: "In 2002, the total energy consumption in the chemical industry for the manufacture of Plastics Materials and Resins was 1,821 trillion BTU equivalent and an additional 351 trillion BTU equivalent was consumed by the Plastics and Rubber industry (essentially to fabricate products)." (Note: BTU stands for "British thermal unit, and it is the basic unit of energy used to describe the heat value—or energy content—of both fuels and power.)

The total of 2,172 trillion BTU was about 2.5 percent of total U.S. energy consumption in 2002. Add to that the amount of energy equivalent in the feedstock used to make plastics and resins in 2002—66 trillion BTU of natural gas and 1,283 trillion BTU of liquid petroleum gases and natural gas liquids—a total of 1,349 trillion BTU and 1.7 percent of total US energy consumption in 2002. Combined you get 3.5 trillion BTU, over 4 percent.

114   *equivalent of . . . 600 million barrels:* A good converter from BTUs to barrels of oil and kilowatt-hours of energy: www.eia.doe.gov/kids/energyfacts/science/energy_calculator.html.

114   *"rushed out to buy a pure cotton":* Stephen Fenichell, *Plastic: The Making of a Synthetic Century* (New York: HarperBusiness, 1996), p. 4.

116   *Plastic bag . . . can take decades:* A National Park Service analysis of litter es-

timated that plastic bags can take ten to twenty years to decompose. It also says that nylon fabric takes thirty to forty years, a plastic six-pack holder takes one hundred years, and plastic bottles and Styrofoam last "indefinitely": www .nps.gov/crmo/forteachers/activity-3b.htm. Note that photodegradable forms of polyethylene and polypropylene have been around for decades and are coming into wider use: www.degradable.net.

116   *glut our landfills:* Bryan Walsh, "The Truth About Plastic," *Time,* July 21, 2008. Among the facts in this article: "The United States produced 28 million tons of plastic waste in 2005—27 million tons of which ended up in landfills."

116   *sully our oceans with . . . waste:* Alan Weisman, "Polymers Are Forever: Alarming Tales of a Most Prevalent and Problematic Substance," *Orion,* May/June 2007. It describes "terrible tales of sea otters choking on polyethylene rings from beer six-packs . . . of a green sea turtle in Hawai'i dead with a pocket comb, a foot of nylon rope, and a toy truck wheel lodged in its gut."

118   *a quick primer on polymers:* This primer on polymers was inspired by a great overview of oil inputs in plastics by advice columnist Umbra on the Web site Grist.org, where I am a regular contributor, www.grist.org/advice/ask/2007/03/14/plastics.

118   *American engineers found a way:* There were plenty of European scientists working on early plastics, but the first synthetic plastics were commercialized in America by Belgian-born chemist Leo Heindrick Baekeland, who immigrated to New York, www.chemheritage.org/classroom/chemach/plastics/baekeland.html.

119   *1 gallon of oil yields 3 pounds:* A report from the nonpartisan research group the Pacific Institute (http://www.pacinst.org/) looks at the example of PET plastic and estimates that it takes 83,000 MJ to make one metric ton of PET, which converts to 13.5 barrels. Multiply that times 42 gallons per barrel and you get 570 gallons of oil to make 2,204 pounds of PET resin. Divide 569.67 by 2,204 and you get that it takes .258 gallons of oil to make one pound of PET resin. So in this case it's between 3 and 4 pounds being made from a gallon equivalent of oil. Also see this study of oil use in the European plastics industry: www.lca.plasticseurope.org/download/pp.zip.

119   *65 million tons a year:* From *Global Polyethylene Market Analysis and Forecasts to 2020,* a report by Global Markets Direct, published March 2009.

120   *commonly used in medicine:* Much of my information on plastics in medicine came from an interview with Professor Steven McCarthy, a professor of plastics engineering at University of Massachusetts, Lowell, who specializes in medical plastics.

126   *emerging field of "green plastics":* E. S. Stevens, *Green Plastics: An Introduction*

to the New Science of Biodegradable Plastics (Princeton, New Jersey: Princeton University Press, 2001).

127   *"Thin as tissue but hard to tear"*: From an essay by James Piani, marketing employee of the DuPont Cellophane Company in *DuPont Magazine*, December 1923.

128   *"The boom was on"*: "Transparent," *New Yorker*, December 24, 1949.

128   *"You're the purple light"*: Quoted in Fenichell, *Plastic*, p. 106. Permission to use quote granted by Cole Porter Musical and Literary Property Trusts.

129   *"civilization as we know it today"*: "Transparent," *New Yorker*.

129   *"At the rate this peekaboo"*: Fenichell, *Plastic*, p. 116.

129   *"fine as the spider's web"*: Editorial in the *Industrial and Engineering Chemistry Magazine*, November 1938.

129   *64 million pairs of nylons*: Dr. G. P. Hoff, director of DuPont Nylon Research, Akron, Ohio, "Nylon: Its Development and Uses," June 23, 1943. From DuPont archives.

129   "an entirely new arrangement": Quoted in Ann and Patrick Fullick, *Chemistry for AQA* (Portsmouth, New Hampshire, 2001), p. 49.

130   *"When the armed services faced shortages"*: Jeffrey Meikle, *American Plastic: A Cultural History.* (New Brunswick, New Jersey: Rutgers University Press, 1995), p. 125.

130   *130 million pounds . . . in 1938*: See the following 2007 production statistics: American Chemistry Council, "PIPS Resin Statistics Summary: 2007 vs. 2006, Production, Sales, and Captive Use," December 28, 2007. See www .americanchemistry.com/s_acc/sec_policyissues.asp?CID=996&DID=6072.

130   *a quarter of all global plastics*: According to this PlasticsEurope report, global plastics production reached 230 million tons (460 billion pounds) in 2005. It's expected to hit 304 million tons (608 billion pounds) by 2010. See this article from Plastic News at www.plastemart.com/plasticnews_desc.asp?news _id=9761&P=P.

130   *"lighthearted aesthetic indulgence"*: Meikle, *American Plastic*, p. 126.

130   *"War Makes Gimcrack Industry"*: Headlines quoted ibid., p. 126.

131   *Earl Tupper was a farm boy*: Much of the Earl Tupper bio material is from Alison Clarke's *Tupperware!: The Promise of Plastic in 1950s America* (Washington D.C.: Smithsonian Institute Press, 1999), pp. 8–19.

131   *"help dispel the discontentment"*: Ibid., p. 10.

131   *"ham inventor and Yankee trader"*: "Tupperware," *Time*, September 8, 1947.

132   *cover of* BusinessWeek: "How Brownie Wise Whoops Up Sales," *BusinessWeek*, April 17, 1954.

132   *"now dressed in white gloves"*: PBS, American Experience, *Tupperware!* docu-

mentary, based on Clarke's book. See online materials: www.pbs.org/wgbh/amex/tupperware/filmmore/index.html.

132 *cultish motivational tactics:* "How Brownie Wise Whoops Up Sales," *Business-Week.*

133 *"bulwark against communism":* Simon Jeffery, "Tupperware Hits 50," *Guardian,* November 35, 1999.

133 *own special look and feel:* Meikle, *American Plastic,* p. 181.

133 *decency, frugality, and indulgence:* Clarke, *Tupperware!,* pp. 112–18.

134 *produced its hearts for $20,000:* William Broad, "Dr. Clark's Heart: A Story of Modern Marketing as well as Modern Medicine," *New York Times,* March 20, 1983.

134 *"what God can grow":* Quoted in Fenichell, *Plastic,* p. 330.

134 *Pop Art a movement:* Meikle in *American Plasic,* pp. 223–241.

135 *"plastic was the emblem":* Carter Ratcliff, "Of Redemption and Damnation: On Mark Rothko," *Tate Etc.,* Issue 14, www.tate.org.uk/.

135 *"Just look at the surface":* John Updike, "Andy Warhol," *Rolling Stone,* May 15, 2003.

135 *"entered middlebrow folklore":* Meikle, *American Plastic,* p. 180.

136 *"objects that could never be built":* Ibid., p. 230.

136 *"Some people in the industry":* "Plastics: A Market You Can Mold," *Business-Week,* June 16, 1956.

136 *Betsey Johnson to Pierre Cardin:* Fenichell, Plastic, p. 293–97.

136– *"witchery the chemist performs":* Ruth Carson, "Plastic Age," *Collier's,* July 19,
137 1947.

137 *"We gave away our freedom":* Noman Mailer, "The Big Bite," *Esquire,* April 1963.

138 *"the faceless plastic surfaces":* Ibid.

138 *21,000 tons of toxic waste:* Love Canal facts from Eckardt C. Beck, "The Love Canal Tragedy," *EPA Journal,* January 1979, www.epa.gov/history/topics/lovecanal/01.htm.

138 *"not so much to serve":* Barry Commoner, *The Poverty of Power: Energy and the Economic Crisis* (New York: Knopf, 1976).

139 *most widely used materials:* From the American Chemistry Council. See a brief history of plastics, www.americanchemistry.com/s_plastics/doc.asp?CID=1102&DID=4665.

139 *"it's getting difficult":* From an interview with Keith Christman, senior director of packaging for the Plastics Division of the American Chemistry Council.

139 *25 billion plastic water bottles:* From the Container Recycling Institute's "Beverage Market Data Analysis" for 2008, www.container-recycling.org.

139   17 *million barrels of oil*: The Pacific Institute report "Bottled Water and Energy: Getting to 17 Million Barrels," published December 2007 in conjunction with the Earth Policy Institute and Container Recycling Institute.

139   500 *billion plastic bags*: Plastic bags statistics from Carolyn Sayre, "Just Say No to Plastic Bags," *Time*, April 9, 2007.

140   *concerns about "endocrine disruptors"*: A good overview of endocrine disruptors in plastics by the Rollins School of Public Health at Emory University can be found here: www.sph.emory.edu/PEHSU/html/exposures/endocrine.htm#Q1.

140   *bisphenol A (BPA)*: Tara Parker-Pope, "A Hard Plastic Is Raising Hard Questions," *New York Times*, April 22, 2008.

141   *less than 7 percent*: The EPA reports that of the 30.7 million tons (or 61.4 billion pounds) of plastics discarded in 2007, about 6 percent were recycled. See "Municipal Solid Waste in the United States," U.S. EPA, 2008.

142   *"the Big McDonough Idea"*: Florence Williams, "Prophet of Bloom," *Wired*, October 2002.

144   *"hierarchy of substances is abolished:"* Roland Barthes, *Mythologies* (New York: Noonday Press, 1972).

5. Cooking Oil

147   *fifth-biggest producer*: From the United States Department of Agriculture 2007 Census of Agriculture, state profile of Kansas. I calculated level of production in terms of the total value of the livestock and agriculture produced, not the number of units produced.

147   *90 percent of the state*: See federal statistics on the state of Kansas here: www.fedstats.gov/qf/states/20000.html.

147   *wheat, corn, soybeans*: A full portfolio of Kansas's agriculture industry can be found here: www.nass.usda.gov/Statistics_by_State/Ag_Overview/AgOverview_KS.pdf.

148   *most plentiful harvests*: 2007 and 2008 were two of the most plentiful harvests in the last one hundred years, according to the USDA's National Agricultural Statistics Service, www.nass.usda.gov/QuickStats/Create_Federal_Indv.jsp.

149   6.2 *billion pounds*: From the International Fertilizer Industry Association (http://www.fertilizer.org/): The U.S. uses 3.13 million tons of N as ammonia for direct application per year, which equals 6.2 billion pounds.

149   *"40 percent of our nation's crop production"*: Interview with Rosemary O'Brien of the Agriculture Energy Alliance, www.agenergyalliance.com.

149   *produces over $110 billion:* "Farm Income and Costs: Value-added to the U.S. Economy by the Agricultural Sector via the Production of Goods and Services, 2005–2009," USDA, Economic Research Service.

150   *fertilizers . . . roughly a quarter:* Danielle Murray, "Oil and Food: A Rising Security Challenge," Earth Policy Institute, May 9, 2005.

150   *10 percent of annual U.S. energy:* Martin C. Heller and Gregory A. Keoleian,"Life Cycle–Based Sustainability Indicators for Assessment of the U.S. Food System," Center for Sustainable Systems, School of Natural Resources and Environment, University of Michigan, December 2000, css.snre .umich.edu/css_doc/CSS00-04.pdf.
       Note that the estimates for how much energy is consumed by America's food system vary greatly, and I tried to pick numbers in the middle. The EIA says that only about 2.2 percent of all U.S. energy consumption goes to agriculture and food production. See Table 35: "Food Industry Energy Consumption," www .eia.doe.gov/oiaf/aeo/supplement/supref.html. This number did not account for fertilizer use, among other factors considered in the Michigan study. Meanwhile, the United Nation's Food and Agriculture Organization has estimated that fully 19 percent of America's fossil fuel usage goes into our food system. See Jodi Ziesemer, "Energy Use in Organic Food Systems," FAO, August 2007, p. 5.

150   *greenhouse gases from eating:* Gidon Eshel and Pamela Martin, "Diet, Energy and Global Warming," University of Illinois Department of the Geophysical Sciences, May 2005, http://geosci.uchicago.edu/~gidon/papers/nutri/nutri3.pdf.

151   *140 pounds of nitrogen:* Wen Yuan Hwang, "Impact of Rising Natural Gas Prices on U.S. Ammonia Supply," USDA Economic Research Service, August 2007, www.ers.usda.gov/publications/WRS0702/wrs0702.pdf. The report says that about 138 pounds of nitrogen are needed per acre of corn for fertilizer.

151   *the equivalent of 92 gallons:* David Pimentel, *Biofuels, Solar and Wind as Renewable Energy Systems: Benefits and Risks* (Boston: Springer Publishing, 2009), p. 376.

151   *"forty-five thousand items":* Much of the information in the two paragraphs describing corn by-products is from Michael Pollan, *The Omnivore's Dilemma: A Natural History of Four Meals* (New York: Penguin Press, 2006).

151   *quarter of all the corn . . . ethanol:* Tom Doggett, "Ethanol to Take 30 Pct of US Corn Crop in 2012," Reuters, June 12, 2007. See also: "USDA Agricultural Predictions to 2017," Interagency Agricultural Projections Committee, USDA, February 2008, pp. 22 and 25.

152   *half goes to animal feed:* See "Crop Production 2008 Summary," National Agricultural Statitics Survey, USDA, January 2008, http://usda.mannlib.cornell .edu/usda/current/CropProdSu/CropProdSu-01-12-2009.pdf.

154   *86 million acres of farmland:* "Crop Production Summary 2008," National Agricultural Statistics Survey, USDA.

158   *"Sunlight nourishes the grasses":* Michael Pollan, "Farmer in Chief," *New York Times Magazine,* October 12, 2008.

159   *"The main peculiarity":* Quoted in Egbert Tellegen and Maarten Wolsink, *Society and Its Environment* (Amsterdam: Gorden and Breach Science Publishers, 1998).

159   *European scientists recognized:* David E. Fisher and Marshall Jon Fisher, "The Nitrogen Bomb," *Discover,* April 2001.

160   *Haber came up with an answer:* Biographical materials on Haber came from the following: *Nobel Lectures, Chemistry 1901–1921* (Amsterdam: Elsevier Publishing Company, 1966); Vaclav Smil, *Enriching the Earth: Fritz Haber, Carl Bosch, and the Transformation of World Food Production* (Cambridge, Massachusetts: MIT Press, 2001); and Daniel Charles, *Master Mind: The Rise and Fall of Fritz Haber, the Nobel Laureate Who Launched the Age of Chemical Warfare* (New York: HarperCollins, 2005).

160   *three hundred American households:* This calculation assumes that 138 pounds of nitrogen are needed per acre of corn for fertilizer, and 33 million BTU of natural gas are needed to produce 1 ton of ammonia. Given these numbers, about 9.5 billion BTU would be the amount of energy needed to produce the necessary fertilizer for a 4,000-acre farm. The average American household uses roughly 10,000 KWH, annually, which converts to 36,371,920 BTU.

160   *"Airplanes, nuclear energy":* Smil, *Enriching the Earth,* p. xiii.

161   *"Haber actually insisted":* Dan Charles, "The Tragedy of Fritz Haber: Nobel Laureate Transformed World Food Production, War," National Public Radio, July 11, 2002.

162   *"Nitrogen fixation teaches":* Charles, *Master Mind,* p. 79.

163   *"expansion hit its limits":* Richard Manning, "The Oil We Eat: Following the Food Chain Back to Iraq," *Harper's,* February 2004.

163   *"The scale, severity and duration":* "Green Revolution: Curse or Blessing?" International Food Policy Research Institute, 2002, www.ifpri.org/pubs/ib/ib11.pdf.

164   *"He probably sleeps four hours":* Quoted in Mark Stuertz, "Green Giant," *Dallas Observer,* December 5, 2002.

164   *"No. No. That can't be, Margaret.":* Leon Hesser, *The Man Who Fed the World* (Dallas, Texas: Durban House Publishing Company, 2006), p. 3.

164   *"it's better to fill you head":* Ibid., p. 8.

165   *"I saw how food":* Ibid., p. 202.

165   *"[O]ften sick with diarrhea":* Ibid., p. 42.

165 *"powered its forward thrust"*: Norman Borlaug, December 11, 1970, Nobel Prize acceptance speech, see www.nobelprize.org.

165 *edge of bloody battles*: Gregg Easterbrook, "A Forgotten Benefactor of Humanity," *Atlantic Monthly*, January 1997.

165 *HYV seeds were bred*: Information from the report "Green Revolution: Curse or Blessing?" International Food Policy Research Institute, 2002.

166 *"I call it the Green Revolution"*: Speech by William S. Gaud to the Society for International Development, Washington D.C., March 8, 1968.

166 *Scientists have serious concerns*: For a comprehensive overview of environmental concerns related to industrial farming and the Green Revolution, see Pollan, "Farmer-In-Chief," and Lester R. Brown, *Outgrowing the Earth: The Food Security Challenge in an Age of Falling Water Tables and Rising Temperatures* (New York: W.W. Norton & Company, 2005).

167 *doubled between 1970 and 1975*: "Green Revolution: Curse or Blessing?" International Food Policy Resarch Insitutute, Washington, D.C., 2002.

167 *"My good friend Norman"*: Introduction to Hesser, *The Man Who Fed the World*, p. i.

168 *"he drops his 'g's'"*: James Risser and George Anthan, "Why They Love Earl Butz," *New York Times*, June 13, 1976.

168 *incomes shot up . . . 60 percent*: According to The USDA's Economic Research Service, average farm incomes were $9,822 in 1971 and $16,461 by 1976.

168 *exports almost tripled*: USDA Economic Research Service data, www.ers.usda .gov/data/fatus/DATA/XMScy1935.xls.

168 *"breadbasket to the world"*: "Farming's Golden Challenge," *Time*, September 10, 1973.

168 *"Beef plants . . . shutting down"*: William Robbins, "Unhappily, Less Really is More," *New York Times*, August 5, 1973.

169 *"Adapt or die, resist and perish"*: Risser and Anthan, "Why They Love Earl Butz."

169 *ties to three large agribusiness*: Ibid.

169 *6.3 million . . . to roughly 2.2 million*: Data from the USDA's Economic Research Service: http://www.ers.usda.gov/AmberWaves/June05/DataFeature/.

170 *Twenty years ago . . . $23 for food*: Earl Butz, "Farmer as the Good Guy," *New York Times*, April 15, 1972.

170 *"That is marvelous"*: Quoted in *King Korn*, directed by Aaron Woolf, Mosaic Films Incorporated, released October 2007.

170 *Europe spends 18 percent*: "Food Safety and Quality in Europe," Food and Agriculture Organization of the United Nations, Twenty-fourth FAO Regional Conference for Europe, May 2004.

170   *developing world . . . 50 percent*: Ronald Trustle, *Global Agricultural Supply and Demand: Factors Contributing to the Recent Increase in Food Commodity Prices*, Economic Research Service, USDA, Rev. ed. July 2008. www.ers.usda.gov/Publications/WRS0801/WRS0801.pdf.

170   *"Just stop for a minute"*: Quoted in Wendell Berry, *The Unsettling of America*, (San Francisco: Sierra Club Books, 1996), p. 99.

170   *"last bastion of patriotism"*: Quoted in Risser and Anthan, "Why They Love Earl Butz."

171   *$48 billion to $175 billion*: Data from the USDA's Economic Research Service, www.ers.usda.gov/Data/FarmBalanceSheet/50stbsht.htm.

171   *lost nearly 200,000 farms*: Data from the National Agriculture Statistics Service, www.nass.usda.gov/Statistics_by_State/Kentucky/Publications/Annual_Statistical_Bulletin/B2008/Pg017.pdf.

171   *banks foreclosed on dozens*: From my interview with Ken McCauley.

172   *An algae bloom occurs*: For more on algae blooms and nitrogen contamination in the environment see Tom Hornton and Heather Dewar, "Nitrogen's Deadly Harvest," *Newsday*, September 26, 2000.

172   *soil quality erodes*: Jason McKenney "Artificial Fertility: The Environmental Costs of Industrial Fertilizers," in *The Fatal Harvest Reader: The Tragedy of Industrial Agriculture*, Andrew Kimball, ed. (Sausalito: Foundation For Deep Ecology, 2002), p. 121.

173   *35 percent by 2030*: "Fertilizer Use by 2015 and 2030," UN Food and Agriculture Organization, Rome, 2000, ftp://ftp.fao.org/agl/agll/docs/barfinal.pdf.

174   *"only a fifth of the total energy"*: Pollan, *Omnivore's Dilemma*, p. 183.

174   *"sinking sea of petroleum"*: Ibid., p. 184.

175   *Reformers want to see*: Recommendations from my interview with Tom Philpott.

176   *small comfort in . . . hyperengineered*: Donald Barlett and James Steele, "Monsanto's Harvest of Fear," *Vanity Fair*, May 2008. Also see: Tom Philpott: "GM-OH, NO!" Grist.org, November 13, 2008, www.grist.org/article/gm-oh-no.

177   *First Lady Michelle Obama*: Michael Pollan, "A Food Revolution in the Making from Victory Gardens to White House Lawn," on NRDC's OnEarth blog, www.onearth.org/node/1061.

6. Chain of Fuels

179   *miles to my chopping block*: This is a rough estimate, given that the food labels don't indicate the exact region of origin, just the nation of origin. I picked

cities somewhat at random for the calculation. From Nashville it is roughly 5,050 miles to Curico, Chile; 1,400 miles to Guanaja, Honduras; 4,800 miles to Milan, Italy; 7,500 miles to Shanghai, China; and 1,100 miles to Monterrey Mexico—in total,19,850 miles.

180 *produce . . . travels about 1,500 miles:* Rich Pirog, "Checking the Food Odometer: Comparing Food Miles for Local versus Conventional Produce," *Leopold Letter,* Leopold Center for Sustainable Agriculture, Iowa State University, July 2003, www.leopold.iastate.edu/pubs/staff/ppp/food_chart0402.pdf.

181 *imported more than 25 billion:* From USDA's Agricultural Marketing Service's annual "Market News and Transportation Data," www.ams.usda.gov/.

181 *more than half of our fresh fruit . . . nearly a quarter:* Ibid. See further related statistics from the University of California, Davis: http://migration.ucdavis .edu/rmn/more.php?id=1139_0_5_0.

181 *Walmart effect:* Charles Fishman, *The Walmart Effect* (New York: Penguin, 2006).

183 *nine times less energy:* Christopher L. Weber and H. Scott Matthews, "Food-Miles and the Relative Climate Impacts of Food Choices in the United States," *Environmental Science and Technology Journal,* American Chemical Society, 2008.

183 *85 percent . . . 15 percent is delivered by train:* Richard Pirog, "Food, Fuel, and Freeways," Leopold Center For Sustainable Agriculture, Iowa State University, 2001.

183 *roughly 3 billion miles:* Note that this number, from Weber and Matthews' study in Environmental Science and Technology refers to the delivery of more than just produce, it also includes these categories: red meat, oils/sweets/condiments, dairy products, chicken/fish/eggs, cereals/carbs, beverages and other miscellaneous products.

184 *6 billion pounds of bananas:* "Fresh-Market Bananas: Background Statistics and Information," USDA Economic Research Service, June 9, 2008.

185 *Whole Foods . . . 17 percent:* From a 2007 email interview with Karen Christensen, Global Produce Coordinator of Whole Foods Market.

185 *Tesco . . . "carbon labeling":* Eoin O'Caroll, "Tesco Tries Out Carbol Labels," *Christian Science Monitor,* May 3, 2008. See also: www.carbon-label.com/ index.htm.

186 *Walmart . . . going local:* All facts in this passage are from my interview with Matt Kissler, Walmart's director of sustainability. See also: Jon Gambrell, "Walmart Branches Out into Locally Grown Produce," Associated Press, July 1, 2008.

187 *artificially heating such indoor farms:* Alison Smith and Paul Watkiss, "The Validity of Food Miles as an Indicator of Sustainable Development," Depart-

ment for Environment, Food, and Rural Affairs (DEFRA) in the United Kingdom, July 2005.

188 *Sara Lee has to source:* Alexiei Barroneuiveo, "Globalization in Every Loaf," *New York Times,* June 16, 2007.

188 *A typical Nokia cell phone:* Pranshu Singhal, "Lifecycle Integrated Analysis: Environmental Issues of Mobile Phones," Nokia Company Report, April 2005.

189 *billions of phones:* There were 1.15 billion phones sold in 2007, and 1.22 billion units sold in 2008, according to Gartner research, www.gartner.com.

190 *three-quarters of the toys:* Joseph Kahn and Ralph Blumenthal, "Made in the USA? Not Anymore," *New York Times Upfront,* April 26, 2004.

190 *"several other countries contributed to . . . Barbie":* Rone Tempest, Robert C. Feenstra, and Alan M. Taylor, *International Trade* (New York: Worth Publishers, 2008); and "Barbie and the World Economy," *Los Angeles Times,* September 22, 1996.

191 *"We need markets":* William H. Chafe, *The Unfinished Journey: America Since World War II* (London: Oxford University Press, 2003), p 70.

193 *"We were at the end of two decades":* Economist Herbert Stein quoted in Daniel Yergin and Joseph Stanislaw, *The Commanding Heights: The Battle for the World Economy* (New York: Simon and Schuster, 1998).

193 *"Only by reducing the growth":* John M. Berry, "Risky Reagan Remedy Attacks Economic Ills on All Fronts at Once," *Washington Post,* February 19, 1981.

193 *"a means to an end":* Milton Friedman, "Freedom's Friend," *Wall Street Journal,* June 11, 2004.

193 *experienced 500 percent growth:* A comprehensive overview of transportation deregulation and its impact on commerce can be found in this Department of Transportation report, *Freight Management Operations: From Economic Deregulation to Safety Regulation.* http://ops.fhwa.dot.gov/freight/theme_papers/final_thm8_v4.htm.

194 *"We don't need to build walls":* Bill Clinton, *Between Hope and History: Meeting America's Challenges for the 21st Century* (New York: Random House, 1996), pp. 33–34.

194 *GDP shot up 35 percent:* U.S. Department of Commerce Bureau of Economic Analysis, "National Income and Product Accounts Table (1.1.6. Real Gross Domestic Product), Chained Dollars," March 26, 2009.

195 *"world was full of small manufacturers":* Marc Levinson, *The Box: How the Shipping Container Made the World Smaller and the World Economy Bigger* (Princeton, New Jersey: Princeton University Press, 2006), p. 3.

196 *"He wouldn't be able to sit still":* Quoted ibid., p. 43.

196 *"never one to be carried away":* Pamela Sherrid, "Captain Courageous," *Forbes,* October 24, 1983.

196 *McLean sold eggs:* Bio material from "Malcolm McLean," obituary in the *Times* (London), May 30, 2001.

197 *"I'd like to sink that":* Quoted in Brian J. Cudahy, "The Containership Revolution: Malcolm McLean's 1956 Innovation Goes Global," *TR News,* Transportation Research Board, Sept.–Oct. 2006.

197 *merchandise imported . . . doubled:* U.S. Department of Commerce, Bureau of Economic Analysis, U.S. International Transactions Accounts Data, Table 2b U.S. Trade in Goods. March 18, 2008.

198 *"Chinese fabric . . . American cotton":* Levinson, *The Box,* p. 266.

199 *"when U.S. customs authorities":* Ibid.

199 *Zara . . . had restocked:* Thomas L. Friendman, *The World is Flat,* 3rd ed. (New York: Picador/Farrar, Straus and Giroux, 2005), p. 154.

199 *"chicken and fish . . . crossing":* Larry Rohter, "Shipping Costs Start to Crimp Globalization," *New York Times,* August 3, 2008.

200 *"Even as Americans lament":* Mark Levinson, "Freight Pain," *Foreign Affairs,* vol. 87, no. 6, November/December 2008.

200 *"The cost of shipping a 40-foot":* Rohter, "Shipping Costs Start to Crimp Globalization."

201 *"A lot of our supply chain design":* Quoted in Jonathan Birchall and Elizabeth Rigby, "Oil Costs Force P&G to Rethink Supply Network," *Financial Times,* June 26, 2008.

201 *Kimberly Clark . . . reengineered:* Justin Lahart, "U.S. Retools Economy, Curbing Thirst for Oil," Associated Press, August 12, 2008.

201 *Home Depot . . . sourcing:* Claudia H. Deutsch, "For Suppliers, the Pressure Is On," *New York Times,* Novemeber 7, 2007.

201 *Emerson . . . transferred the production:* Timothy Appel, "Stung By High Transportation Costs, Factories Bring Jobs Home Again," *Wall Street Journal,* June 18, 2008.

201 *costs jumped more than $35,000:* "Few Options as Energy Costs Soar for Small Businesses," Reuters, May 27, 2008.

202 *"Look at this lead pencil":* Milton's pencil commentary aired on his television show *Free to Choose,* www.youtube.com/watch?v=d6vjrzUplWU, in which he focused on the work of Leonard Read, founder of the libertarian think tank Foundation for Economic Education—specifically Read's essay "I Pencil," a story on the complex origins of a pencil, www.econlib.org/library/Essays/rdPncl1.html.

7. Short Circuits

205   *home to 50 million people:* "August 14 Blackout," Department of Energy, Office of Electricity Delivery and Energy Reliability, www.oe.netl.doe.gov/hurricanes_emer/blackout.aspx.

206   *The accident resulted:* A step-by-step account of the causes of the 2003 Northeastern blackout can be found in "Final Report on the August 14, 2003 Blackout in the United States and Canada: Causes and Recommendations," Department of Energy, Conducted by the U.S.-Canada Power System Outage Task Force, April 2004.

208   *electricity prices . . . tripling:* Interview with Kurt Yeager, former president and CEO of the Electric Power Research Institute (EPRI) and executive director of the Galvin Electricity Intiative.

209   *$700,000 per minute:* The city lost an estimated total of $1 billion in twenty-four hours, divided by 1,440 minutes in that time period.

210   *"reaches into your":* Phillip Schewe, *The Grid* (New York: Henry Holt, 2001) p. 1.

211   *300,000 miles of high-voltage:* The statistics on the existing grid came from a comprehensive report by the Department of Energy: "The Smart Grid: An Introduction: Exploring the Imperative of Revitalizing America's Electric Infrastructure," Department of Energy, www.oe.energy.gov/DocumentsandMedia/DOE_SG_Book_Single_Pages.pdf.

211   *5.2 million miles:* From the Edison Electric Institute, www.eei.org.

211   *increasingly prone to . . . interruptions:* See Kurt Yeager, *Perfect Power* (New York: McGraw-Hill, 2009). See also Brad Foss, "Frustrations Grow over Lack of Electric Reliability Standards," *USA Today*, August 9, 2004.

211   *50 percent more electricity:* Gary Caruso, "Long-term Outlook for Energy Markets," Energy Information Administration, November 21, 2005, slide 20.

212   *"superpower with a third-world grid":* See this PBS *Frontline* interview with Bill Richardson from April 10, 2001, www.pbs.org/wgbh/pages/frontline/shows/blackout/interviews/richardson.html. See also: Chris Suellentrop, "Why This Was a First World Blackout," Slate.com, August 15, 2003, www.slate.com/id/2087036.

212   *"not much smarter than sewage pipes":* Interview with Yeager.

212   *infrastructure is . . . 40 years:* Ibid.

212   *$150 billion annually:* From "The Smart Grid: An Intoduction," DoE.

212   *5 to 10 percent . . . $12 billion:* Interview with Yeager.

213   *a third of all greenhouse:* "Inventory of U.S. Greenhouse Gas Emissions and Sinks, 1990–2007," Environmental Protection Agency, April 15, 2009.

213   *80 percent*: "Emissions of Greenhouse Gasses Report," Energy Information
       Administration, December 3, 2008, Table 11.

213   *more than fifty years old*: M. Granger Morgan, "Don't Grandfather Coal Plants,"
       Science Magazine, November 17, 2006.

214   *80,000 miles of cable*: The facts on New York City's electricity grid came from
       Con Ed spokesperson Chris Olert.

217   *"the biggest and most responsible"*: Quoted in Jill Jonnes, *Empires of Light:
       Edison, Tesla, Westinghouse and the Race to Electrify the World*, (New York:
       Random House, 2004), p. 81.

217   *Born in Milan, Ohio*: Much of the Edison bio material from Neil Baldwin, *Edison:
       Inventing the Century* (New York: Hyperion, 1995). And William Adams Simonds,
       Edison: His Life, His Work, His Genius (New York: Read Books, 2007).

217   *"Genius is 1 percent"*: M. A. Rosanoff, "Edison in His Laboratory," *Harper's*,
       September 1932.

217   *"clear, cold, and beautiful"*: Quoted in Jonnes, *Empires of Light*, p. 57.

218   *"a thousand—aye, ten thousand"*: "Edison's Newest Marvel," *New York Sun*,
       September 16, 1878.

218   *The day had come*: Details on Edison's demonstration day and the early growth
       of the electricity industry: Richard Munson, *From Edison to Enron: The Busi-
       ness of Power and What It Means for the Future of Electricity* (Westport: Prae-
       ger Publishers, 2005); Maury Klein, *The Power Makers: Steam, Electricity,
       and the Men Who Invented Modern America* (New York: Bloomsbury Press,
       2008), ch. 8; Paul Israel, *Edison: A Life of Invention* (New York: John Wiley
       & Sons, 1998), pp. 206–207; and Jean Strouse, *Morgan: American Financier*
       (New York: Random House, 1999), p. 233.

219   *"Edison was vindicated"*: "Edison's Illuminators," *New York Herald*, September
       5, 1882.

219   *"light was soft, mellow"*: "Edison's Electric Light: The Times' Building Illumi-
       nated by Electricity," *New York Times*, September 5, 1882.

219   *200 customerss . . . 3,500 bulbs*: Numbers on Edison's customers and bulbs
       from Klein, *The Power Makers*, p. 172.

219   *more than a million bulbs*: From my interview with Munson.

220   *less than 5 percent of lamps*: Ibid.

221   *power to rural America*: Facts in this section from my interview with Munson
       and his book *The Power Makers: The Inside Story of America's Biggest Business—
       and Its Struggle to Control Tomorrow's Electricity* (New York: Rodale Press,
       1985), Chap. 6 (not to be confused with Klein's book of the same title).

222   *the companies refused*: Robert Caro, *The Years of Lyndon Johnson: The Path to
       Power* (New York: Alfred A. Knopf, 1982), pp. 516–519.

222   *GE spent nearly $50 million*: Daniel Nye, *Electrifying America: Social Meanings of a New Technology, 1880–1940* (Cambridge, Massachusetts: MIT Press, 1990), p. 270.

222   *"Any Woman" campaign*: Slogans and ad copy quoted ibid., pp. 270–71.

223   *"Come up from slavery"*: Quoted ibid., p. 271.

223   *"toiled in a nineteenth-century"*: William Edward Leuchtenburg, *Franklin D. Roosevelt and the New Deal, 1932–1940* (New York: Harper & Row, 1963), p. 157.

223   *"Cold figures do not measure"*: Quoted in Nye, *Electrifying America*, p. 304.

223   *"they are the bulwark"*: Quoted in Caro, *Path to Power*, p. 520.

224   *"at night on a hillside"*: Leuchtenburg, *Franklin D. Roosevelet*, p. 158.

224   *1,045,000 in 1953*: Malcolm Jones Jr., "Air Conditioning," *Newsweek*, vol. 130, Winter Special Issue, 1997, p. 42.

224   *Television came of age*: Many of these examples drawn from Munson's *From Edison to Enron*, pp. 77–79.

225   *"Millions of women lived"*: Betty Friedan, *The Feminine* Mystique (New York: W. W. Norton, 2001), p. 136.

225   *female consumer force*: For more on this concept see Ruth Schwartz Cowan's essay "The Industrial Revolution in the Home: Household Technology and Social Change in the 20th Century," *Technology and Culture* magazine, January 1976.

225   *"women's . . . workforce jumped"*: Abraham Mossia and Steven Hipple, *Trends in Labor Force Participation in the United States*, Bureau of Labor Statistics, October 2006, p. 36.

225   *"52 hours a week . . . to 45"*: Valerie Ramey, *"A Century of Work and Leisure,"* University of California, San Diego, and National Bureau of Economics Research, May 2006.

225   *people employed in housekeeping jobs*: See data from "Historical Statistics of the United States, Colonial Times to 1970," US Census Bureau: www2 .census.gov/prod2/statcomp/documents/CT1970p2–01.pdf.

226   *"a power plant . . . typewriter"*: Quoted in the essay "Frontiers," in *MANAS Journal*, Volume XXXIII No. 12, March 19, 1980.

227   *"Some river! Chocolate-brown"*: "The Cities: The Price of Optimism," *Time*, August 1, 1969.

227   *"Pollution was rampant"*: Interview on NPR's WPCN, June 22, 1999.

228   *"initiative away from the Democrats"*: J. Brooks Flippen, *Nixon and the Environment* (Albuquerque: University of New Mexico Press, 2000).

228   *"reason Earth Day worked"*: Keith Schneider, "Gaylord Nelson, Founder of Earth Day, Is Dead at 89," *New York Times*, July 4, 2005.

229   *"industry is out of control"*: Richard Morgan and Sandra Jerabek, "How to Challenge Your Local Electric Utility: A Citizen's Guide to the Power Industry," Environmental Action Foundation, 1974.

230   *"Project Independence"*: "A Consumer's Guide to the Issues in the Restructuring of Electric Utilities," Consumer Energy Council of America Research Foundation, http://www.liheapch.acf.hhs.gov/dereg/cecarf.htm, also see this DoE timeline of energy milestones: www.energy.gov/about/timeline1971-1980.htm.

231   *"Every act of . . . conservation"*: From Jimmy Carter's "Crisis of Confidence" speech on July 15, 1979. http://www.pbs.org/wgbh/amex/carter/filmmore/ps_crisis.html.

231   *"a mosquito bite on an elephant's fanny"*: When oil expert and author of *The Prize* Daniel Yergin was a young professor at Harvard Business School, he edited a book entitled *Energy Future: Report of the Energy Project at the Harvard Business School*, where this quote from the World Oil editor appears. See: http://www.energybulletin.net/node/36930.

233   *"The waste heat discarded"*: Amory Lovins, "More Profit for Less Carbon," *Scientific American,* September 2005.

234   *impacts of coal:* See this Sierra Club summary of the environmental impacts of coal: www.sierraclub.org/coal/dirtytruth/frame-mining.asp.

235   *electronics . . . 10 percent:* "Electricity Consumption by End Use in U.S. Households, 2001," *EIA Household Electricity Reports, United States,* Table US-1.

235   *twenty-seven hours a month:* Nielsen Annual Report 2008.

235   *Plasma TVs . . . consume as much:* Peter Ostendorp and Suzanne Foster, "Televisions: Active Mode Energy Use and Opportunities for Energy Savings," *Issue Paper,* ed., National Resources Defense Council, March 2005, p. 31.

235   *10 percent . . . wasted:* Alan Mier, "Standby Power," Lawrence Berkeley Laboratory, http://standby.lbl.gov/standby.html.

235   *the "smart grid":* For more details on the Smart Grid see: "The Smart Grid: An Introduction," DoE. See also Ira Flatow, "What's a Smart Grid?" *Science Friday,* NPR, February 13, 2009. See also: Steve Silberman, "The Energy Web" *Wired,* July 2001.

237   *5,000 energy providers . . . 124 million:* Industry statistics from the Edison Electric Institute communications office.

237   *29 percent growth:* From the EIA's Annual Energy Outlook 2009, www.eia.doe.gov/oiaf/aeo.

238   *"The mind can not conceive"*: Quoted in Elbert Hubbard, *Little Journeys to the Homes of the Great* (New York: Wm. H. Wise & Co., 1916), p. 267. Reprinted online: www.gutenberg.org/files/12933/12933-h/12933-h.htm#THOMAS_A_EDISON.

8. Earth, Wind, and Fire

243   *"distinctive blend of technology"*: John Kouwenhoven, *The Beer Can by the Highway* (Baltimore: Johns Hopkins University Press, 1988), p. 17.

245   *fund the Swift-boat ad campaign*: Wayne Slater, "For Pickens, Denver's a Breeze," *Dallas Morning News*, August 28, 2008.

245   *"Pickens Plan"*: Further information on Pickens, his plan, and his wind project can be found at www.pickensplan.com.

246   *a defector, a turncoat*: Steven Milloy, "Is T. Boone Pickens 'Swiftboating' America?" FoxNews.com, July 24, 2008, www.foxnews.com/story/0,2933,390821,00.html.

248   *He credits his father*: Pickens bio material from his memoir: *Boone* (New York: Houghton Mifflin, 1987). And Pickens's *The First Billion Is the Hardest: Reflections on a Life of Comebacks and America's Energy Future* (New York: Crown Publishing Group, 2008).

249   *$700 billion annually*: Pickens's number, which is based on $125 a barrel of oil, has been widely debated. The following article reports that the U.S. spent roughly half that—$328 billion—on imported oil and petroleum by-products in 2007: H. Josef Hebert, "Unfriendly Oil Import is Wildly Inflated," Associated Press, October 16, 2008.

251   *Pickens's grand vision . . . roadblocks*: Elizabeth Souder, Pickens paring down wind farm project," *Dallas Morning News*, July 6, 2009.

251   *plunged . . . 97 percent*: Joe Carroll and Mark Chediak, "Pickens Reduces Energy Investments, Holdings Fall 97%," *Bloomberg News*, February 17, 2009.

251   *"the price is going back up"*: Quoted in Paul Davidson, "Pickens Relishes Role as Leader in Push for Alternative Energy," *USA Today*, April 15, 2009.

251   *"I haven't changed anything"*: Ibid.

252   *14 and 18 miles per hour*: See the National Renewable Energy Laboratory's maps of America's wind resource potential: www.nrel.gov/gis/wind.html.

252   *Enough solar energy beats*: From the Solar Energy Industries Association. See NREL's maps of America's solar resources: www.nrel.gov/gis/solar.html.

252   *"This scheme of combustion"*: Quoted in Elbert Hubbard, *Little Journeys*, p. 285.

253   *betting . . . economic survival*: Makower, Joel. *Strategies for the Green Economy: Opportunities and Challenges in the New World of Business* (New York: McGraw-Hill Companies, 2008). See also: Rob Pernick, *The Clean Tech Revolution: Discover the Trends, Technologies and Companies to Watch* (New York: HarperCollins, 2009).

253 *$47.5 billion in wind:* From the Global Wind Energy Council. http://www
.gwec.net/.

253 *six of the world's ten:* "International Wind Energy Development World
Market Update," BTM Consult, 2009, www.btm.dk. According to Angelika
Pullen of Global Wind Energy Coincil, "the largest turbine manufacturers are based in Denmark (Vestas, Siemens), Germany (Enercon, Nordex),
the US (GE) and Spain (Gamesa, Acciona). India's Suzlon now also has
a market share of around 10 percent, and the Chinese manufacturers are
starting to expand into the international markets."

253 *90 percent of solar:* From the Solar Energy Industries Association: http://www
.seia.org/galleries/pdf/Solar_manufacturing_tax_credit.pdf.

253 *$70 billion over . . . ten years:* From a summary of the legislation by Speaker
of the House Nancy Pelosi, www.speaker.gov/newsroom/legislation?id=0273
#energy.

253 *"big step down the road":* "Obama's Remarks on Signing the Stimulus Plan,"
CNN.com, February 17, 2009, www.cnn.com/2009/POLITICS/02/17/obama
.stimulus.remarks/.

254 *(DARPA) created . . . technologies:* Mitch Waldrop, "DARPA and the Internet
Revolution," DARPA, www.darpa.mil/Docs/Internet_Development_200807
180909255.pdf.

254 *"'green' is green":* For more on G.E.'s decision to go green, see the feature
I wrote for *Vanity Fair*, "G.E.'s Green Gamble," www.VF.com, July 10,
2006.

255 *"the green coal baron":* Clive Thompson, "The Green Coal Baron," *New York
Times Magazine*, June 22, 2008. Also see my interview with Jim Rogers:
"Rogers and Me," Grist.org, April 4, 2007.

256 *CO_2 . . . of New Hampshire:* Walmart's 2007 emissions were 20.2 million
metric tons of CO_2 equivalent. New Hampshire's 2005 emissions were 21.21
CO_2, www.epa.gov/climatechange/emissions/state_energyco2inv.html.

256 *"As I got exposed":* From my interview with H. Lee Scott in April 2008, www
.msnbc.msn.com/id/12316725/.

257 *$500 million . . . $8 billion:* "Clean technology venture investment reaches
record $8.4 billion in 2008 despite credit crisis and broadening recession,"
Cleantech Group, January 6, 2009, http://cleantech.com/about/pressreleases/
010609.cfm.

258 *"The field of green-tech":* "Venture Capitalist Taps Green Technology," Associated Press, April 12, 2006.

258 *"saving the future of civilization":* For more Khosla commentary see: http://
www.grist.org/article/little9.

258   *"RE<C"*: See my Grist.org interview with Google's Green Energy Czar, Bill Weihl, www.grist.org/article/weihl/.

259   *zero-energy home*: For more information on zero-energy homes see: www1 .eere.energy.gov/buildings/goals.html. My articles on the Harmony Heights development have appeared in Grist.org: "Little Solar Homes for You and Me," October 7, 2003, and Mother Earth News: "Super Solar Homes That Everyone Can Afford," January 2005.

264   *first photovoltaic cells*: See the first Bell Laboratory paper on solar panels: D. M. Chapin, C. S. Fuller, and G. L. Pearson, "A New Silicon *p-n* Junction Photocell for Converting Solar Radiation into Electrical Power," *Journal of Applied Physics*, May 1954.

264   *two central challenges*: For an overview of the solar industry's challenges and opportunities see: Fred Krupp and Miriam Horn, *Earth: The Sequel: The Race to Reinvent Energy and Stop Global Warming* (New York: W. W. Norton & Company, 2008), chapter 2.

264   *the cost to install solar*: See "Tracking the Sun: The Installed Costs of Photovoltaics in the U.S. from 1998–2007," Lawrence Berkeley National Laboratory, February 2009.

265   *product called the Sunflower*: E-mail exchange with Bill Gross.

265   *Miasolé . . . thin films*: Krupp and Horn, *Earth: The Sequel*, pp. 33–36.

265   *"nanoink" can be "printed"*: Josh Wolfe, "Solar Power Heats Up with Nanotechnology," *Forbes*, September 2009. Also see: Michael Moyer, "The New Dawn of Solar," *Popular Science*, 2007, www.popsci.com/popsci/flat/bown/2007/green/item_59.html.

266   *"technology . . . save humanity"*: Joseph Romm, "The Technology That Will Save Humanity," Salon.com, April 14, 2008. For an overview of CSP also see: www.nrel.gov/learning/re_csp.html.

268   *"20 percent . . . by 2030"*: U.S. Department of Energy. "Wind Energy Could Produce 20 Percent of U.S. Electricity by 2030," *Energy Efficiency and Renewable Energy*, July 2008.

268   *3 to 6 percent . . . from wind*: This and the Texas statistics from "2008 Annual Rankings Report," American Wind Energy Association, April 2008.

269   *bioengineer viruses . . . batteries*: See Krupp and Horn, *Earth: The Sequel*, pp. 232–234.

270   *"innovation known as geothermal"*: Joseph Romm, "Hot Rocks are Rockin Hot," www.climateprogress.org, May 23, 2008.

270   *best geothermal resources*: "Geothermal Energy: Energy from the Earth's Core" *Energy Kid's Page*, Energy Information Administration, July 2008.

270   *at least 15 times that*: U.S. Department of the Interior, U.S. Geological Survey.

"Assessment of Moderate and High-Temperature Geothermal Resources of the United States," Fact Sheet 2008-3082.

270 *45 and 75 degrees:* U.S. Department of Energy. "Energy Savers: Your Home: Geothermal Heat Pumps," *Energy Efficiency and Renewable Energy*, February 2009.

271 *"a thousand times more dense":* Quotes in the ocean-energy section from my interview with Hendrick and his book, cowritten with Jay Inslee, *Apollo's Fire: Igniting America's Clean-Energy Economy* (Washington: Island Press, 2008), pp. 189–191.

271 *coastlines . . . up to a quarter:* California Energy Commission. "Ocean Energy," http://www.energy.ca.gov/oceanenergy/index.html.

272 *"has rigged up 13 machines":* Jane Spencer, "While You're at It, Why Not Generate a Little Electricity?" *Wall Street Journal*, March 1, 2007.

272 *ReRev.com is now making elliptical:* "Universities Generate Electricity in the Gym," Associated Press, May 17, 2009.

273 *vaporizing trash:* "Florida County Plans to Vaporize Landfill Trash," Associated Press, September 9, 2006.

273 *"poop power":* Brian Handwerk, "Device Uses Sewage Bacteria to Produce Electricity," *National Geographic News*, March 1, 2004.

275 *a peaceful protest:* See: www.capitolclimateaction.com/.

275 *candlelight prayer vigil:* "Coal plants put on hold," News 8 Austin, February 21, 2007.

275 *plunged by more than half:* EIA forecasted 100 gigawatts new coal in Annual Energy Outlook of 2008. In 2009 they are forecasting 42 gigawatts. See: www.eia.doe.gov/oiaf/aeo/.

276 *nuclear power . . . promoted as "clean":* For more on nuclear see "The Future of Nuclear Power: An Interdisciplinary MIT Study," Massachusetts Institute of Technology, 2003, http://web.mit.edu/nuclearpower.

279 *price on carbon dioxide:* For an overview of how a carbon cap-and-trade system works, see the congressional testimony of David Doniger, Climate Center Policy Director of the Natural Resources Defense Council, on the "American Clean Energy and Security Act of 2009," April 24, 2009, http://www.nrdc.org/globalWarming/aces/files/glo_09042401.pdf.

279 *reconnecting to the past:* See Joseph Romm's "The Technology That Will Save Humanity." Also see Romm's "Winds of Change," Salon.com, May 15, 2008; DoE, "A History of Geothermal Energy in the United States," *Geothermal Technologies Program*, November 2006.

9. Autopia

285   *"commitment to the electrification"*: Mike Spector, "Detroit Car Show Hit by Industry's Gloom: Like the Auto Makers, Annual Event Is Downsized; Nissan Won't Attend, GM Cancels Fashion Show," *Wall Street Journal*, January 22, 2009.

285   *"speed the acceptance"*: Martin LaMonica, "Big Three Plug Electric Cars at Auto Show," CNET Tech News, January 11, 2009.

285   *"seeking treatment for wounds"*: Quoted in Greg Hitt and Mathhew Dolan, "Big Three Plead for Aid," *Wall Street Journal*, November 19, 2008.

285   *"Any superfluous show business"*: Martin LaMonica, "Big Three Plug Electric Cars at Auto Show."

286   *the sizable barriers:* John Stewart, "Toyota, Experts See Plug-In Car Trouble, Electric Reality Check," *Popular Mechanics*, October 2, 2008.

288   *40 percent of America's total oil:* "EPA's Fuel Economy Programs," EPA, October 2007.

288   *"better from a global warming standpoint"*: David Sandalow bears this point out in his book *Freedom from Oil: How the Next President Can End the United States' Oil Addiction* (New York: McGraw-Hill, 2008).

    Further data on the global warming benefits of electric cars can be found here: "Executive Summary: Environmental Assessment of Plug-In Hybrid Vehicles," Electric Power Research Institute, 2007.

    See also: Brendan Koerner, "The Electric Vehicle Acid Test," Slate.com, December 11, 2007.

289   *battery . . . 400 pounds:* Details on the GM battery can be found here: http://media.gm.com/volt/eflex/docs/battery_101.pdf.

289   *introduction to the Chevrolet Volt:* "GM's Chevy Volt Plug-In Concept—How it Works," *Popular Mechanics*, animated diagram, www.popularmechanics.com/automotive/new_cars/4215492.html.

289   *over 100 miles per gallon:* It's difficult to tally fuel economy for an electric car that could conceivably never go to a gas station. Even the EPA is confused about how to calculate the fuel economy of these newfangled vehicles. Using the EPA's standard formulas to calculate fuel economy, the Volt averages over 100 mpg, though the agency thinks that its standard test should be revised for plug-ins. See: http://reviews.cnet.com/8301-13746_7-10037173-48.html.

291   *meet in Gremban's . . . garage:* a full account of these car-hacking sessions can be found on www.calcars.org.

291   *"This does not make Toyota happy":* Danny Hakim, "A Car That Plugs into the Wall," *New York Times,* April 2, 2005.

291   *Dozens of orders came in:* A chronological list of plug-in hybrid conversions can be found here: www.calcars.org/where-phevs-are.

293   *hydrogen-car future:* Ulf Bossel, "Does a Hydrogen Economy Make Sense?" *Proceedings of the IEEE* (94)10: 1826, October 2006.

293   *half of all cars in Brazil:* "Brazil Energy Data Statistics and Analysis," Energy Information Administration, www.eia.doe.gov/emeu/cabs/Brazil/Oil.html.

293   *price of oil skyrocketed so did the price of corn:* Michael Rosenwald, "The Rising Tide of Corn," *Washington Post,* June 15, 2007.

293   *practically and affordably scaled up:* James B. Meigs, "The Ethanol Fallacy," *Popular Mechanics,* February 2008.

293   *food shortages spread:* See: *Press Conference by United Nations Special Rapporteur on Right to Food,* Dept. of Public Information, News and Media Division, http://www.un.org/News/briefings/docs/2007/071026_Ziegler.doc .htm. See also: Peter Robison, "Ethanol's Boom Holds Hidden Costs: Higher Food Prices," *Bloomberg News,* February 9, 2007. See also: Keith Johnson, "E-corn-omics: Ethanol vs. Gasoline vs. Food," *Wall Street Journal,* April 9, 2009.

293   *not all biofuels have green benefits:* "The Truth about Ethanol," *Clean Vehicles,* Union of Concerned Scientists, December 7, 2007.

294   *cut carbon . . . up to 60 percent:* Adam J. Liska and Haishun S. Yang, "Improvements in Life Cycle Energy Efficiency and Greenhouse Gas Emissions of Corn-Ethanol," *Journal of Industrial Ecology,* 2009.

294   *1 million units in ten years:* "Worldwide Prius Sales Top 1 Million Mark," Toyota Press Release, May 15, 2008, www.toyota.co.jp/en/news/08/0515.html.

295   *counteract Toyota's surging popularity:* The backstory of GM's plunge into the electric car in: Jonathan Rauch, "Electro-Shock Therapy," *Atlantic,* July/August 2008.

295   *"electricfication . . . is inevitable":* Bob Lutz, "The Road Ahead for Cars," *Newsweek,* April 25, 2008.

295   *"live green or die."* David Welch, "Live Green or Die," *BusinessWeek,* May 15, 2008.

295   *all-electric past:* Iain Carson and Vijay Vaitheeswaran, *Zoom: The Global Race to Fuel the Car of the Future* (New York: TwelvePublishers, 2007), p. 112.

295   *40 percent powered by steam:* From "Automobile History," in "Early Electric Automobiles," Brittanica Online Encyclopedia.

295   *targeted at female consumers:* See these images of early electric-car advertisements targeted at women: www.autolife.umd.umich.edu/Gender/Scharff/

Waverly_Electric.htm, and http://upload.wikimedia.org/wikipedia/commons/c/c7/1910_Waverley_Coupe.GIF.

296 *strides in battery development:* "In Search of the Perfect Battery," *Economist*, March 6, 2008. See also: M. Armand and J. M. Tarascon, "Building Better Batteries," *Nature* 451, February 7, 2008.

297 *Enter . . . lithium-ion:* For more on this technology, see: Dave Chameides, "Watt the Deal with Lithium-Ion Batteries?" Edmunds.com, August 21, 2007.

297 *"ignite and burn up grandma":* Quoted in Norihiko Shirouzu, "Race to Make Electric Cars Stalled by Battery Problems: GM, Toyota Seek Ways to Snuff Out Fire Risk," *Wall Street Journal*, January 11, 2008.

298 *triple to $2.3 billion:* "In Search of the Perfect Battery."

298 *Toyota changed its plan:* Bengt Halvorson, "Li-ion Not Ready for Prius," *BusinessWeek*, June 18, 2007.

298 *"The software will say, 'Okay'":* Quoted in Richard Truett, "Safety First: A lot is riding on the batteries that will power tomorrow's hybrid vehicles," *Automotive News*, November 19, 2007.

302 *intelligent grid system:* See: Brendan I. Koerner, "Power to the People: Seven Ways to Fix the Grid, Now," *Wired*, March 23, 2009.

304 *"power-generation capacity trapped":* Quoted in "Building the Energy Internet," *Economist*, March 11, 2004.

305 *most impressive display:* Warren Brown, "China Could Charge Ahead in the U.S. Market," *Washington Post*, January 18, 2008.

305 *China, BYD Auto:* For pictures and descriptions of the company's latest innovations see the blog ChinaCarTimes: http://www.chinacartimes.com/category/byd-auto/.

305 *"We are confident of exporting":* Quoted in Li Fangfang, "Automakers Are Plugging into Green Energy," *Business Daily Update*, February 9, 2009.

306 *middle class wants cars:* Lee Schipper and Wei-Shiuen Ng, "Rapid Motorization in China," *World Resources Institute*, October 19, 2004.

306 *16 million cars:* Lisa Margonelli, *Oil on the Brain* (New York: Doubleday, 2007) p. 265.

306 *52 million . . . to up to 660 million:* M. Wang, H. Huo, L. Johnson, and D. He, "Projection of Chinese Vehicle Growth: Oil Demand and CO_2 Emissions Through 2050," Energy Systems Division, Argonne National Laboratory, and the Energy Foundation, December 2006.

306 *road-building boom:* Calum MacLeod, "China's Highways Go the Distance," *USA Today*, January 1, 2006.

306 *six times the rate of the United States:* Margonelli, *Oil on the Brain*, p. 270.

306   *twenty of the thirty most polluted*: "China Quick Facts," World Bank, www
        .worldbank.org.

306   *restricting the country's GDP growth*: Margonelli, *Oil on the Brain*, p. 272.

306   *importing nearly half of its petroleum*: "China Energy Data," Energy Informa-
        tion Administration, http://tonto.eia.doe.gov/country/excel.cfm?fips=CH

306   *from suppliers in Russia*: "Country Analysis Briefs: China," EIA, August 2006.

307   *McKinsey & Co. report*: Quoted in Emma Graham-Harrison, "Electric Cars
        Are a Big Chance for China," Reuters, October 29, 2008.

307   *"We've built this in three years"*: Quoted in "Cars in China," *Economist*, June 2,
        2005.

307   *Chery Automobile*: "Chery Unveils First EV," Edmunds 'Inside Line,' February
        23, 2009. www.edmunds.com.

308   *plans to construct a new smart grid*: Fu Chenghao, "China Gets Smart on
        Power Supply," *Shanghai Daily*, June 6, 2009.

308   *don't struggle with . . . nostalgic attachment*: Margonelli, *Oil on the Brain*,
        p. 275.

308   *government of Israel . . . sanctioned*: Steven Erlanger, "Israel Is Set to Promote
        the Use of Electric Cars," *New York Times*, January 21, 2008.

309   *Shai Agassi's big idea*: "Business Bosses and the Environment," *Economist*,
        February 28, 2008. Also see: Tom Friedman, "Texas to Tel Aviv," *New York
        Times*, July 27, 2008.

309   *"an open, standards-based network"*: From a CBS News Sunday Morning inter-
        view with Shai Agassi, conducted by David Pogue, March 15, 2009.

309   *"The mission is to end oil"*: Daniel Roth, "Driven: Shai Agassi's Audacious Plan
        to Put Electric Cars on the Road," *Wired*, August 18, 2008.

309   *"massive disruption"*: See the Deutsche Bank report quoted by WiredNews
        .com, www.wired.com/autopia/2008/04/deutsche-bank-1.

310   *"Saudis don't control the sun"*: Steven Erlanger, "Israel Is Set to Promote the
        Use of Electric Cars."

312   *"add a minimum of $8,000"*: Joseph Romm, "The Car of the Future Is Here,"
        Salon.com, January 22, 2008.

313   *equivalent of 75¢ per gallon*: Jerry Mader, *"Battery Powered Vehicles: Don't Rule
        Them Out,"* Transportation Energy Center, University of Michigan, Novem-
        ber 16, 2006. Also see: "PEV Frequently Asked Questions," Plug-in Hybrid
        Electric Vehicles, Duke Energy.

314   *already-existing . . . "right of way"*: Richard Lacayo, "Re-Instating the Inter-
        state," from "What's Next 2009: 10 Ideas Changing the World Right Now,"
        Time, March 12, 2009, http://www.time.com/time/specials/packages/article/
        0,28804,1884779_1884782_1884764,00.html.

10. City, Slicker

318   *last revolution in architecture*: See Paul Goldberger, *The Skyscraper* (New York: Knopf, 1981).

319   *roughly three-quarters of the electricity*: Lew Pratsch, "Yes, You Can Build a Zero-Energy Home Today," Department of Energy Efficiency and Renewable Energy Program, April 28, 2009.

319   *40 percent of global energy*: "Buildings and Climate Change: Status, Challenges, and Opportunities," United Nations Environment Programme, 2007. See also: Hal S. Knowles, III, "Realizing Residential Building Greenhouse Gas Emissions Reductions," Program for Resource Efficient Communities/ School of Natural Resources and Environment, University of Florida, June 2008.

319   *"greater than total carbon"*: Anthony W. King, Lisa Dilling, Gregory P. Zimmerman, "United States Climate Change Science Program Synthesis and Assessment Product 2.2," Oak Ridge National Laboratory, May 2007.

319   *30 percent of . . . materials*: Materials consumption and waste production statistics from "EPA Green Buildings" data compilation, *Greening EPA*, U.S. EPA.

319   *"In 2030, about half of the buildings"*: Arthur C. Nelson "Toward a New Metropolis: The Opportunity to Rebuild America," Virginia Polytechnic Institute and the Brookings Institution, December 2004, p. 5. The report reads: "The nation had about 300 billion square feet of built space in 2000. By 2030, the nation will need about 427 billion square feet of built space to accommodate growth projections. About 82 billion of that will be from replacement of existing space and 131 will be new space. Thus, 50 percent of that 427 billion will have to be constructed between now and then."

321   *solar-powered planned communities*: Pratsch, "Yes, You Can Build a Zero-Energy Home Today."

321   *five thousand new low-income green homes*: "Habitat for Humanity International and the Home Depot Foundation Announce National Green Building Effort," press release, March 20, 2008.

322   *"ice batteries"*: For more details on ice cooling visit the Web site of Calmac, the company that manufactures and maintains the ice batteries: www .calmac.com/.

324   *cement production . . . 4 percent*: Thomas A. Boden, "National Cement Production Estimates: 1950–2004," Carbon Dioxide Information Analysis Center at Oak Ridge National Laboratory, 2005.

325   *"nature deficit disorder"*: Richard Louv, *Last Child in the Woods: Saving Our Children from Nature Deficit Disorder* (New York: Workman Publishing, 2008).

325   *"buildings reflect no understanding"*: For more on this concept see Stephen R. Keller, Judith H. Heerwagen, Martin L. Mador, *Biophilic Design: The Theory, Science, and Practice of Bringing Buildings to Life* (Hoboken, New Jersey: John Wiley & Sons, 2008).

326   *"biomimicry"*: Janine M. Benyus, *Biomimicry: Innovation Inspired by Nature* (New York: HarperCollins, 1998).

326   *examples of nature-inspired design*: From Joel Makower, "The Blossoming of Biomimicry," Two Steps Forward (blog), September 24, 2008.

326   *"The goal, in essence"*: Mark Gunther, "Buildings Inspired by Nature," *Fortune*, October 3, 2008.

327   *infinitely reused and recycled*: William McDonough, *Cradle to Cradle: Remaking the Way We Make Things* (New York: Melcher Media, 2006), p. 15. See also: www.mcdonough.com.

327   *"Growing up we were always taught"*: Quoted in Brian Dumaine, *The Plot to Save the Planet: How Visionary Entrepreneurs and Corporate Titans Are Creating Real Solutions to Global Warming* (New York: Crown Business, 2008), p. 94.

330   *60 percent of . . . commercial projects*: "Global Green Building Trends," Research and Analytics, McGraw-Hill Construction, September 19, 2008.

330   *70 percent of home buyers*: "SmartMarket Report: The Green Home Consumer," McGraw-Hill Construction, 2008.

331   *LivingHomes*: Virtual tour at www.wired.com/promo/wiredhome.

331   *"people who buy organic"*: Christopher Palmeri, "The Greenest House on the Planet," *BusinessWeek*, September 11, 2006.

333   *Chicago's Merchandise Mart*: Howard Cincotta, "Merchandise Mart Sets 'Green' Standard for Existing Buildings: Prestigious LEED Design Award Recognizes Old as Well as New Structures," www.america.gov, March 28, 2008.

333   *No detail . . . was too small*: Adam Aston, "Past Is Prologue: How Merchandise Mart Transformed Itself from a Relic into a Certified Energy-Efficient Marvel—and a Model for Other Outdated Buildings Around the U.S.," *BusinessWeek*, March 19, 2008.

334   *$100 million green retrofit*: Ilaina Jonas, "Empire State Building to Go 'Green,' Save Millions," Reuters, April 6, 2009.

335   *energy-efficient mortgage*: Shaheen Pasha, "Go Green with Your Mortgage: Energy Efficient Mortgages May Help Homebuyers Qualify for a Better House and Save on Energy Costs," CNNMoney.com, November 2, 2005. For

more information on energy-efficient mortgages see: Bob Tedeschi, "Mortgages: Saving Both Money and Energy," *New York Times,* September 10, 2006. See also: Jay Romano, "Your Home; Ways to Whittle Your Energy Bill," *New York Times*, January 13, 2008. And: Sara Schaefer Munoz, "Going Green to Save Some Green: Lenders Push Mortgages with Discounts and Credits for Energy-Efficient Upgrades," *Wall Street Journal,* September 12, 2007.

335   *Pat and Mynette Theard:* Case study in "Energy Efficient Mortgage Homeowner Guide," Homes and Communities: U.S. Department of Housing and Urban Development, November 29, 2001.

336   *"One homeowner doesn't make":* Quoted in the Rocky Mountain Institute documentary, *High Performance Building, Perspecitves and Practice,* http://bet .rmi.org/video.

338   *7 units per acre today to 11 units per acre:* Reid Ewing, Keith Bartholomew, Steve Winkelman, Jerry Walters, and Don Chen, *Growing Cooler: The Evidence on Urban Development and Climate Change* (Washington, D.C.: Urban Land Institute, 2008).

338   *$4 million . . . to power its streetlights:* This is a rough estimate based on the city of Raleigh, North Carolina, from data supplied by the mayor's office.

338   *"We're the ones building":* From my article "The Revolution Will Be Localized," Grist.org, July 14, 2005.

338   *snapshot of the mayoral green strategies:* Some of these examples from Tom Murphy, John Miller, and Uwe S. Brandes, "On the Front Lines of Positive Change," Urban Land Institute, February 2005.

340   *hubs for new mixed-use developments:* Bryan Walsh, "Recycling the Suburbs," from "What's Next 2009: 10 Ideas Changing the World Right Now," *Time,* March 12, 2009.

341   *In Singapore . . . Hong Kong:* From the International Association of Public Transport, www.uitp.org.

341   *5 percent of all commutes:* U.S. Census Bureau, "Most of Us Still Drive to Work—Alone: Public Transportation Commuters Concentrated in a Handful of Large Cities," *U.S. Census Bureau News* press release, June 13, 2007.

342   *U.S. households are located within even a mile:* Tara Bartee, "Public Transit in America, Analysis of Access Using the 2001 National Household Travel Survey," Center for Urban Transportation Research, University of South Florida, February 2007.

342   *"Given the settlement patterns":* Quoted in David Roberts, "Alternative Energy Guru Reflects on Policy, Fuels," July 27, 2007.

342   *$8 billion to the development of mass transit:* Dave Obey, "Summary: American

Recovery and Reinvestment," Conference Agreement, Committee on Appropriations, U.S. House of Representatives, February 13, 2009.

343 *The streets of Strasbourg:* Examples in this passage are from my interview with Robert Cervero, transportation expert at the University of California, Berkeley.

344 *planner and futurist Mitchell Joachim:* See Joachim's Web site, www.archinode .com.

345 *"Made from neoprene":* Tom Vanderbilt, "Mitchell Joachim: Redesign Cities from Scratch," *Wired,* September 22, 2008. See also: "The Best Inventions of the Year," *Time,* 2007.

345 *bicycles known as Vélib':* John Ward Anderson, "Paris Embraces Plan to Become City of Bikes," *Washington Post,* March 24, 2007.

346 *Jacobs . . . promoted a vision:* Jane Jacobs, *The Death and Life of Great American Cities* (New York: Modern Library, 1993).

347 *population to grow from 310 million:* Jennifer Cheeseman Day, "Population Profile of the United States," *National Population Projections,* U.S. Census Bureau, July 8, 2008.

347 *"Whenever and wherever":* Jacobs, *Death and Life of Great American Cities,* from Jacobs's forward in the Modern Library edition.

11. Fresh Greens

349 *half were from this neighborhood:* From my interviews with Tom Darden and Ninth Ward residents.

350 *about three-quarters of the families:* Ibid.

350 *Make It Right Foundation:* For more info see: www.makeitrightnola.org. See also: Bill Sasser, "New Orleans' Lower Ninth Ward Stirs and Rebuilds," *Christian Science Monitor,* November 5, 2008. See also: Malcolm Jones and Cathleen McGuigan, "Toward a New New Orleans," *Newsweek,* May 5, 2008.

351 *"zero waste" principles:* See: www.zerowaste.org.

351 *Pitt launched the foundation:* For more on Brad Pitt's vision behind Make It Right, see Gerald Clarke, "Brad Pitt Makes It Right In New Orleans," *Architectural Digest,* January 2009.

354 *larger social and environmental trend:* I wrote about pollution impacts on low-income communities in "Not in Whose Back Yard?" *New York Times Magazine,* September 2, 2007.

354   *four power plants, two prisons:* Data on the South Bronx industrial facilities
        from Sustainable South Bronx. See: www.ssbx.org.

355   *several times the national average:* "Asthma Facts, Second Edition," New York
        City Child Asthma Initiative from the Department of Health and Mental
        Hygiene, May 2003.

355   *roughly half of the low-income:* See the *Toxic Wastes and Race at Twenty,* report
        by the United Church of Christ Commission for Racial Justice, 2007, www
        .ucc.org/justice/environmental-justice/pdfs/introduction.pdf.

355   *80 percent more likely:* See "More Blacks Live with Pollution" Associated
        Press, December 16, 2008. Also see: Robert D. Bullard, "More Blacks
        Overburdened with Dangerous Pollution," Environmental Justice Resource
        Center, www.ejrc.cau.edu/BullardAPEJ.html.

355   *Lead-poisoning rates:* Data from my article, "Not in Whose Back Yard?" *New
        York Times Magazine,* September 2, 2007.

355   *30 percent of America's public schools:* See: "Many U.S. Public Schools in 'Air
        Pollution Danger Zone,'" University of Cincinatti Academic Health Center,
        August 20, 2008, http://healthnews.uc.edu/news/?/7358/.

356   *burgeoning environmental justice:* See: "Poverty & the Environment: On the
        Intersection of Economic and Ecological Survival," www.grist.org/article/
        pate.

357   *Green For All:* See this essay by Green For All CEO Phaedra Ellis-Lamkins,
        "Green-Collar Jobs: Equal Pay for Equal Work," Huffingtonpost.com, April
        28, 2009. Also see: www.greenforall.org.

357   *green jobs advocate Van Jones:* See Elizabeth Kolbert, "Greening the Ghetto,"
        New Yorker, January 12, 2009.

359   *"green-collar" economy:* See Van Jones, *The Green Collar Economy* (New York:
        HarperCollins, 2008).

361   *$1.7 billion:* Democracy Alliance data on environmental group funding and
        spending quoted in Mark Hertsgaard, "Green Grows Grassroots," *Nation,*
        July 13, 2006, www.thenation.com/doc/20060731/hertsgaard.

363   *President Theodore Roosevelt:* See Douglas Brinkley, *The Wilderness Warrior*
        (New York: HarperCollins, 2009).

363   *"protect the health and well-being":* For more on Republicans for Environmen-
        tal Protection see: www.repamerica.org.

363   *33 percent of its electricity:* See "Governor Schwarzenegger Advances State's
        Renewable Energy Development," http://gov.ca.gov/press-release/11073/, and
        "Governor Establishes World's First Low Carbon Standard for Transportation
        Fuels," press release, http://gov.ca.gov/index.php?/fact-sheet/5465/.

363   *Sunshine State utilities . . . 20 percent:* "For Renewables, No Time to Wait,"

St. Petersburg Times, April 21, 2009, www.tampabay.com/opinion/editorials/article993795.ece.

363   *greenest metropolises by 2030*: See PlaNYC: www.nyc.gov/html/planyc.

363   *curb government subsidies*: Snowe teamed up with Democratic Senator Feinstein of California to introduce a bill that would repeal drilling subsidies for oil companies. See: http://feinstein.senate.gov/05releases/r-incentive.htm.

364   *"We represent 40 percent"*: From my interview with Richard Cizik, "Cizik Matters," Grist.org, October 5, 2005, http://www.grist.org/article/cizik.

365   *"a new way to love your neighbor"*: From Ball's commentary at "Powering the Planet: Energy For the Long Run," Aspen Institute Environment Forum, March 25–28.

365   *"We focus on tangible results"*: See the Interfaith Power and Light Web site: www.theregenerationproject.org/About.htm.

365   *"inspire a vision for the future"*: See the Jewish Climate Initiative Web site: www.jewishclimateinitiative.org.

365   *"Producing the [water] bottles"*: From the sermon of Thomas Kleinert at the Vine Street Christian Church in Nashville, Tennesee. Full sermon here: http://thomas-kleinert.blogspot.com/2008/11/royal-vocation.html.

366   *"Profound and powerful forces"*: From Bill Clinton's first inaugural address, delivered January 20, 1993.

367   *"We can create millions of jobs"*: For details on the Obama administration's Green Jobs Initiative visit Department of Labor, Employment and Training Administration, www.doleta.gov/Business/.

367   *"consequences that will last for centuries"*: From Steven Chu's speech accepting his nomination. See also inteview of Chu by Stephen Power, *Wall Street Journal*, "We've Got to Do This," February 6, 2009, http://online.wsj.com/article/SB123393841471357455.html.

367   *"Worried about the economy"*: From Jackson's commentary at "Powering the Planet: Energy for the Long Run," Aspen Institute Environment Forum, March 25–28.

368   *first 100 days*: See Kate Sheppard, "100 Days: Obama's Big Green Dream," Grist.org, April 29, 2009.

369   *"We've got to stop crying"*: From Jimmy Carter's "Crisis of Confidence" speech, July 15, 1979, www.pbs.org/wgbh/amex/carter/filmmore/ps_crisis.html.

376   *"The greatness of America lies"*: Alexis de Tocqueville, *Democracy in America* (New York: Penguin Classics, 2003), p. 113.

Bibliography

Ambrose, Stephen E. *Eisenhower: The President*. New York: Touchstone, 1990.

Audretsch, David B. *The Entrepreneurial Society*. Oxford: Oxford University Press, 2007.

Baldwin, Neil. *Edison: Inventing the Century*. New York: Hyperion, 1995.

Barthes, Roland. *Mythologies*. New York: Noonday, 1972.

Benyus, Janine M. *Biomimicry: Innovation Inspired by Nature*. New York: HarperCollins, 1998.

Berry, Wendell. *The Unsettling of America*. San Francisco: Sierra Club, 1996.

Bone, John Herbert Aloysius. *Petroleum and Petroleum Wells: A Complete Guidebook*. Philadelphia: J. P. Lippincott, 1865.

Brinkley, Douglas. *Wheels for the World: Henry Ford, His Company, and a Century of Progress*. New York: Penguin, 2003.

Bronson, Rachel. *Thicker Than Oil: America's Uneasy Partnership with Saudi Arabia*. Oxford: Oxford University Press, 2006.

Brown, Lester. *Seeds of Change: The Green Revolution and Development in the 1970s*. New York: Praeger, 1970.

Caro, Robert. *The Years of Lyndon Johnson: The Path to Power*. New York: Knopf, 1982.

Carson, Iain, and Vijay Vaitheeswaran. *Zoom: The Global Race to Fuel the Car of the Future*. New York: Twelve, 2007.

Carson, Rachel. *Silent Spring*. Fortieth Anniversary Edition. New York: Houghton Mifflin Harcourt, 2002.

Chanda, Nayan. *Bound Together: How Traders, Preachers, Adventures, and Warriors Shaped Globalization.* New Haven, Conn.: Yale University Press, 2007.

Charles, Daniel. *Master Mind: The Rise and Fall of Fritz Haber, the Nobel Laureate Who Launched the Age of Chemical Warfare.* New York: HarperCollins, 2005.

Chernow, Ron. *Titan: The Life of John D. Rockefeller, Sr.* New York: Random House, 1998.

Clarke, Alison J. *Tupperware: The Promise of Plastic in 1950s America.* Washington, D.C.: Smithsonian Institute Press, 1999.

Clark, James Anthony, and Michel T. Halbouty. *Spindletop: The True Story of the Oil Discovery That Changed the World.* New York: Random House, 1952.

Commoner, Barry. *The Poverty of Power: Energy and the Economic Crisis.* New York: Knopf, 1976.

Davenport, E. H. and S. R. Cooke. *The Oil Trusts and Anglo-American Relations.* New York: Macmillan, 1924.

Deffeyes, Kenneth S. *Hubbert's Peak: The Impending World Oil Shortage.* Princeton, N.J.: Princeton University Press, 2008.

DeGolyer, E. L., and L. W. MacNaughton. *Twentieth Century Petroleum Statistics.* Ann Arbor: University of Michigan Press, 1945.

Dumaine, Brian. *The Plot to Save the Planet: How Visionary Entrepreneurs and Corporate Titans Are Creating Real Solutions to Global Warming.* New York: Crown Business, 2008.

Eddy, William A. *F.D.R Meets Ibn Saud.* Vista, Calif.: Selwa, 2005.

Eisenhower, Dwight. *At Ease: Stories I Tell to Friends.* Garden City, N.Y.: Doubleday, 1967.

Engler, Robert. *The Politics of Oil: A Study of Private Power and Democratic Directions.* New York: Macmillan, 1961.

Ewing, Reid, Keith Bartholomew, Steve Winkelman, Jerry Walters, and Don Chen. *Growing Cooler: The Evidence on Urban Development and Climate Change.* Washington, D.C.: Urban Land Institute, 2008.

Feenstra, Robert C., and Alan M. Taylor. *International Trade.* New York: Worth, 2008.

Fenichell, Stephen. *Plastic: The Making of a Synthetic Century.* New York: Harper Business, 1996.

Friedan, Betty. *The Feminine Mystique.* New York: Norton, 2001.

Friedman, Thomas L. *Hot, Flat, and Crowded: Why We Need a Green Revolution—And How It Can Renew America,* New York: Farrar, Straus and Giroux, 2008.

———. *The World Is Flat,* 3rd ed. New York: Picador / Farrar, Straus and Giroux, 2005.

Flink, James J. *The Automobile Age*. American Council of Learned Societies. Cambridge, Mass.: MIT Press, 1988.

Galvin, Robert, and Kurt Yeager. *Perfect Power: How the Microgrid Revolution Will Unleash Cleaner, Greener, and More Abundant Energy*. New York: McGraw-Hill, 2009.

Gardner, Bruce L. *American Agriculture in the Twentieth Century: How It Flourished and What It Cost*. Cambridge, Mass.: Harvard University Press, 2002.

Goddard, Stephen B. *Getting There: The Epic Struggle between Road and Rail in the American Century*. Chicago, Ill.: University of Chicago Press, 1996.

Goodell, Jeff. *Big Coal: The Dirty Secret behind America's Energy Future*. Boston, Mass.: Houghton Mifflin, 2006.

Gross, Daniel. *Greatest Business Stories of All Time*. New York: Wiley, 1996.

Halberstam, David. *The Fifties*. New York: Villard, 1993.

Hamill, Thomas, and Paul T. Brown. *Escape in Iraq: The Thomas Hamill Story*. Accokeek, Md.: Stoeger, 2005.

Hawken, Paul, Amory Lovins, and L. Hunter Lovins. *Natural Capitalism: Creating the Next Industrial Revolution*. Boston, Mass.: Little, Brown, 2000.

Heinberg, Richard. *The Party's Over: Oil, War, and the Fate of Industrial Societies*. Gabriola Island, B.C.: New Society, 2003.

Hesser, Leon. *The Man Who Fed the World: Nobel Peace Prize Norman Borlaug and His Battle to End World Hunger*. Dallas, Tex.: Durban House, 2006.

Howell, Mark D. *Moonshine to Madison Avenue: A Cultural History of the NASCAR Winston Cup Series*. Bowling Green, Ohio: Bowling Green State University Popular Press, 1997.

Hubbard, Elbert. *Little Journeys to the Homes of the Great*. New York: Wm. H. Wise, 1916. Reprinted online: http://www.gutenberg.org/files/12933/12933-h/12933-h.htm#THOMAS_EDISON.

Inslee, Jay, and Bracken Hendricks. *Apollo's Fire*. Washington, D.C.: Island, 2008.

Israel, Paul. *Edison: A Life of Invention*. New York: Wiley, 1998.

Jacobs, Jane. *The Death and Life of Great American Cities*. New York: Modern Library, 1993.

Jones, Van. *The Green Collar Economy*. New York: HarperCollins, 2008.

Jonnes, Jill. *Empires of Light: Edison, Tesla, Westinghouse, and the Race to Electrify the World*. New York: Random House, 2004.

Kellert, Stephen R., Judith H. Heerwagen, and Martin L. Mador. *Biophilic Design: The Theory, Science, and Practice of Bringing Buildings to Life*. Hoboken, N.J.: Wiley, 2008.

Klare, Michael T. *Blood and Oil: The Dangers and Consequences of America's Growing Petroleum Dependency*. New York: Holt, 2004.

Klein, Maury. *The Power Makers: Steam, Electricity, and the Men Who Invented Modern America*. New York: Bloomsbury, 2008.

Knowles, Ruth Sheldon. *The Greatest Gamblers: The Epic of American Oil Exploration*. Norman: University of Oklahoma Press, 1978.

Korda, Michael. *Ike: An American Hero*. New York: HarperCollins, 2007.

Kouwenhoven, John Atlee. *The Beer Can by the Highway: Essays on What's "American" about America*. Baltimore, Md.: Johns Hopkins University Press, 1988.

Krupp, Fred, and Miriam Horn. *Earth: The Sequel—The Race to Reinvent Energy and Stop Global Warming*. New York: Norton, 2008.

Kunstler, James H. *The Geography of Nowhere: The Rise and Decline of America's Man-Made Landscape*. New York: Simon & Schuster, 1993.

Lacey, Robert. *The Kingdom: Arabia and the House of Sa'Ud*. New York: Avon, 1981.

Leahy, William D. *I Was There: The Personal Story of the Chief of Staff to Presidents Roosevelt and Truman, Based on His Notes and Diaries Made at the Time*. New York: McGraw Hill, 1950.

Leuchtenburg, William Edward. *Franklin D. Roosevelt and The New Deal 1932–1940*. New York: Harper and Row, 1963.

Levinson, Marc. *The Box: How the Shipping Container Made the World Smaller and the World Economy Bigger*. Princeton, N.J.: Princeton University Press, 2006.

Louv, Richard. *Last Child in the Woods: Saving Our Children from Nature Deficit Disorder*. New York: Workman, 2008.

Makower, Joel. *Strategies for the Green Economy: Opportunities and Challenges in the New World of Business*. New York: McGraw-Hill, 2008.

Margonelli, Lisa. *Oil on the Brain*. New York: Nan A. Talese/Doubleday, 2007.

Marolda, Edward J. *Shield and Sword: The United States Navy and the Persian Gulf War*. New York: Crown, 2001.

McDonough, William. *Cradle to Cradle: Remaking the Way We Make Things*. New York: Melcher Media, 2006.

McKenney, Jason. "Artificial Fertility: The Environmental Costs of Industrial Fertilizers," in Andrew Kimball, ed., *The Fatal Harvest Reader: The Tragedy of Industrial Agriculture*. Sausalito, Calif.: The Foundation for Deep Ecology and Island Press, 2002.

McNichol, Dan. *The Roads That Built America: The Incredible Story of the U.S. Interstate System*. New York: Sterling, 2006.

Meikle, Jeffrey L. *American Plastic: A Cultural History*. New Brunswick, N.J.: Rutgers University Press, 1995.

Miller, Aaron David. *Search for Security: Saudi Arabian Oil and American Foreign Policy*. Chapel Hill: University of North Carolina Press, 1980.

Miller, John Anderson. *At the Touch of a Button*. Schenectady, N.Y.: Mohawk Development Service, 1962.

Morris, Edmund. *The Rise of Theodore Roosevelt*. New York: Modern Library, 1979.

Mowery, David C., and Nathan Rosenberg. *Paths of Innovation: Technological Change in Twentieth Century America*. Cambridge: Cambridge University Press, 1998.

Munson, Richard. *From Edison to Enron: The Business of Power and What It Means for the Future of Electricity*. Westport, Conn.: Praeger, 2005.

———. *The Power Makers: The Inside Story of America's Biggest Business—And Its Struggle to Control Tomorrow's Electricity*. New York: Rodale, 1985.

Nye, David E. *Electrifying America: Social Meanings of a New Technology, 1880–1940*. Cambridge, Mass.: MIT Press, 1990.

Paarlberg, Don. *The Agricultural Revolution of the Twentieth Century*. Ames: Iowa State University Press, 2000.

Painter, David. *Oil and the American Century: The Political Economy of U.S. Foreign Oil Policy, 1941–1953*. Baltimore, Md.: Johns Hopkins University Press, 1986.

Payne, Robert. *Report on America*. New York: John Day, 1949.

Pernick, Rob, and Clint Wilder. *The Clean Tech Revolution: Discover the Trends, Technologies, and Companies to Watch*. New York: HarperCollins, 2009.

Pickens, Thomas Boone. *Boone*. New York: Houghton Mifflin, 1987.

———. *The First Billion Is the Hardest: Reflections on a Life of Comebacks and America's Energy Future*. New York: Crown, 2008.

Pimentel, David. *Biofuels, Solar, and Wind as Renewable Energy Systems: Benefits and Risks*. Boston, Mass.: Springer, 2009.

Pollan, Michael. *The Omnivore's Dilemma: A Natural History of Four Meals*. New York: Penguin, 2006.

Porter, Michael I. *The Competitive Advantage of Nations*. New York: Free Press, 1998.

Pyle, George. *Raising Less Corn, More Hell: The Case for the Independent Farm and Against Industrial Food*. New York: Public Affairs, 2005.

Reilly, Michael F. *Reilly of the White House*. New York: Simon & Schuster, 1947.

Roberts, Paul. *The End of Oil: On the Edge of a Perilous New World*. Boston, Mass.: Houghton Mifflin, 2004.

Rose, Mark H. *Interstate: Express Highway Politics, 1939–1989*. Knoxville: University of Tennessee Press, 1990.

Roosevelt, Elliott. *As He Saw It*. New York: Duell, Sloan, and Pearce, 1946.

Roosevelt, Theodore. *Presidential Addresses and State Papers*. New York: Review of Reviews, 1909.

Ross, Eric B. *The Malthus Factor: Poverty and Population in Capitalist Development*. London: Zed, 1998.

Sachs, Jeffery D. *Common Wealth: Economics for a Crowded Planet.* New York: Penguin, 2008.

Sampson, Anthony. *The Seven Sisters: The Great Oil Companies and the World They Made.* New York: Viking, 1975.

Sandalow, David. *Freedom from Oil: How the Next President Can End the United States' Oil Addiction.* New York: McGraw-Hill, 2008.

Shah, Sonia. *Crude: The Story of Oil.* New York: Seven Stories, 2004.

Simon, Linda. *Dark Light: Electricity and Anxiety from the Telegraph to the X-Ray.* Orlando, Fla.: Harcourt, 2004.

Simonds, William Adams. *Edison: His Life, His Work, His Genius.* New York: Read Books, 2007.

Simmons, Matthew R. *Twilight in the Desert: The Coming Saudi Oil Shock and the World Economy.* Hoboken, N.J.: Wiley, 2005.

Sinclair, Upton. *Oil!* Berkeley: University of California Press, 1926.

Smil, Vaclav. *Enriching the Earth: Fritz Haber, Carl Bosch, and the Transformation of World Food Production.* Cambridge, Mass.: MIT Press, 2001.

———. *Feeding the World: A Challenge for the Twenty-First Century.* Cambridge, Mass.: MIT Press, 2000.

Solberg, Carl. *Oil Power.* New York: Mason/Charter, 1976.

Stegner, Wallace. *Discovery! The Search for Arabian Oil.* San Francisco, Calif.: Selwa, 2007.

Stevens, Eugene S. *Green Plastics: An Introduction to the New Science of Biodegradable Plastics.* Princeton, N.J.: Princeton University Press, 2001.

Strouse, Jean. *Morgan: American Financier.* New York: Random House, 1999.

Tarbell, Ida M. *The History of the Standard Oil Company.* New York: Norton, 1966.

Thompson, Neal. *Driving with the Devil: Southern Moonshine, Detroit Wheels, and the Birth of NASCAR.* New York: Three Rivers, 2006.

Tocqueville, Alexis de. *Democracy in America.* New York: Penguin Classics, 2003.

United States Office of Naval Petroleum and Oil Shale Reserves. *Twentieth Century Petroleum Statistics.* Ann Arbor: University of Michigan, 1998.

Wollen, Peter, and Joe Kerr, eds. *Autopia: Cars and Culture.* London: Reaktion, 2002.

Yates, Brock. *The Decline and Fall of the American Automobile Industry.* New York: Houghton Mifflin, 1983.

Yergin, Daniel. *The Prize: The Epic Quest for Oil, Money, and Power.* New York: Simon & Schuster, 2008.

Yergin, Daniel, and Joseph Stanislaw. *The Commanding Heights: The Battle for the World Economy.* New York: Simon & Schuster, 1998.

Index